D0982689

LAW AND SPACE TELECOMMUNICATIONS

For Heather

LAW AND SPACE
TELECOMMUNICATIONS

FRANCIS LYALL

Dartmouth

Published by
Dartmouth Publishing Company Limited
Gower House
Croft Road
Aldershot
Hants GU11 3HR
England

Gower Publishing Company
Old Post Road
Brookfield
Vermont 05036
USA

Printed in Great Britain by
Billing & Sons Ltd, Worcester
Typeset by Inforum Typesetting, Portsmouth

British Library Cataloguing in Publication Data
Lyall, Francis
    Law and space telecommunications
    1. International telecommunication services.
    Legal aspects
    I. Title
    341.7'577

ISBN 1 85521 039 8

Contents

List of Abbreviations

ARABSAT	Arab Satellite Organization
ARPA	(US) Advanced Research Projects Agency
AT & T	American Telephone and Telegraph Company
CCIR	International Consultative Committee on Radio
CCITT	International Telegraph and Telephone Consultative Committee
CEPT	Committee on Postal and Telecommunications Administrations
CERN	European Organization for Nuclear Research
CETS	European Conference of space Telecommunications
CO-COM	Coordinating Committee on Information
COMSAT	Communications Satellite Organization
COPERS	European Preparatory Commission on Space Research
COPUOS	(UN) Committee on the Peaceful Uses of Outer Space
CSAGI	Comite Special de l'Année Geophysique Internationale
EBU	European Broadcasting Union
ECS	European Communications Satellite (programme)
ECU	European Currency Unit
ELDO	European Launcher Development Organization
ESA	European Space Agency
ESC	European Space Conference
ESOC	Earth Station Ownership Committee
ESRO	European Space Research Organization
EUTELSAT	European Telecommunications Satellite Organization
FCC	(US) Federal Communications Commission
GDMS	Global Maritime Distress and Safety System
IBS	INTELSTAT Business Service
ICAO	International Civil Aviation Organization
ICSC	Interim Communications Satellite Committee
IFRB	International Frequency Registration Board
IGY	International Geophysical Year of 1957
IMCO	Intergovernmental Maritime Consultative Organization

IMO	International Maritime Organization (formerly IMCO)
INMARSAT	International Maritime Satellite Organization
INTELSAT	International Telecommunications Satellite Organization
IRU	Indefeasible right of user
ISI	International Satellite Inc.
ITT	International Telegraph and Telegraph World Communications Inc.
ITU	International Telecommunication Union
IWG	Intersessional Working Group (of Plenipotentary Conference)
JISO	Japanese International Joint Users Organization
MPM	(ITU) Multilateral Planning Meeting
NASA	(US) National Aeronautics and Space Administration
OTS	Orbital Test Satellite
PanAmSat	Pan American Satellite
PTT	Post, Telegraph and Telephone Administration
RCA	Radio Corporation of America
RFP	Request for Proposals
SCORE	Signal Communications Orbital Relay Experiment
SIG	(US) Senior Interagency Group on International Communications and Foreign Policy
TDMA	Time-division multiple access
TRU	Transponder right of user
UN	United Nations
UNDP	United Nations Development Programme
UNESCO	United Nations Educational, Scientific and Cultural Organization
URSI	International Scientific Radio Union
WMO	World Meteorological Organization
WWI	Western Union International Inc.

Preface

In 1965 I wrote my LL.M thesis on 'Law and Space Telecommunications'. The years since have seen swift development both in the field of international telecommunications and in their legal regulation. We have made considerable progress towards what Marshall McLuhan dubbed the 'Global Village', although we have not yet learned fully to cope with the volume and pace of communications that space has made possible. The events of Black Monday in October 1987 were compounded, if not caused, by the telecommunicative interrelationships of the world's stock markets. On the other side, it cannot be doubted that the swift availability of TV pictures for Viet Nam at the breakfast tables of American helped change government policy in that arena.

Swiftness of change and development has made the production of this book awkward, not to say difficult. The publishing process itself takes time. The basic text is based on material available to me at May 1988, but that material has no uniform cut-off. The availability of data has been affected as competition in the telecommunications market has led various companies and agencies to impose a confidentiality which was not previously present. This is irritating to a researcher, but quite understandable. It does mean, however, that uncertainty is compounded. As far as possible Chapters 1–8 take account of developments to 31 December 1987, and in some cases later than that. The publishers have permitted me to make revisions at the copy-editing stage (very late December 1988–January 1989), and I have recast Chapter 9 to include major developments.

As a 'snapshot' of matters which are in motion or in flux, the book is necessarily untrue to reality, but nevertheless may aid understanding of the subject. Hokusai's 'The Wave' looks unlike a wave, but tells you much about waves. I hope that this book will provide a similar service – though I hope also that reviewers will not avail themselves of the easy jibe just provided.

These pages contain some hard statements about US attitudes within the field of telecommunications regulation. It may be that I have got some things wrong. But I would state here that it is the very openness of the US political system which allows one to see, for example in the FCC (Federal Communications Commission) proceedings and Congressional Hearings, things which in other countries are shrouded in secrecy. The current attitudes of various countries and PTTs (Post, Telegraph and Telephone Administrations) to INTELSAT, for example, are not unlike those of the USA. Documentation in their cases is lacking.

The many people I have talked to in the UK, in Europe and North America about telecommunications have been, almost without exception, helpful. Some expressly wish not to be named, and others have indicated reluctance. It is fairest, therefore, to thank everyone anonymously, but that does not imply that my thanks are a mere formality. I have been enormously helped. I apologize to those who found my questions immature and ignorant, and to those who will see from these pages that I have not understood what they were saying. To some I must say that, in the face of other evidence, I have been unable fully to accept their points of view.

But I will give some public thanks. Various of my friends, knowing my love of science fiction, were greatly amused when in 1963–64 I went to study at the Institute of Air and Space Law at McGill University, Montreal. There Space Law was made interesting by Professor Ivan Vlasic, who also encouraged me in writing a term paper and then the LL.M. thesis on the topic of this book. My greetings and thanks to Ivan for first directing me to this fascinating area.

The British Academy made the book possible by financing, by means of its Small Grants Scheme, research in London, the USA and at the ITU headquarters in Geneva in 1981–82. The Trustees of the Davidson Fund further financed a research trip to New York and Washington in 1986. I am grateful. I thank them for their patience and hope that these bodies will not think their money wasted.

I would thank my colleague Douglas Cusine, for permitting me to use his computer to feed the text to the Faculty Decmate, and Mrs M. Mercer, Mrs A. Walton and Miss S. Kilpatrick for processing the text.

Finally, my family has put up with much. Too often have I disappeared up to the wordprocessor instead of listening, playing, and reading and generally being a good father. As of now, that stops!

Francis Lyall
Department of Public Law
University of Aberdeen
Scotland, UK
January, 1989

1 Introduction, Advent and Context

INTRODUCTION

The development of space telecommunications within the last twenty-five years or so has been swift and has had a dramatic impact on civilization. Had pictures of the previous day's fighting in Viet Nam not been beamed into the breakfast programmes in the USA, would events have turned out as they did?

The swiftness of development is, however, awkward for the author. The 'lead time' for a book is of the order of nine months, and much can happen in that time. Accordingly it is helpful to view space telecommunications with the benefit of history: how the waves have rolled in the past may help us to understand how things are as they are, and how they may develop in the future. Again, the developments do not occur in isolation. There is a political, an economic and a legal context within which events unfold. In telecommunications there is also the constraint of the laws of physics. Some scene-setting therefore has to be done, some of which takes place within the individual chapters. As a preliminary matter, however, this chapter will continue with a description of the advent of the satellite into the telecommunications environment as it had developed down to the 1960s. The problems of that environment explain how useful the satellite facility is, a matter sometimes less known to lawyers. Then the legal context within which the regulation of the new medium occurs will be outlined for the benefit of those without a legal background.

Thereafter, in Chapter 2, we review the US Communications Satellite Corporation, the USA's major initial response to the possibilities of the new facilities. Although occurring within one state, that response has affected all other international developments. The primary and major international response is the International Telecommunications Satellite Organization (INTELSAT) which has established and runs the largest global satellite system. Chapters 3 and 4 concerning it therefore take the largest share of this book. Chapter 5 on the International Maritime Satellite Organization (INMARSAT) deals primarily with maritime communications and is a smaller, though global, organization. Chapter 6 considers Europe's response to space through the European Space Agency (ESA), and, in telecommunications, through a regional organization, the European Telecommunications Satellite Organization (EUTELSAT). Chapter 7 then deals with two smaller responses, that of INTERSPUTNIK on a global basis and that of ARABSAT in a regional dimension.

Radio frequencies undergird virtually all satellite usage for all purposes. The last major portion of the book, Chapter 8, deals with the development of the International Telecommunication Union (ITU), the international agency which has competence in such matters, and whose competence is being extended as states use its fora to discuss the allocation of positions on the geostationary orbital torus. Finally, I attempt a provisional conclusion. But that is a long way away. First, there are technical questions: what is telecommunication? How is it carried out? How have its modalities been so significantly improved since 1957?

ADVENT: COMMUNICATIONS AND THE SPACE AGE

Society has always required telecommunications. The developments of the last half of the twentieth century were, perhaps, unforeseen by many, and there is an argument that their potential has not yet been fully comprehended by regulators and politicians. Yet these developments lie within the general historical pattern, public and private needs, governmental and mercantile interests as well as the wish of the general population to stay in touch with the news and with relatives, all being formative of the demands which technical progress have made easier to meet.

Historically the development of telecommunications[1] has been encouraged and shaped by the needs of society. Communications is necessary for an organized society, and two principal needs – needs of state and needs of trade – are fundamental. The earliest form of telecommunication was information brought by a messenger either orally or by letter, using the main communications routes of kingdoms and empires. Early roads were often built to afford easy transit to troops, but the transfer of information was a useful corollary. Assyrian and Babylonian ruins have yielded many clay tablets on which military, political and economic events were notified to central government. In Roman times, the letters of governors of provinces, and the Imperial replies, were extensively used as source material by such authors as Pliny. Indeed, Roman rule largely rested on a basis of sound and rapidly transmitted information. The same habit of collecting and disseminating news is found in the Middle Ages; the Roman

Catholic Church, for example, established its own information network.

While the state and similar institutions were creating and using their communications networks, a parallel use by traders and bankers grew up. Trading enterprises used the roads to move their goods, but the information on prices and plenty and scarcity that also travelled the trade routes was of obvious interest to entrepreneurs. They, as well as the governments of the day, found it both necessary and desirable to keep informed about what was going on. Such banking houses as the Medici had a network of correspondents throughout Europe and even into the Near East to keep it informed for, then as now, the exclusive command of news and information is an advantage giving both power and the opportunity for wealth to those that have it.

The dissemination of information through communication is also important. Ideology can be spread. The growth of printing and the correlative spread of opinion produced a significant change in society. The ideas of Erasmus and Luther, for example, led to the overthrow of the hegemony of the Roman Catholic Church in northern Europe, with far-reaching effects. The spread of the ideas of Copernicus, Tycho Brahe, Johannes Kepler and Galileo added to the upheaval in religion, leading ultimately to the scientific revolution and in turn to the scientific basis of our own day. By the dissemination of ideas through the printed word society itself was altered. Education made communication available to many: to some it was passively available as they became receivers of information and opinion from a wider world; to others it afforded a larger stage through their ability to make use of the media of the time. The urge to communicate spread throughout society. The development of a public postal system in the mid-nineteenth century was a major step in the establishing of a telecommunications system which was open to the use of many.

So far we have spoken of the communication of information by physical means – a message memorized, a letter carried between writer and recipient, books and, at a later date, newspapers. But information is not a physical matter and it can have immense effects if known and acted upon quickly. The rapidity of information transmission is vastly increased if only the information is transferred, not some physical object containing it. Crude messages can, of course, be passed by simple signals. The bonfires which warned of the coming of the Armada are examples of such. But if the signal can be made more complex, more information can be passed. Drums and smoke signals overcome the limitations of their media by the imposition of a complex pattern, and this is the same technique which is used in modern wire and wireless communications.

The modulation of a medium to pass a message was first most successfully solved by the use of flags, which were used to pass commands among the Roman and Greek galleys. By the time of the Napoleonic Wars such flag codes had become extremely complex.[2] Another flag signalling system is semaphore. Claude Chappe, a Frenchman, then took the concept of the moveable arm from the semaphore system, put such arms on to a sequence of towers, added the use of the telescope, and produced the first reliable long-distance communications facility which he called the 'telegraph'.[3] His method was useless at night or in fog,[4] but the method was feasible and effective. By 1844, just over 50 years from Chappe's first proposals, a network of over 5,000 km linked Paris with 29 French cities. The value of the system in allowing the central government rapidly and accurately to know what was going on throughout the country was readily apparent to other governments, and the system was copied elsewhere. The time was one of flux and dissension, culminating in 1848, 'the Year of Revolutions', and a speedy and reliable method of telecommunications was seen by governments as essential both for security and their preservation.

Even as the usefulness of the visual telegraph was becoming known, its supplanter was beginning to be established. This, the system of the 'electric telegraph' was the result of the efforts of many men, the best known of whom is the American, Samuel Morse. As in the case of the drum or smoke signal, the system involves the modulation of a medium in accordance with a recognizable and pre-arranged pattern. The duration of flow of an electrical current along a wire is arranged so as to conform to a code, allowing the conversion of words into that code, and the decipherment of the pattern at the other end of the wire. The method is reliable and can be operated over long distances without the need of a chain of stations. It is not subject to interruption by darkness or inclement weather and is speedier than the mechanical semaphore system. Finally, the electric telegraph can handle a greater volume of traffic than the Chappe system. It was therefore practicable to open the system for use by businessmen as well as for government messages.[5]

Here again, we can see the needs of state interacting with the needs of commerce. In Europe most telegraph services were nationalized from the beginning (or very soon thereafter), but the telegraph lines were laid down in conjunction with the other main development of the time, the railways. The state wanted rapid telecommunications both for its own needs and to knit the country together. The railways could function in safety and speed only as the positions of trains on the railway system were known. Businessmen wanted to know when goods would arrive, and price levels in other parts of the country. The railway 'corridors' cutting through the countryside permitted the laying of cables without the necessity of negotiations with a succession of landowners. Small wonder that the telegraph was both successful and intimately bound up with rail traffic.

But, of course, abuses did occur. Of these, the most flagrant may have been the emendations made to the 'Ems telegram' by Count Bismarck, which resulted in the Franco-Prussian War of 1870. That war itself reinforced the military usefulness of the telegraph. As so often happens in history, the pressures of war produced technological advance and, in the 1870s, the first multiplex systems and automatic sending and receiving were introduced. Of these, multiplex, allowing the transmission of several messages 'simultaneously', was the most important.

Technical difficulties always arose when a cable had to cross water through damage by the cable rubbing on underwater obstacles and by the 'shorting' effect of the water. Despite these, however, the first submarine cable was laid between Dover and Calais in 1851, making Britain part of a network which soon stretched from Moscow to Italy and the Mediterranean. By 1865 cables had reached Calcutta. In 1866 the Atlantic was bridged, and by 1871 cables had reached Australia. Naturally the international flow of messages increased with the provision of the new facility, with France, for example, stating that in 1880 she had handled 15,864,298 internal messages and 3,393,330 international messages through her facilities.[6] In 1906 a cable was laid from the USA to the Far East, and the frame of the cable system had been completed. There were over two thousand cables in operation, and the process of filling in the interstices continued.[7]

The next major telecommunication system to develop was the telephone. Invented by Alexander Graham Bell in 1876, the telephone was slow in gaining acceptance for a variety of reasons. At first it was technically deficient, the system acquiring too many whistles and clicks to be of value over a long distance. Then again, the system had to compete with the by then popular telegraph, and investors in the telegraph companies were keen to preserve their interests. The one exception was Germany where the state took up the telephone and developed it vigorously. Berlin, for example, had a city-wide telephone system in the early part of this century. The UK was, however, more typical in its attitude. The telegraph had early been made a state monopoly under the supervision of the General Post Office,[8] and a nineteen-year feud between the two telecommunication systems was resolved only when the telephone companies were bought out by the government and added to the Post Office's responsibilities.[9]

In the USA a similar battle was fought between the various private interests involved. The government had established a telegraphic link between Washington and Baltimore in 1844, but when it proved uneconomic soon sold it to private enterprise. Thereafter the US government remained aloof from involvement with the provision of a telecommunications system, and when the telephone arrived on the scene the evolutionary principle of private enterprise was allowed to work itself out with great bitterness in the marketplace and in the courts, until at length the Bell System was established as an entity entirely apart from either government or the telegraph companies. This dispute was to have effects in all subsequent questions of tele-communications in the USA, and still echoes even in the arguments as to the setting-up of the Communications Satellite Corporation.

Internationally, the telephone cable went through the same de-velopments as had the telegraph before it. Like the telegraph, there were problems with the establishment of submarine connections because of shorting and friction but, although the expertise of the tele-graph cable helped with those problems, a more serious problem was lurking. Over long distances the loss of signal quality in telephone cables was such as totally to destroy the usefulness of the system. The interruption of a current was one thing; maintaining the more soph-isticated signal required to transmit speech was something else. Thus, while a short cross-Channel telephone cable was laid in 1891, it was not until 1954 that a long-distance submarine telephone cable link was established (between Aberdeen, Scotland and Bergen, Norway). The first successful transatlantic telephone cable was not laid until 1956, and only in the 1960s were a series of reliable telephone cables laid across the Atlantic and down to late 1988 there were only two major transpacific cables. This long delay is partly explained by the difficulties of solving the problem of signal degradation, partly as a matter of cost, and partly because of the establishment of a reason-ably adequate transatlantic radio-telephony link in 1927.

Radio communication involves the radiation of energy in the form of radio waves. These, emitted from a transmitter at a particular frequency of cycles per second, can be picked up by a receiver tuned to that frequency. The signal can be arranged in a pattern to permit the transmission of information by code. At a more sophisticated level of development, either the amplitude of the wave or its fre-quency can be modulated so as to allow a greater volume of data to be contained within the signal, such that suitable equipment in the receiver can translate the incoming signal into speech or music, or (a later development) into the instructions required to produce a picture on a television screen. Unlike the cable systems of communication, radio is not dependent upon the existence of a physical channel of communication along which messages are passed. Messages can be passed between radio transmitters and receivers which are otherwise unconnected. The message may be sent either 'broadcast', where anyone can pick up the signal with suitable equipment, or the signal may be 'narrowcast' both to limit its reception and sometimes to enhance its clarity at some specific receiver.

The credit for the experiments which laid the foundation for the invention of radio must be shared among many people, but the first patent in the field stands in the name of Guglielmo Marconi in 1896. Many experiments followed to improve the facility, and there were many public demonstrations; even Queen Victoria expressed interest.

The first transatlantic transmission took place in 1901, and by 1914 significant strides had been made. As in the case of the telephone, there were problems with the already established systems, and in retrospect it is quite understandable that the first steps were taken in an arena where these could not operate – at sea. The Marconi Company made full use of its ability to communicate between ships, and ship to shore, even seeking to choke out competing systems by refusing to accept messages from non-Marconi stations. The arrest of Dr H.H. Crippen in 1910 was one event which brought the new medium to public attention. Dr Crippen was identified on board the SS *Montrose* by a radioed description, and police, warned by radio, were able to cross the Atlantic on a faster ship and arrest the fugitive before the *Montrose* could reach territorial waters. Another event which stressed the potential of radio was the loss of the *Titanic*, which had a consequent effect on the London Radio-Telegraph Conference of 1912. The London Conference on Safety of Life at Sea of 1914 required all ships carrying more than 50 passengers to have a radio capable of transmitting over a distance of 100 miles, and this caused a vast increase in the number of ship-borne and land stations.

As the 1870 Franco-Prussian War affected the telegraph, so the 1914–18 War affected the development of radio. Radiotelegraphy was given impetus, and the power of transmitters was greatly increased to provide communications links with farther areas of the globe. One of the first acts of the British government on the outbreak of the war was to rip up the German Atlantic cables, leaving the Germans with radio as their only method of communication with the Americas, and necessitating the setting-up of immensely powerful transmitters south of Berlin. Of course, such signals could be intercepted, and the successful crypto-analysis of the Zimmerman Telegram probably changed the course of history.

Despite the flaw of interceptability, radio made immense advances under the pressures of war. Clarity of signal was added to range, and soon after the First World War radio telephony was available over long distances, including, as noted above, the transatlantic service. Radio also developed as a cultural medium. Commercial radio began in the USA and, in Europe, each country set up its own broadcasting facilities.[10] Long- and short-wave broadcasting also commenced. Television was not far off. However, telephone and telegraphic communication remained the means of telecommunication of business and government.

If one considers the communications arena immediately prior to the availability of satellite communications, one can clearly see why the new modality was sought. There was a great demand for communications facilities from governments. There was also a huge and increasing demand from business for its information requirements and a more limited demand for private communications by individuals. Some also saw a potential for the use of satellites for television transmission for re-broadcast by the ordinary TV stations. The more farsighted dreamed of direct broadcast from a satellite to a home. But the methods of telecommunication available in, say, the mid-1950s, were inadequate for such demands.

To deal first with the cable system, a cable had the major defect that it was fixed. Once a cable is established between two points there is no way that that length of wire and its technical attachments can be used to serve a different set of points. Within a country or a continent, of course, the cabling network makes it feasible to route messages by the fixed cables to different points with a measure of flexibility. That is why the early conventions on radio-telegraphy all require the signatories to ensure that coastal radio stations are linked into the cable network. The high quality of cable transmission was preferable, wherever possible, to reliance on the vagaries of the ether. But where there is a body of water or inhospitable terrain to be crossed, a cable, once laid, despite its repair problems is likely to be the only connection to be made, and once laid is immoveable. Thus, in the early days of the cable facility, one of the primary aims of an attacking army was to gain control over the terminals of the enemy's international cables, or to use naval means to break the cable.[11]

A further problem of the cable system was simply that the cables were configured with the old colonial empires, for it was their interests which were served by their establishment. Thus, in the 1950s it was not unusual for a cable telephone call between reasonably proximate but newly independent countries to have to be routed through either or both of the capitals of the former colonial empires concerned.[12] Although this was intolerable to the newer nations, there was little that could be done. The cost of establishing direct cable links, whether for telegraph or telephone, across the rugged terrain, for example, of Africa, would have been prohibitive. On the other hand, radio links were not satisfactory as reliable alternatives.

Another difficulty of cables in the modern period was that, until the development of the optical fibre cable within the last decade, cables lacked versatility in the facilities which they could provide. The first cables were laid for telegraphic communications and cannot be adapted to other purposes. The later cables, although capable of providing voice links, operated on a restricted waveband which was insufficient to carry television signals and could not be expanded for that purpose.

All this does not imply, however, that cable technology has not been developed very far from its early state. Refined multiplexing techniques permit a great volume of traffic to use a cable 'simultaneously' without apparent loss of quality, and channel splitting has allowed many data links to use one telephone channel. Nonetheless, until the very recent invention of the optical fibre, the cable system increasingly could not cope with the demands that were being put on it.

Radio communication helped solve some of these problems, and aided greatly in meeting demand for services, but it was not a complete solution. Over short distances, very high frequencies could carry many channels providing the clarity and sophistication of signal required for complex messaging, including that needed by television and analogue computer traffic and the like. Microwave relay stations are the obvious manifestation of these abilities. However, the very high frequencies are not reflected back to earth by the ionosphere. VHF and UHF operate in 'line of sight'. Terrestrial radio could therefore help with long-distance telecommunication only as far as the end of the short-wave bands. Further, all radio bands are subject to interference by natural and artificial sources. Weather and sunspots, other radio stations and unsuppressed electrical equipment can spoil radiocommunications. While the ITV provides a mechanism through which man-made interference is lessened and to a degree coped with, it has no jurisdiction over natural interference, in particular that caused by solar flares which can wholly blank out shortwave bands.

In the 1950s then, long-distance communications had to be booked in advance and, when used, were frequently less than acceptable. Radio was not the solution that had been looked for, and cable technology was limited in what it could offer. Nonetheless cable did (and does) provide a suitable means for much telecommunication. One might have thought, therefore, that new higher-capacity cables were the solution, albeit that they did not provide a long-distance means of television transmission to a broadcasting point. That solution was, in fact, being pursued, a major British Commonwealth cable being opened in December 1963. But political difficulties in deciding routes, together with cost (about £3,300 per mile at 1960 prices) did not make new cables an entirely attractive proposition. On the other hand, the world demand for telecommunications facilities was being projected as rising some 15 per cent per year on major routes.[13]

Such were the problems. However, the advent of the satellite changed things drastically, providing a new means of telecommunication, new services and new opportunities. The result has been alterations also in demand patterns and volume. Importantly also, it has allowed each state to become slightly more independent in its use of international communications.

The advent of satellites

In 1945 in a then obscure, but now famous, article 'Extra-terrestrial Relays. Can Rocket Stations Give World-Wide Radio Coverage?',[14] Arthur C. Clarke noted the potential of the geostationary satellite as a solution to the problems of relaying radio signals for long-distance communication needs. That article has given Clarke an honoured place in the history of applied science.[15] Unfortunately, as he has pointed out, he did not patent the idea.[16]

Having been interested in space matters even before the Second World War, Clarke saw that in 1945 the state of knowledge of rocketry, particularly that gained through the German V2 programme, meant that it would soon be possible to launch three synchronous satellites, which, if stationed approximately over Tokyo, Cairo and New Orleans would give effective radio coverage over most of the Earth. He repeated this idea, without development, in 1951 in his book *The Exploration of Space*.[17] In both cases Clarke envisaged the use of manned space stations and, as transistors were unknown when he wrote, foresaw rather bulky equipment.[18] However, not long afterwards transistors were invented and developed, John Bardeen and Walter Battain of Bell Laboratories making the initial invention in December 1947, William Shockley creating the junction transistor in 1949 and G.K. Teal developing the germanium and the silicon transistor.[19] This (and the later echo-cancelling chip) made possible the satellite as we know it. As we now know, technology progressed rapidly, and the geostationary communications satellites of today are performing precisely the function foreseen by Clarke.

The impetus towards the development of space satellites had certainly a scientific element, but it is interesting to find that the provision of new telecommunications facilities very soon became a purpose of equal status with that of scientific inquiry. Whether or not deliberately proposed to influence budget decisions, the practical application of space science, and in particular the provision of telecommunications facilities, definitely made the scientists' proposals more attractive to their governmental and commercial paymasters. Be that as it may, the origin of the satellite programmes lies with the scientific community and the International Geophysical Year, 1957, which itself had historical roots.

In 1882–83, and again in 1932-33, the world's scientists cooperated in what were known respectively as the 'First' and the 'Second Polar Year'. In both Years much information and understanding was gained by studies which were planned and coordinated on a world-wide basis. In 1950, however, a Washington conference of scientists suggested that a fifty-year gap between such events was too long, that a further 'Polar Year' be held in 1957-58, and that thereafter such cooperative efforts should recur at intervals of twenty-five years. In 1952 the International Council of Scientific Unions, having considered this proposal, decided that the confining of the cooperative effort to only the polar regions was undesirable. It felt that, instead, the whole world should be the object of concerted scientific effort. The International Geophysical Year of 1957 (IGY) was therefore decided upon.[20] The IGY was, of course, to have significant polar aspects. Much exploration of both polar regions, and the 'Trans-Antarctica Expedition', did occur during it, and scientific bases, most of which are still operational, were established by many countries.[21] For our purposes, it was the decision to use space satellites during the IGY which is of most importance.

In August 1953, Dr S.F. Singer of the USA proposed, to the Fourth Congress of the International Astronautical Federation meeting in Zurich, that a small satellite be launched to orbit the earth for scientific purposes. Other proposals followed, and the matter was discussed by rocket specialists eventuating in the establishment in August 1954 of Project Orbiter as a joint US Army and Navy venture. Meanwhile Singer repeated his proposal to a meeting of the International Scientific Radio Union (URSI) at The Hague which endorsed his ideas and passed them to the IGY coordinating body, the Comité Special de l'Année Geophysique Internationale (CSAGI). At its Rome meeting in October 1954, CSAGI recommended that the matter be pursued.[22]

In retrospect it is clear that both the USA and the USSR took up the matter, the USA announcing its plans in 1955, and keeping the CSAGI informed as to progress. Despite prompting by the CSAGI and others, the Soviet scientists were more coy about revealing their intentions, other than stating in 1956 that they too would have a satellite programme. The Russians also participated in a discussions as to the establishment of data centres, and on other arrangements for the observing of satellite and the interchange of information on rockets, satellites and the results obtained.[23] However, Soviet intentions became plain when Sputnik I was launched on 4 October 1957, towards the end of a meeting of a special CSAGI working group which was preparing a Rocket and Satellite Manual to facilitate the interchange of information. The President of the CSAGI, who was at the meeting, commented on the success of the IGY satellite programme. He congratulated the Russians present, noting that all had felt, when the Russians had made their intimation of a programme in 1956, that that was a momentous statement. He did not forebear also to note that the

1956 statement had been 'marked by a characteristic reserve'.[24] Other launches followed. Sputnik II with the dog Laika on board, was launched on 3 November 1957. The USA followed with Explorer I (which discovered the Van Allen radiation belts) on 31 January 1958, Vanguard I on 17 March 1958, and (following a failure) Explorer III on 26 March. Sputnik III went up on 13 May 1958, and on 2 January 1959, Lunik I, the first successful deep-space probe (rather than an earth satellite) was launched by the USSR.

We need not follow through the general history of satellites thereafter, but must now turn to communications matters. At one level, of course, all active satellite experiments require radiocommunications for command, telemetry and control, matters which require spectrum space and planning as much as do the relay elements aboard. But we can take these matters for granted until we turn to the functions of the ITU. The actual development of the communications satellite, however, is of importance for us here. Such satellites began to be launched quite soon after the first scientific probes, an indication of the potential that was perceived in satellite communications.

The first broadcast from space, other than the simple 'bleep' of the early satellites and the return of scientific data from experiments, was by the SCORE satellite, (the Signal Communications Orbital Relay Experiment), launched by the US Air Force and the Convair Corporation on 18 December 1958, the Atlas missile taking it into an orbit ranging between 110 and 920 miles. This primitive experiment was designed to demonstrate the feasibility of voice and teletype transmission and reception at high frequencies between the satellite and ground stations. SCORE operated as a delayed repeater, recording information sent to it from one ground station and relaying it on command to another, although it was also capable of serving as a real-time communication link. Using SCORE, a Christmas message from President Eisenhower was broadcast to the world on 19 December 1958, the first time a voice message was received from outer space. Other experiments were carried out until 31 December, and the satellite decayed on 21 January 1959.

It was some eight months before other communications experiments followed. Courier, another delayed repeater and real-time link, worked for two and a half weeks in October 1960 and was designed to handle communications successively in its orbit between four ground stations at up to 360,000 words per station. In Project Lofti the US Navy demonstrated the use of low-frequency signals using a low orbit satellite configuration. However, the commercial future lay with Telstar, Relay and Syncom.

The Telstar series of communications satellites was an important development, both technically and commercially. Its commercial importance will be dealt with in another chapter: suffice it here to note that the Telstar series was conceived, procured, launched and operated by a private commercial company, the American Telephone and Telegraph Company (AT&T). Technically, the series was a significant advance. Telstar I was launched into a 593/3,503 mile elliptical orbit on 10 July 1962, and, with a lapse of five weeks after four months, was operational until late February 1963. This was not, however, continuous operation, for the satellite had a limited power supply and was usable only for some 10 per cent of the time. Nonetheless, successful tests were conducted of a variety of communications media including telephone, telegraph, and data transmission at high and low speeds, as well as both black and white and colour television. This confirmed the potential of satellites for use for a variety of telecommunications modes.[25] Telstar II was heavier than Telstar I by some five pounds, the electronic apparatus within it having been given greater shielding against radiation. Launched on 7 May 1963, it was placed in a higher orbit than its predecessor, and with its apogee some 3,000 miles farther out in space.

Chronologically by launch date, though not by planning, the Relay series of satellites (a NASA project), come after Telstar, the first Relay being launched after the first Telstar. Relay I weighing 172 lbs was launched on 13 December 1962 into an 819/4,612 miles orbit, and was in part devoted to experiments other than telecommunications. Relay II, launched on 1 January 1964 had a perigee of about 1,298 miles.

Implicit in the figures quoted for the orbits of the Telstar and the Relay satellites is the argument for the synchronous satellite, which Clarke had clearly stated in his 1945 article.[26] Simply put, it is that the higher its orbit the more widespread the ground stations which can simultaneously use a satellite, thus permitting real-time communication between them. The corresponding disadvantages of the low-orbit satellite are many. First, the satellite has to be tracked by a ground station as it appears over the horizon and crosses the sky to its 'setting'. Second, a satisfactory global satellite system requires the permanent interconnection of its ground stations. There must therefore always be at least one satellite mutually visible between successive ground stations to allow signals to be sent round the world. In a low orbital system that would mean that any one ground station would always have to be tracking at least two satellites, one 'east' and one 'west'. Third, such a global system would therefore require a minimum of six satellites in service to provide continuity of visibility for all the ground stations, and the lower the satellite orbits, the more satellites would be required. Fourth, the satellites would have to be constantly tracked by movable dish aerials. At least three dish aerials would therefore be required, one tracking the 'east' satellite, one tracking the 'west' satellite, and one preparing to acquire the next satellite in the orbital chain as it rises above the horizon. Indeed, some would consider that a secure service would require each ground station to have a fourth aerial dish ready and waiting to act as a 'spare' in case of problems among the operational three. There would also be problems of the technological sophistication (particularly at the ground station). Finally, in all the above elements, the low-orbit system will run into greater problems of cost than the synchronous system.

Provided that you can get the satellites into the correct orbit Clarke's proposed synchronous satellites eliminate many of these problems. A satellite in eastbound equatorial orbit at a height of some 22,300 miles (35,000 km) will remain virtually stationary with respect to points on the surface of the Earth, and will be visible between latitudes 81.5 North and South which cover almost all the inhabited parts of the globe. Three synchronous satellites placed at a mutual separation of 120° longitude would provide global coverage and permit the operation of a global satellite communications network. One synchronous satellite placed appropriately over the Atlantic could provide telecommunications for an area within which, in the early 1960s, some 90 per cent of the world's telephones were to be found. Such a satellite was seen as able to supplement, if not replace, the complex of cables which then carried much of the international traffic in that area, particularly on the transatlantic route.

The first attempt to place a satellite into a synchronous orbit, Syncom I, was made on 14 February 1963. The placing seems to have been attained, but at the final 'nudge' into position, all telemetry from the sattelite ceased. Syncom II was launched on 26 July 1963 into a synchronous but non-geostationary orbit, inclined at some 33° to the Equator, with the result that the sattelite appeared to describe a figure of eight in the sky as the Earth rotated under it. Its main purpose however, was, to be used as a test-bed for acquiring expertise in placing and moving satellites in the geostationary orbit. It was a success in that as well as in its telecommunications aspect.

Syncom III was launched on 19 August 1964, and was 'walked' into a geostationary orbit roughly over the International Date Line the following day. Its equatorial inclination was less than one half of a degree, a very creditable achievement. The satellite was used for training ground crew and to test new equipment both on the ground and in space. At the request of the US State Department, it was also used by the then newly created Communications Satellite Corporation to relay live television transmission from the Tokyo Olympic Games to the USA, although Clarke notes that, ironically, most networks did not want to upset their previously arranged schedules and hence did not take the historic transmissions.[27]

Thereafter technical progress continued. This can be best documented through the developments of the spacecraft used by

INTELSAT, the major provider of international telecommunications by satellite.[28]

The first in the INTELSAT series, Intelsat I, more familiarly known as 'Early Bird' was launched on 6 April 1965. By later standards it was a small piece of equipment. Without its antennae deployed it was just over 2 ft (0.7m) wide and just under 2 ft (0.6m) high. Its capacity was 240 voice circuits or one television channel and it was incapable of serving more than one pair of stations at a time. Positioned over the Atlantic, it was serviceable for more than double its design life of 18 months and, apart from its use for telecommunications traffic, it provided much data useful in the design and development of its successors. Most importantly its successful insertion into orbit, stationkeeping and functioning proved that a global satellite system based on geostationary satellite links was technically feasible.

The first Intelsat II launch in October 1966 failed to attain geostationary orbit, although the satellite, 'Lani Bird', provided a service between Hawaii and the US mainland for a period. The second launch in the series in January 1967 was successful. The same height as its predecessor the Intelsat II was twice its width and resembled a squat cylinder. Like its predecessor it also provided 240 voice channels or one TV channel, but had one significant improvement. Many groundstations could use it simultaneously since it was capable of multipoint access as have been all subsequent developments.

The Intelsat III series, begun in 1968, was a set of satellites each of which could accommodate 1,500 circuits or four TV channels, or combinations of these providing TV capacity without interruption of telephone signal. It also could handle any of, or any combination of, telephone, telegraph, television, high-speed data and facsimile services. The same width as its immediate predecessors, the series found space for the necessary hardware inside a height of just over 3 ft (1m).

The next series began in 1971. Intelsat IV and its modification, IVA, was extremely successful. The 12 transponders of an Intelsat IV satellite provided a capacity of 4,000 voice circuits with a two TV channel ability. The satellite itself was almost 8 ft wide when undeployed and just over 17 ft high. Intelsat IV-A satellites, which began to be launched in 1975, were the same width but 22 ft high. Each had 20 transponders providing 6,000 circuits and two TV channels. The Intelsat V series, begun in 1980, provides 12,000 circuits plus two TV channels and achieves its higher capacity by signal polarization and spot-beaming. In addition it carries L-band packages providing maritime communications (taken up by INMARSAT). Its modification, Intelsat V-A, first launched in 1985 has an average capacity of 15,000 two-way voice channels plus two TV channels. Its technical capabilities include modifications to serve the needs of newly introduced INTELSAT business services, and domestic leased services. Even so, the Intelsat V and VA series are slightly smaller than their predecessor, being only six feet six inches (2m) wide and 20 ft 6 ins (6.4m) high. This, in part, is because of the introduction of nickel hydrogen batteries.

For the future, Intelsat VI, first scheduled for launch in 1987, though subject to delay through problems with both the Ariane launch service and the NASA shuttle, is a much larger satellite. The satellite will be 11 ft 6 ins wide (3.6m) and 17 ft (5.3m) high. It will have an average capacity of 30,000 two-way voice circuits and three TV channels. Directional beaming will permit the more efficient use of radio frequencies, and the provision of services in small ground areas. INTELSAT VII, which is proceeding into its contract stage, will be even more advanced.

Two things remain to be said. First, the design lifetimes of international communications satellites have risen from 18 months in the case of Intelsat I, through three years for Intelsat II and five years for Intelsat III to the seven years for each of Intelsats IV to V-A. The Intelsat VI series has a design life of 10 years. However, in each case, the design life has been exceeded, which has had significant economic benefits for INTELSAT. It has permitted the actual substitution of satellites to be taken more slowly than was expected back in the early days of INTELSAT, and has therefore allowed successive satellites to accommodate the technical progress made through several years' research. Some early problems have therefore been 'jumped' by emergent technology. Second, there has been a significant advance in the use made of the radio frequency spectrum. Intelsat I used 50 MHz; Intelsat VI will use 3,300 MHz. Frequency usage has also been made more efficient through polarization of the signals, through the design of antenna including steerable antennae, and through spot-beaming. New wavelengths are also in use. 6/4 GHz and 14/11 GHz bands are now commonly used, providing quality of signal with decreased interference problems. However, that is not to say that either the problems of frequency assignment and usage or those of orbital locations have been solved.

Such have been the developments within the INTELSAT family of satellites. Similar progress has been made by the other organizations which later entered the field of satellite telecommunications. Many countries now have their own domestic satellite telecommunications services, some wholly reliant on INTELSAT. The USA is, naturally, best served by a sophisticated and efficient set of satellites designed and built by various supplier companies and operated by a number of different communications providers.

In short, technical developments have made synchronous communications satellites reliable vehicles for an efficient and economic telecommunications service. They therefore provide a good answer to many of the problems of international telecommunications, to say nothing of domestic needs where traditional telecommunications links are difficult. Of course, satellites are not necessarily a complete answer to all telecommunications problems. In appropriate cases, terrestrial microwave links can be very economic. Furthermore, within the last few years significant developments have been made in glass technology. Fibre optic cables can now provide a signal bandwidth which can accommodate television signals, the lack of which was a major deficiency of the wire cable. But even if there is a spectacular resurgence in cables, satellites are here to stay as a major telecommunications medium. Their advantages secure their position.

But, of course, the new technology brought problems, many of them legal. Questions of the allocation of radio frequency for satellite services is a matter which has been given to one of the oldest international organizations, the ITU. Its machineries and procedures have also been found suitable for dealing with many of the problems of the use of orbits and orbital positions. We shall consider such questions later. The other major question is how best to organize the provision of a satellite service. International services require international arrangements. In addition, most countries have found it more convenient and economic to buy a domestic satellite service from an external source rather than establish their own domestic satellite system for their limited needs. So far, the tendency has been for such countries to go to those who provide the international services, buying both international and domestic service from them. That makes more acute the problem of organizing the international system. Should it be a business or a service? How can states gain a voice within the organization which seeks to meet their needs in a businesslike manner? Can states participate in it or are they to be left as customers only? Various solutions have been found to these problems of organization, but all of them lie within a legal context.

THE LEGAL CONTEXT

The development of the technical and organizational facilities by which international telecommunications are carried out did not occur in a legal vacuum. International communications is an area which is governed by rules and, in discussing telecommunications, it is necessary to remember that these rules exist. Regrettably, some seem to think that whatever they wish is legally possible: it is not so. Law and agreement restricts what can be done.

A country's rules of law are for that country to decide, except to the extent that it has restricted its freedom under international law and to the extent that some acts (for example, genocide) are never lawful.

Similarly, how a country organizes its telecommunications is a matter for its decision. Logically, telecommunications could be organized as a monopoly owned and operated by an individual, or as a private enterprise, as a state enterprise, or as a hybrid between state and private enterprise and with or without competition. Different countries have made different choices as to their internal telecommunications arrangements. We cannot pay detailed attention to any of them, with the exception of the USA. Because of its domination of the satellite business and the size of its market share, how the USA has organized telecommunications and the decisions it makes on national and international services are of acute concern to international organizations formed as an orderly response to emergent needs. Accordingly Chapter 2 deals with the Communications Satellite Corporation (COMSAT) and in measure with the US Federal Communications Commission (FCC). The FCC is also relevant in the discussion of INTELSAT. Apart from that specially justified treatment, domestic telecommunications arrangements cannot be allocated word space here, even though domestic arrangements do impinge on international matters.

The law which governs the arrangements for international telecommunications arrangements is then a matter for the municipal law of that country.

The law which governs the arrangements for international telecommunications is a different matter. Logically, and broadly speaking, there are four aspects to this law: contract, general international law, space law, and the law and practice of international telecommunications.

Contract

First, there is a strictly contractual element. Some aspects of the international arrangements for telecommunications are governed by the terms which are agreed between the participants in an international telecommunications network. Telecommunications entities can enter into a contract which they agree is to be construed according to the law of contract of a particular legal system. Thus private companies in the USA and the UK contract as to the establishment and operation of a transatlantic cable link.

Of course, it is never as simple as that. No telecommunications entity has entire freedom to contract. International telecommunications is too important a matter for a state to let a national company have a free and unfettered hand in international agreements. Governmental concern is manifested either through some formal body such as the US Federal Communications Commission, or through some other more direct governmental intervention. In any event, no Foreign Office or State Department is without a section having some responsibility for the communications field. But even with that governmental element, there is nonetheless a strong contractual element in the working of such bodies as INTELSAT and INMARSAT. This is particularly found in the operating agreements of both organizations, which are agreements between telecommunications entities each nominated for its interest by a Party to the intergovernmental agreement.

General international law

At the opposite pole from ordinary contract, international communications involve, by nature, relationships between states and hence international law.

International law contains the rules and norms regulating the relationships and conduct of states and any other entities which are recognized as subjects of international law by the other subjects of international law. Non-state subjects of international law are bodies such as the international organizations whose constituent documents agreed by states either expressly or impliedly give them 'personality' in international law. Most of the international telecommunications organizations start as interim associations of states and telecommunications operators and, as such, are without legal personality. Normally the permanent arrangements give legal personality, and hence the new organizations are made subjects of international law. The advantage of this is that they thereby become entitled to hold rights and exercise duties, and can therefore hold property, contract in their own names and so on. The United National is the best known of such bodies, although the other international organizations with which we will deal are other examples.

Unlike other legal systems, international law has no legislature. The United Nations has its General Assembly but that is a body which debates, discusses and adopts resolutions. It does not decree law. There is no body which makes international law in the way that parliaments, congresses and councils make municipal laws. There is an International Court with its seat at The Hague which takes decisions on cases brought by states and other recognized international persons (see below), but it is hampered. Unlike a municipal court, the International Court's jurisdiction is not compulsory but depends on a state's willingness to make itself subject to its jurisdiction, either generally or subject to conditions and restrictions. A state can agree to submit itself to the jurisdiction of the International Court for one particular case only. When deciding a case the International Court applies international law, unless the parties to the dispute agree that the Court may come to a decision *ex aequo et bono*. In that event the Court gives a decision to its perception of what is the right decision in the matter, irrespective of what the law might indicate. (As yet no case has been formally decided on this basis.)

What, then, is the source of these rules and norms referred to as being contained in international law?

Article 38 of the Statute of the International Court of Justice is often looked to as a statement of the sources of international law.[29] It lists the legal material which the Court is to apply in deciding a case as:

1. international conventions, whether general or particular, establishing rules expressly recognized by the contesting states;
2. international custom as evidence of a general practice accepted as law;
3. the general principles of law recognized by civilized nations;[30]
4. judicial decisions [though the International Court is not bound by precedent] and the teachings of the most highly qualified publicists [jurists and commentators] of the various nations as subsidiary means for the determination of rules of law.

From these sources, manifested through treaty and formal statements, through cases and state practice, there has grown a body of law regulating the relationships between states which we call international law.

The list in article 38 sets treaties as first in the hierarchy of sources of law. In the arena of telecommunications, treaties and agreements are very important: they form a body of rules to which states have formally agreed. The Law of Treaties – that is, the rules by which treaties are entered into and obeyed – is undergirded by principles of customary international law. These have themselves been reduced to formal statement in an international treaty, the Vienna Convention on the Law of Treaties, 1969.[31] The whole of the Law of Treaties is of relevance for us since it prescribes how a treaty is entered into, the right to make reservations as to parts of an agreement (which affect their applicability to the state making the reservation), the entry into force of a treaty, the canons of interpretation and so on.[32]

However, in the light of developments within the USA regarding international satellite telecommunications, two particular elements should be highlighted. The first is the principle *pacta sunt servanda*, which means that obligations undertaken by a state are to be fulfilled or performed with the utmost good faith.[33] The second is that a treaty is binding until it is lawfully abrogated (or denounced by a party which wishes to leave the treaty), or until there is a fundamental change in the circumstances with which the treaty was designed to deal.[34] This last is expressed in the Latin tag that the treaty binds *rebus sic stantibus,* but that tag has absolute referents, and its standard is not met simply by a change of mind on the part of a party which now perceives its interest to lie in a different direction. These elements of the Law of Treaties, *pacta sunt* and *rebus sic,* do not sit squarely with the USA's willingness to license systems separate from the INTELSAT system for the provision of international satellite telecommunications services from the USA.[35] The perception of international law seems to differ in the USA from that found elsewhere, a difference which appears to owe much to the attitude of Americans to their internal or municipal law. For too many US business lawyers, anything regulatory is something which can be changed, overcome or circumvented as necessary and without any respect for its intrinsic status as law. International law cannot be approached in that manner. In fairness, international law is not so thought of by the better lawyers and by many of those trained in that field, but the problem is that too many of those active in telecommunications seem not to know about international law, and approach it as merely US law applicable to international transactions. They use law to suit their interests, and seek to circumvent (or subvert) it as they consider their interests require. It is appealed to or not depending upon the desired result. This is worrying. International agreements are treaties, not negotiating stances.

The other major element of international law is custom. Custom is a practice engaged in by states in the belief that it is legally required. Apart from the rules it contains about treaties, an element of Customary International Law important for this book is the principle that a state is a sovereign entity and, as such, is not subject to constraint by other states. That principle is the basis of a state's right to regulate its own telecommunications. Another element is the principle (loosely expressed) that a state must not cause harm to another state. The concept of 'harmful interference' is intrinsic to the regulation of the use of the radio frequency spectrum through the agency of the International Telecommunication Union (ITU).

Space law

Third, there is a body of international law relating to the use of space, which interacts with, and to a degree now forms the context within which, the law relating to international telecommunications by satellite must live. What is now called 'space law' sets parameters within which telecommunications has to operate.

Despite the caveat above about its lack of legislative powers many of the major international treaties and agreements (as well as certain influential principles) which govern Space Law have been developed through the United Nations.[36] As indicated above, the International Geophysical Year of 1957 both showed the potential of space and ensured that developments in space technology would continue. Although there were earlier indications of a legal interest elsewhere, the IGY attracted attention within the United Nations to the need for a legal framework for the exploration and use of space. Within six weeks of the launch of Sputnik 1 the UN adopted GA Res. 1148 (XII) (14 November 1967), the first of a long sequence of space-oriented Resolutions. This Resolution, directed towards questions of disarmament, mentioned that studies should be made to ensure that objects sent into outer space should be exclusively for peaceful and scientific purposes. In the train of moves by both the USSR and the USA, the USA proposed in 1958 the establishment of an Ad Hoc Committee intended to prepare a programme of international cooperation in the peaceful uses of outer space. GA Res. 1348 (XIII) of 13 December 1958, stressed the need for international cooperation in space and that outer space should be used for peaceful purposes only. More importantly for this book, the Resolution went on to establish an Ad Hoc Committee on the Peaceful Uses of Outer Space. In 1959 by GA Res. No. 1472 (XIV) this became a permanent Committee of the General Assembly, Assembly Committee 105,[37] the Committee on the Peaceful Uses of Outer Space (COPUOS).

It is unnecessary here to review in detail the whole work of the Committee. Suffice it to say that it has done valuable work, particularly in its early years. As originally constituted it had a membership of 18, before being constituted as a permanent committee with a membership of 24. Over the years that number has increased: 28 in 1961, 37 in 1973, 47 in 1977, 48 in 1978 (with the addition of China) and 52 in 1980. Such an increase is to be regretted. States do want to be involved at the initial stages of formulating policy and law, particularly because many of the newer nations are convinced that present international law favours the older states. In addition, the expansion of COPUOS has perils. Conversation with a variety of persons who have served on COPUOS, and reading the Proces Verbaux and Summary Record, show the problem developing. In part, it is a matter of group dynamics. The smaller body was more cohesive: the participants knew each other well, respected each other's competence and could discuss informally together. Partly because certain matters were easier to agree, progress was considerable. But even conceding that point, the Committee is too big. Too many members have little knowledge of either the technical or the legal elements involved. Some seem incompetent to deal with the remit of COPUOS. Too many representatives appear only for one year, make political speeches for an audience other than that in front of them, and then depart for other more prestigious or lucrative appointments. As this has run in step with the greater difficulties of the areas under the Committee's consideration, it has had an adverse effect on the work of both the Committee and the General Assembly. Until recently, COPUOS worked on the basis of consensus, and the Committee's personnel problems resulted in COPUOS making little progress on many matters in the 1980s. A break with that basis in 1985 has merely showed the wisdom of the earlier principle.[38]

That said (and it had to be),[39] COPUOS has been responsible for many important developments in space law, proposing first, principles and then treaties to the community of nations through the agency of the General Assembly. Of these, the major treaties are: the Outer Space Treaty, 1967; the Agreement on the Rescue and Return of Astronauts, and of Objects Launched into Outer Space, 1968; the Convention on International Liability for Damage Caused by Space Objects, 1972; the Convention on Registration of Objects Launched into Outer Space, 1975; and the Moon Treaty, 1979. Obviously not all of these have major relevance for this book, but together they provide a matrix of law relating to space and activities in space within which the law relating to space telecommunications finds its context.

The 1967 Treaty on Principles Governing the Activities of States in the Exploration and Use of Outer Space Including the Moon and Other Celestial Bodies built on and expanded the Declaration of Legal Principles Governing the Activities of States in the Exploration and Use of Outer Space drafted by COPUOS which was unanimously adopted by the General Assembly in 1963 (GA Res. 1962 (XVIII)). That Declaration itself had roots in such documents as GA Res. 1721 (XVI) of 20 December 1961 on 'International Cooperation in the Peaceful Uses of Outer Space' and the similarly titled GA Res. 1802 (XVII) of 19 December 1962. In particular, GA Res. 1721 (XVI) specifically commended two principles to states for their guidance in the exploration and use of outer space: first, that international law, including the UN Charter applies to outer space and celestial bodes; and, second, that outer space and celestial bodes are free for exploration and use by all states in conformity with international law and are not subject to national appropriation.[40] This was agreed in the belief that 'the exploration and use of outer space should be only for the betterment of mankind and to the benefit of states irrespective of the stage of their economic or scientific development'.[41] In relation to telecommunications, the Assembly went on to state its belief that 'communication by means of satellites should be available to the nations of the world as soon as practicable on a global and nondiscriminatory basis'[42] – words which influenced, first, the US Communications Satellite Act of 1962 and then, as we shall see, the principles of INTELSAT and form a substratum for its operations.

The next major step in the development of space law was adoption of the Declaration of Legal Principles Governing the Activities of States in the Exploration and Use of Outer Space drafted by COPUOS and unanimously adopted by the General Assembly in 1963 (GA Res. 1962 (XVIII)). There is argument as to whether a UN declaration unanimously adopted forms international law, but there is strong evidence that it does form 'instant custom' because it constitutes a formal declaration of a state as to its views on the matters covered. However, irrespective of such niceties, the main elements of the 1963 Declaration appear in true treaty form in the 1967 Treaty on Principles Governing the Activities of States in the Exploration and Use of Outer Space Including the Moon and Other Celestial Bodies.[43] Certain of its provisions must be quoted:

Article I

The exploration and use of outer space, including the moon and other celestial bodies, shall be carried out for the benefit and in the interests of all countries, irrespective of their degree of economic or scientific development, and shall be the province of all mankind.

Outer space, including the moon and other celestial bodies shall be free for exploration and use by all States without discrimination of any kind, on a basis of equality and in accordance with international law and these shall be free access to all areas of celestial bodies.

Article II

Outer space, including the moon and other celestial bodies, is not subject to national appropriation by claims of sovereignty, by means of use or occupation, or by any other means.

Article III

States Parties to the Treaty shall carry on activities in the exploration and use of outer space, including the moon and other celestial bodies, in accordance with international law, including the Charter of the United Nations, in the interest of maintaining international peace and security and promoting international cooperation and understanding.

Article IV goes on to prohibit the placing of nuclear or other weapons in space and to proscribe the establishment of military bases on celestial bodies. Astronauts are the envoys of mankind and are to be assisted (Art. V), and objects found are to be returned to their home states (Art. VIII).[44] States are internationally liable for national activities in outer space, and are required to establish that activities carried out by agencies under their jurisdiction conform with the Treaty (Art. VI).[45] States are liable for damage caused by objects they launch into space (ART. VII).[46] States must maintain a system of registration of objects which they so launch and retain jurisdiction over them (Art. VIII).[47] Article IX stipulates for international cooperation, the avoidance of contamination and the avoidance of harmful interference with the activites of other states – this last being capable of interpretation for radio purposes. Finally, of major contextual relevance to this book, Article XIII clearly contemplates that not only may a state use and explore outer space so may also a group of states, or an international organization. The same international law applies to all such endeavours.

So, in summary, the development of international telecommunications by satellites does not, and has not, taken place in a legal vacuum. There is now a considerable body of space law which affects what is done. It may also be noted that some of that law existed in a more general, often inchoate but sometimes in specific, form before the United Nations got to work. The question of harmful interference, for example, was long a matter of concern in radio and was coped with through the ITU. What is notable, however, is that through the United Nations there have been strong, and partly successful, attempts to influence and constrain the freedom of the technologically advanced nations. It cannot now be thought that 'those who can, do' to the exclusion of others. That space is to be used for the benefit of all is a principle which can be drawn into discussions as to future developments. That space is part of the 'common heritage of mankind' is another formulation of the same point, highlighted in, for example, the deliberations of the Second United Nations Conference on the Peaceful Uses of Outer Space held in Vienna in 1982.[48] The background papers and official documents of that conference as well as such as those of the Nairobi Conference of the International Telecommunication Union, held also in autumn 1982,[49] show that there has been a change in attitude and in international expectation. Various international agreements as well as the more aspirational formulations of the UN now state that some of the resources of the Earth and near space form part of the common heritage, of mankind with a correlative that the exploration and use of such must be carried out for the benefit of all mankind.[50]

Such notions put a constraint upon states' activities and interact with the other purposes and interests in furtherance of which countries associate together to in such bodies as INTELSAT and INMARSAT and show up in the statements of aspiration in the Preambles to their constituent documents. Indeed, matters have now largely passed to these other institutions. Matters progress too rapidly to fit the UN timetable, and the difficulties of the expanded COPUOS have also affected the UN's authority and leadership in all space matters, not just in telecommunications questions. To many working in the telecommunications field, to write as I have just done of 'the authority and leadership of the UN' is at best idealistic, certainly gravely mistaken, and at worst quite out of touch with reality. Telecommunications policy and law lies elsewhere.

The law of space telecommunications

The last part of law governing international telecommunications is the international law which has grown up specifically to deal with matters of telecommunications. The content of this aspect forms the bulk of this book. The ITU has particular roles in the technical matters of radio frequency allocation, use and registration, orbital positions and rate setting, all of which we will deal with separately. Then there are the organizations and entities which exist to provide an international service, such as INTELSAT, INMARSAT, EUTELSAT and INTERSPUTNIK and the Arab Satellite Organization (ARABSAT) which all have legal personality in international law. The law which relates to the working of these organizations is partly a matter of agreement expressed in the constituent documents of these bodies. The decisions of their governing bodies and decision-making processes are law within that organization for the purposes of that organization. They are also, however, governed by general principles of international law relating to the interpretation of, and compliance with, the terms of international agreements.

Certainly, as stated at the end of the last section, telecommunications policy and law lies now outside the UN. But these matters are still governed by law. They do not lie outside either national or international law. Telecommunications requires a legal basis in order to function, in order that agreements are respected and complied with, in order that the international organizations which operate the satellites facilities exist and work, in order that radio frequencies are properly used for agreed purposes and that orbital positions are secure. To an extent, this can be characterized as mutual cooperation: the interesting thing is that it works as well as it does, and it does so by being not merely cooperation, but by being based upon reliable law. Only in that way is there certainty, and therefore safety, in the investment of the huge sums of money which modern international satellite telecommunications require.

NOTES

1   Sc. Greek : 'tele', = at a distance + communications.
2   See, for example, Captain Th. Lynn, *An Improved System of Telegraphic Communication*, 2 vols. (London, 1818). Captain Frederick Marryat, the novelist, was responsible for a major revision of the British Navy's flag codes prior to his

retirement in 1830; see his tomb at Langham, Norfolk, UK, which summarizes his code.

3   G.A. Codding, *The International Telecommunication Union: An Experiment in International Cooperation* (Leiden: E.J. Brill, 1952) p. 2.

4   Cf. the use made of this deficiency in Dumas's *The Count of Monte Cristo.*

5   The history of the telegraph in the USA was different, the requirements of business coming into play at an earlier stage.

6   Codding, op. cit., p 9.

7   See S.A. Garnham and R.L. Hadfield, *The Submarine Cable,* (London: Sampson Low, Marston and Co., 1934); Willoughby Smith, *The Rise and Extension of Submarine Telegraphy.* (London: J.S. Virtue and Co., 1919).

8   The Telegraph Acts, 1868–69.

9   The Telephone Transfer Act, 1911. Curiously the telephone system in Hull remained in the hands of a separate company.

10 E. Barnouw, 'A Tower in Babel', in *A History of Broadcasting in the US* (New York: Oxford UP, 1966). (See also his *The Golden Web: 1933 to 1953* (1968) and *The Image Empire: from 1953* (1970) for later developments.) Also, A. Briggs, *The History of Broadcasting in the U.K.,* 4 vols., (London: Oxford UP, comprising *The Birth of Broadcasting* (1961); *The Golden Age of Wireless* (1965); *The War of Words* (1970); *Sound and Vision* (1979).

11 See S.A. Garnham and R.L. Hadfield, op. cit., ch. XIV, 'Cable and Cable Ships at War'.

12 See ibid., ch. XVI, 'The Cable Systems of the World'; cf. also the regularly published ITU maps of the present cable networks.

13 R.T. Nichols, 'Submarine Telephone Cables and International Telecommunications', RAND Corporation, Memorandum RM–3472–RC (February 1963); ''International Telecommunications' in the *Prospectus of the Communications Satellite Corporation* at 22 (reprinted III International Legal Materials, 395); W.F. Hilton, 'Communication Satellite Systems Suitable for Commonwealth Telecommunications' in L.J. Carter (ed.), *Communications Satellites* (London: Academic Press, 1961) p. 95 ff.

14 *Wireless World,* October 1945, pp. 303–8.

15 Clarke has gone on to become one of the major writers of science fiction. His scripting of the film *2001* brought his talents to a wider audience and his involvement in the exploitation of the new scientific developments for the benefit of the less developed countries also reflects high credit upon him. He remains a popular figure among those who now are designing and operating systems which, in the 1940s would have been science fiction.

16 A.C. Clarke, 'A Short Pre-History of Comsats, or: How I Lost a Billion Dollars in my Spare Time' in *Voices from the Sky* (London: Gollancz, 1966, and many paperback editions). As he notes, he was unlikely to have gained a patent in 1945, and under US law of that time it would have expired just as COMSAT was established, but the notion intrigues.

17 A.C. Clarke, *The Exploration of Space* (London: Temple Press, 1951; Harmondsworth: Penguin Books, 1958).

18 Cf. George O. Smith, *The Complete Venus Equilateral* (New York: Ballantine, 1976) which collects Smith's stories about interplanetary relay stations. In his Introduction to that edition of Smith's stories which were written in the early 1940s and first published in book form in 1947 (New York: Prime Press), Clarke acknowledges that Smith's stories may have subconsciously influenced him in developing his own revolutionary concept (Introduction, at p. x).

19 J. Millman, *Micro-electronics,* (New York: McGraw-Hill, 1978) pp. xx–xxvii; C. Weiner, 'How the Transistor Emerged', 1973 *IEE Spectrum,* pp. 24–33. In 1956 Bardean, Brattain and Shockley shared the Nobel Prize for Physics.

20 For this, and what follows, see *Documents on International Aspects of the Exploration and Use of Outer Space*, Report prepared for the Committee on Aeronautical and Space Sciences, US Senate. 88th Congress, 1st Session, Doc. no. 18. 9 May 1963. Narrative Text at pp. 2–4 and relative portions of the Documentary Annex; and, *International Cooperation and Organisation for Outer Space*, Staff Report [by Mrs Eilene Galloway, Legislative Reference Service, Library of Congress prepared for the Committee on Aeronautical and Space Sciences, US Senate. 89th Congress, 1st Session, S. Doc. no. 56, 12 August 1965. Part IV 'International Scientific Community', pp. 353–426, and particularly the sections: A. 'International Council of Scientific Unions (ICSU) and the International Geophysical Year (IGY)' at pp. 353–9: B. 'Organization of the Comité Special de l'Année Geophysique Internationalé (CSAGI), pp. 359–64: C. 'US Organisation for the IGY and Earth Satellites', pp. 364–6: and D. 'IGY Rocket and Satellite Program', pp. 366–73.

21 The exploration of the Antarctic was to lead to the Antarctic Treaty of 1959, a significant development in international law: Antarctic Treaty, 402 UNTS 71; (1961) UKTS 97, Cmnd 1535; 12 UST 794, TIAS 4780.

22 'In view of the great importance of observations during extended periods of time of extraterrestrial radiations and geophysical phenomena in the upper atmosphere, and in view of the advanced state of present rocket techniques, the CSAGI recommend that thought be given to the launching of small satellite vehicles, to their scientific instrumentation, and to the new problems associated with satellite experiments, such as power supply, telemetering, and orientation of the vehicle.' *Annals of the International Geophysical Year*, vol. IIA, p. 171 (Pergamon Press, London). Similar resolutions were also adopted by organizations such as the International Union of Geodesy and Geophysics. See *International Cooperation and Organization for Outer Space*, above, at pp. 366–7.

23 *International Cooperation and Organisation for Outer Space*, op. cit., at p. 367–70.

24 ibid., at p. 370.

25 W.J. Bray and F.J.D. Taylor, 'Preliminary Results of the Project TELSTAR Communications Satellite Demonstration and Tests: 10–27 July 1962' *Post Office Engineering Journal*, 55 October 1962, p. 147.

26 Arthur C. Clarke, 'Extra-terrestrial Relays. Can Rocket Stations Give World-Wide Radio Coverage?', *Wireless World*, October 1945, pp. 303–8.

27 A.C. Clarke, 'A Short Pre-History of Comsats . . .', op. cit.

28 A fuller statement with much more technical data is E. Podraczky and J.N. Pelton, 'INTELSAT Satellites' in J.R. Alper and J.N. Pelton, (eds)., *The INTELSAT Global Satellite System*, vol. 93, 'Progress in Astronautics and Aeronautics' (New York: American Institute of Aeronautics and Astronautics, Inc., 1984) pp. 95–133.

29 Cf. C. Parry, *The Sources and Evidences of International Law*, (Manchester: Manchester UP, 1965).

30 Cf. B. Cheng, *General Principles of Law as Applied in International Tribunals*, (London: Stevens, 1953; Cambridge: Grotius Publications, 1988).

31 The Vienna Convention on the Law of Treaties, 1969, (1980) UKTS No. 58, Cmnd. 7964; (1969) 8 ILM 679; (1969) 63 AJIL 875. By Res. 3233 (XXIX) the General Assembly of the UN invited all states to become parties to this treaty – an unusual statement. The USA is not yet a party to the Vienna Convention.

32 Cf. Lord McNair, *The Law of Treaties*, (Cambridge: Cambridge UP, 1961).

33 Cf. Vienna Convention on the Law of Treaties, art. 26.

34 Cf. Vienna Convention on the Law of Treaties, art. 62.

35 The USA is not a party to the Vienna Convention, but the principles indicated are matters of customary international law binding on the USA.

36 See Carl Q. Christol, *The Modern International Law of Outer Space* (New York and Oxford: Pergamon Press, 1982) (excellent bibliographic apparatus in addition to

its text); Nicolas M. Matte, *Aerospace Law* (Toronto: Carswell Company; London: Sweet and Maxwell, 1969).

37 AC. 105 is the code distinguishing its documentation.

38 Cf. E. Galloway, 'Consensus as a Basis for International Space Cooperation', (1978) *Proceedings of the Twentieth Colloquium on the Law of Outer Space*, p. 108.

39 Cf. Christol, op. cit., pp. 17 19.

40 UNGA Res. 1721 (XVI) s. A1.

41 UNGA Res. 1721 (XVI) s. A. Preamble.

42 UNGA Res. 1721 (XVI) s. D.

43 Treaty on Principles Governing the Activities of States in the Exploration and Use of Outer Space Including the Moon and Other Celestial Bodies (1968), 610 UNTS 205; (1968) UKTS 10, Cmnd. 3519; 18 UST 2410, TIAS 6347; 6 ILM 386; 61 AJIL 644.

44 These obligations are expanded by the Agreement on the Rescue of Astronauts, the Return of Astronauts and the Return of Objects Launched into Outer Space, 22 April 1968, 672 UNTS 119; (1969) UKTS 56, Cmnd. 3997; 19 UST 7570, TIAS 6559; 7 ILM 151; (1969) 63 AJIL 382.

45 Cf. the (UK) Outer Space Act 1986, which implements a system for the UK under which the UK can discharge these obligations in respect of companies and persons under its jurisdiction. F. Lyall ''The Outer Space Act 1986' (1987) *Scots Law Times*, pp. 137–40.

46 Expanded by the Convention on International Liability for Damage Caused by Space Objects, 29 March 1972 (1974) UKTS 16, Cmnd. 5551; 24 UST 2389, TIAS 7762; (1971) 10 ILM 965; (1971) 66 AJIL 702.

47 Expanded by the Convention on Registration of Objects Launched into Outer Space, 14 January 1975, (1978) UKTS 70, Cmnd 7271; TIAS 8480; (1975) 14 ILM 43, (1979) 18 ILM 891.

48 See N. Jasentuliyana and R. Chipman (eds), *International Space Programmes and Policies: Proceedings of the Second United Nations Conference on the Exploration and Peaceful Uses of Outer Space (UNISPACE), Vienna, Austria, August 1982*, (Amsterdam, New York, Oxford: North Holland, 1984).

49 See the discussion of the ITU, Chapter 8 below.

50 See Christol, op. cit., pp. 251–4, 277–83, 286–312. See also Carl Q. Christol. 'The Common Heritage of Mankind Provision in the 1979 Agreement Governing the Activities of States on the Moon and other Celestial Bodies', *International Lawyer*, no. 14, 1980, pp. 429–83; B. Cheng, 'The Moon Treaty: Agreement Governing the Activities of States on the Moon and Other Celestial Bodies within the Solar System other than the Earth, December 18, 1979', (1980) 33 *Current Legal Problems*, no. 33, 1980, pp. 213–37, esp. at pp. 220–8. I intend to discuss the 'common heritage' notion in a book on space law.

2 The Communications Satellite Corporation

INTRODUCTION

The most significant national response to the coming of satellite telecommunications is the Communications Satellite Corporation (COMSAT). In its train other companies in the USA and elsewhere have been formed to exploit the new facility, but the Corporation is unique. It is not a general company working with all modes of telecommunications, but is a specialized 'carrier's carrier', given various privileges but also put under remarkable constraints by its enabling legislation and constituent documents.

COMSAT was set up in the USA under the Communications Satellite Act of 1962 to be the US participant in the new international space telecommunications projects. It came into being as the result of certain choices as well as compromises between various interests. As the first organization specifically devoted to matters of communications by satellite, its very existence significantly affected the whole ensuing international developments, notably through its important part in the negotiations leading to the formation and subsequent operations of the International Telecommunications Satellite Organization (INTELSAT) and similarly also through its involvement with the later International Maritime Satellite Organization (INMARSAT).

When space communications began, the ITU was already in existence. One of its major functions was the allocation of radio frequencies and hence it was able to absorb within its duties that of dealing with radio frequencies for the new facilities. But the ITU was not an operational organization, and there was no entity then providing telecommunications services which could easily be adapted to the requirements of the new technology. New skills and expertise had to be gained, and adjustment to the new medium took time. Technology had to be developed, made economic and then applied. The telecommunications entities of the time were skilled only in radio broadcasting and in cable technology.

To understand the present space telecommunications arrangements, we have to cast our minds back to the early 1960s and the
constraints on vision and foresight which then operated, although it
serves no good purpose to look back from the vantage-point of 25
years' development and second-guess what was then decided. Decisions were then taken on the basis of the knowledge, opinion and
prejudice of the time, as well as on such invariant elements as the
wish of states to control their own telecommunications networks and
their interconnection with the networks of other states. Not all the
factors affecting the decisions of the 1960s were well-founded, but
they were effective and go a considerable way towards explaining
modern telecommunications arrangements.[1] Certain of these factors
still play a role in current decision-making.

In a simple logic, the entities which might have entered into the
field of space communications were as various as those which might
have taken part in any business enterprise.[2] An individual might as
well have participated as a state. International law had nothing to say
on the matter, and at that time there were no constraints, not even the
limited constraints of the various Space Treaties. Even now these
constraints are minimal, mainly operating through the Outer Space
Treaty itself, the Registration Convention and the Damages Convention. Broadly speaking, these require states to exercise supervision
over space activities by persons and entities within their jurisdictions,
and permit private and even individual use of space.[3]

But there were other constraints tending to restrict the numbers
and kinds of entities which wished to take part in the development of
space communications. On the one hand there were the matters of
cost and risk. Few companies could muster the capital required for
such an enterprise. Few had the technology. Fewer were willing to
take on the enormous risks which would be attendant upon setting up
the reliable system necessary for commercial operations. The technology was largely untried. Additionally, there was a simple principle
which can be seen at work in all states, that a government wants
to exercise a degree of control over the telecommunications facilities
within its jurisdiction, either loosely through a structure such as the
US Federal Communications Commission, or more directly through
making the provision of communications a governmental responsibility, as in the case of France. Therefore, even where private enterprise indicated a willingness to get involved, states wanted to keep a
check on the new telecommunication facilities.

However, whether through direct involvement or through close
association, there were advantages to having a state involvement in
the development of the new enterprise. First, there were no private
launch facilities. Launching by Western companies could be carried
out only through NASA which was strictly a US government enter-

prise, or through ELDO (the European Launcher Development Organization) which had a bad track record.[4] State involvement made the procurement of launch facilities that much easier. Second, a unified approach to the whole matter of system choice and technical specifications was also likely to be made much easier if states played a controlling role. In hindsight we can observe that a major strength of the INTELSAT system is that it provides a global standard for space telecommunications. INTELSAT is not a unique system, so there is not a single global standard, but at least one exists, and the alternative systems are not equally competitive with it. One may contrast the difficulties which purchasers have had in video-recorders, where it has taken years for competition between rival systems to settle down. Without the state involvement in the 1960s, COMSAT, and hence INTELSAT, might not have come into being. But this book is not in the business of 'alternate worlds'.

Another less often mentioned advantage of state involvement is that state or state-linked enterprises do not have the same difficulty in getting their projects financed as they would had they been required to compete for finance in the open market. As an absolute statement this is, of course, false, for a state does have to pay some attention to the international money market in its determination as to major expenditures, but it is an element to be remembered. As it happened, when COMSAT was floated in 1964 it was a financial success, but its success was principally derived from the 'glamour' of the endeavour. Nor was it unimportant that the enterprise's financial risks had been significantly spread through the negotiations for the Interim Agreements for INTELSAT. Businessmen and legislators considering how best to go about space communications in, say, 1961 considered the state interest and state linkage important in the financial viability of the satellite system.

The then existing private companies did not have these advantages. AT & T had financed the development and construction of the Telstar series of satellites and the relative ground stations. It had paid NASA for the launch vehicles and for the launch facilities required.[5] However, it seemed clear at the time that no commercial company could afford to establish and then run a complete commercial satellite network on its own.

As indicated in Chapter 1, the development of space satellites began as part of the scientific programme for the International Geophysical Year, 1957. Satellites were originally proposed for geodetic purposes, but the potential of satellites for other purposes including telecommunications was quickly perceived and attracted an international concern which grew directly proportionate to the success of satellite programmes. The exploration and use of space was put on the agenda by the technical advances triggered by the IGY.[6]

Thus in 1958 by GA Res. 1348 (XIII) and in 1959 by GA Res. 1472 (XIV), the UN General Assembly noted the importance of the opening of space to exploration and use, and established the Committee on the Peaceful Uses of Outer Space. In 1961 the General Assembly made it known that it considered that: '. . . communication by means of satellite should be available to the nations of the world as soon as practicable on a global and non-discriminatory basis' (GA Res. 1721 (XVI) Part D). The following year it emphasized: '. . . the importance of international cooperation to achieve effective satellite communications which will be available on a world-wide basis' (GA Res. 1802 (XVII) Part E 3). Clearly, however, despite such registering of hope and concern, these objectives could not be achieved by the community of nations, nor within the structures of the UN.

As matters stood in 1961–62, it was the USA which had been preeminent capability in such matters, and it was willing to make the effort.

The USA had been substantially stung by the Soviet launch of Sputnik in 1957.[7] Various commercial enterprises had been encouraged to embark on research programmes and there had been legislative and organizational development within government. Thus, in February 1958, an Advanced Research Projects Agency (ARPA) was created within the US Department of Defence and, amongst other things, started to study the feasibility of a satellite communications facility. This became Project Advent. In September 1959 the ARPA's management responsibility for satellite communications was transferred to the US Army which, as we have seen, was responsible for Project Advent until its termination in 1962.[8] Subsequently, developments of satellite communications within the military sphere have continued, but these lie outside the ambit of this book. Suffice it to say that advances made within a military context do not necessarily always remain confined to that sphere.

Other developments required a new agency. By the National Aeronautics and Space Act 1958 (Public Law 85–568, 72 Stat. 426, 42 USC 2451) (the Space Act 1958) the USA established a National Aeronautics and Space Council under the chairmanship of the Vice-President in order to survey and coordinate national space endeavours (sec. 201; 42 USC 2471).[9] More importantly, s.201 of the Space Act (42 USC 2472–3) established the National Aeronautics and Space Administration (NASA), an entity which has been highly important in subsequent history. The USA's overarching policy was stated in the 1958 Act as being that 'activities in space should be devoted to peaceful purposes for the benefit of all mankind' (s. 102 (a)), words which were later to be reflected in UN pronouncements on a variety of space matters. An ancillary purpose (although an outsider may perhaps consider its status as being high) was 'the preservation of the role of the United States as a leader in aeronautical and space science and technology and in the application thereof to the conduct of peaceful activities within and outside the atmosphere' (s. 102 (c) (5): 42 USC 2451(c) (5)). Thus, by the time the UN was making its seminal Declarations of 1961 and 1962 referred to above, NASA had been in existence and was already engaged in various important space programmes.

Although the 1958 Space Act had not specifically identified space communications as an area of endeavour for NASA, some of its programmes were directed to the development of technology, expertise, and the gathering of information which would be useful for the creation of an active satellite communications system. Projects Courier, Telstar, Relay and Syncom all had either NASA involvement or were NASA projects.[10] Indeed, had the Communications Satellite Act of 1962 not made provision for the establishment of COMSAT, NASA itself might well have gone ahead to provide a synchronous satellite telecommunications system on its own. Only a few synchronous satellites are needed for such a service, and that could easily have been included within the agency's research programme.[11]

However, the provision of communications facilities was an area in which there were (and are) dogmatic difficulties. The UN sought the non-discriminatory provision of a global communications system. However, in the USA and in its international telecommunications services, the provision of communications facilities were in the hands of private commercial entities. It would have run completely counter to the whole concept of American enterprise that a US department of state or federal agency should have been allowed, as a matter of purely public enterprise, to create a satellite communications network responsive to global needs in the way that the UN had hoped. Indeed, before the UN Declarations, President Eisenhower had indicated how things would go, and that private enterprise would have a role in the development. By a Statement on Communications Satellites of 31 December 1960, he had said:

> To achieve the early establishment of a communications satellite system which can be used on a commercial basis is a national objective which will require the concerted capabilities and funds of both Government and private enterprise and the cooperative participation of communications organisations in foreign countries.[12]

This meant that NASA research and data would be made available in the search for a viable communications satellite system, but that the provision of the service would be on a commercial basis and private enterprise would be involved.

President Kennedy, who took over the presidency in January 1961, continued that policy. In the famous Special Message to Congress on

Urgent National Needs of 25 May 1961 he set the objective of putting a man on the moon by the end of 1969 (an objective met by the successful Apollo Program). That was the first special task for which he asked for funding beyond that which he had already requested for space projects. Second, he asked for funding for a rocketry project which has since foundered. Then, before turning to satellites for meteorological purposes, President Kennedy asked for an additional US$ 50 million to accelerate 'the use of space satellites for world-wide communications'.[13]

Eight weeks later came Kennedy's Statement on Communication-Satellite Policy of 24 July 1961 which proved to be formative of much later legal and political development. In the Statement President Kennedy invited 'all nations to participate in a communication satellite system'. In that system: 'Private ownership and operation of the US portion of the system is favoured' provided that certain policy considerations were met. These considerations included the availability, at the earliest possible date, of both new and expanded services and the extension of the system to provide global coverage; foreign participation through ownership of the system or otherwise to be made possible; the non-discriminatory use of, and equitable access to, the system by authorized carriers; effective competition in equipment acquisition and in the operation of the system; compliance with anti-trust legislation; and the development of an economic system, the benefits to be reflected in overseas rates. Governmental responsibilities were also laid down in the Statement. These included the conducting and encouraging of research; conducting or supervision of international agreements and negotiations; control of US spacecraft launching; use of the system for government purposes except where unique government needs indicated otherwise; assuring effective use of the radio spectrum and the shutting-down of satellites when required for effectiveness and efficiency. In addition, government was to 'provide technical assistance to newly developing countries in order to help attain an effective global system as soon as possible', and to examine with other countries the 'most constructive role' for the UN, including the ITU, in international space communications. All government agencies were to help attain these objectives.[14] In the July 1961 Statement Kennedy can therefore be seen to have softened his predecessor's commercial thrust and indicated that, as well basic commercial considerations, other broad public interest objectives would be required to be met. Further, the Statement could be construed as indicating that private ownership and operation might be made conditional upon the safeguarding of these extended objectives.

Meanwhile, the US government agency with responsibility for the administration of telecommunications had been also attempting to clarify future policy. The Federal Communications Commission (FCC), alive to its responsibilities under the Communications Act 1934, had instituted an inquiry into the 'Administrative and Regulatory Problems Relating to the Authorization of Commercially Operable Communications Satellites' on 3 April 1961. The First Report resulting from that inquiry concluded that a joint venture of the existing US international communications common carriers would be the most effective way of ensuring the orderly development of a commercial satellite system; and the FCC later set up an Ad Hoc Committee of Carriers and instructed it to report on how best such a joint enterprise could be undertaken. The FCC also called for procurement for any new system to be carried out by competitive bidding, and for the prior regulatory approval of specifications for all the equipment which would be required in the new system. Both private enterprise and the protection of the market from cartel were therefore shown to be desirable in the eyes of one important US government agency, and that was the one which had telecommunications expertise and responsibility.[15] In addition, the FCC also began to take action on other matters such as the questions of authorized users of such a system, the requirements and obligations to be imposed upon the owners and operators of ground stations, and matters of radio frequency use.[16]

THE COMMUNICATIONS SATELLITE ACT 1962

Three Senate and two House bills sought to establish a legislative framework for developments in space telecommunications. The first, S. 2650, broadly adopted the FCC's view, which had emerged from its 'Inquiry into the Administrative and Regulatory Problems' of the new facility (Docket 14024) outlined above. A private corporation owned by the US international communications common carriers was envisaged, with governmental supervision limited to roles for NASA and the FCC. NASA would have been responsible for launches which would be reimbursed at cost by the corporation. In addition, it would have had a consultancy role on technical specifications, and would have coordinated its communications research and development with the corporation. The FCC would have carried out its functions as usual in all communications matters, requiring the connectivity of the new system with the existing telecommunications networks, ensuring competitive bidding for contracts within the business, and setting a fair rate of return on investment. In S. 2650 therefore, although it was clear that no existing communications common carrier was to be allowed to go it alone, private enterprise was encouraged with relatively little governmental interference, though with a degree of governmental help.

The second Senate bill, S. 2814, was drawn on the lines of the Kennedy Statement of July 1961, proposing a private corporation with ownership not restricted to existing common carriers, but with a large element of public ownership and restriction of the voting powers and dividend revenue of carrier shareholders. Although it foresaw roles for NASA and the FCC not dissimilar to those of the earlier bill, it stipulated an increased element of government supervision designed to achieve broad public interest purposes. In particular, the President had a supervisory role in the planning and execution of a national space communications programme, and in international negotiations.

For the sake of completeness it can briefly be noted that the other bills followed on similar lines. The third Senate bill, S. 2890, would have given ownership of the facility to the government, and House bills, HR 9907 and 10772 proposed alternative ownership arrangements, including governmental participation and with different measures of governmental responsibilities. These, however, did not progress since a compromise was reached between the most of the proponents of the two main Senate bills, HR 11040 containing what is substantially the content of the Communications Satellite Act 1962 (31 August 1962, 76 Stat. 419; Public Law 87–624; 47 USC 701–44 (as consolidated to 1970)). The Act has been subsequently amended, notably to permit COMSAT to participate in the INMARSAT organization as the United States' Signatory to the INMARSAT Operating Agreement, but its broad thrust remains what it was in 1962.

It should not, however, be assumed that the compromise between the two major bills was easily achieved, nor that the passage of the eventual compromise bill was smooth. Many House and Senate Committees held hearings, the matter was extensively discussed in the ephemera of the time, and the argument was bitter on both sides of the question. Those who opposed any private ownership in the new system were as convinced they were right, and opposed to other views, as those who would have the matter entrusted to unfettered private ownership, and therefore in effect to one or, more likely, to a consortium of the then existing communications common carriers.

Nevertheless, the compromise has worked to the satisfaction of those who would have excluded private ownership. Others are less pleased. In fact, as hinted above, in a sense the argument has grumbled on over the years, and the advocates of private enterprise and limited (if any) governmental supervision have emerged once more with the onset of deregulation in the telecommunications field. Now, as we shall see, there is a two-pronged attack within telecommunications, the one aimed at COMSAT itself which, if fully effective, would render the Corporation at least emasculated, and at worst redundant; the other on the international arrangements for the global telecommunications satellite system in the creation of which COMSAT was a major actor, and which we will deal with in our consideration of INTELSAT.

The Communications Satellite Act 1962, to give it its short title (s. 101), an 'Act to provide for the establishment, ownership, operation and regulation of a commercial communications satellite system, and for other purposes' (Public Law 87–624; 76 Stat. 419, 47 USC 701–44), was passed on 31 August 1962. It authorized the creation of COMSAT, in order that it should bring into being the USA's policies in the area of satellite communications. The Declaration of Policy and Purpose which is contained in s. 102 is therefore important, and contains clear echoes and resonances of the 1961 Kennedy Statement.[17] Briefly, the policy was the establishment, in cooperation with other countries and as expeditiously as possible, of a commercial satellite telecommunications system, and its exension as rapidly as possible to give global coverage (s. 102(a)). The communications needs of the USA and other countries were to be met and thereby contribution was to be made to the national goal of international peace and understanding (s. 102(a)). In the expansion of the system to provide global coverage, attention was to be paid to the provision of facilities for the less well developed countries as well as to the needs of the more highly developed (s. 102)(b)). Efficient and economic use was to be made of the radio spectrum and the benefits of the new technology were to reflect through, both in quality of service and in the charges made for it (s. 102(b)).

As noted, although the Act contains duties and powers for other governmental agencies independent of the new entity's area of responsibilities, the main vehicle for this Policy and Purpose is COMSAT which is authorized by Title III of the Act. To this end, it was given monopoly status in the provision of international services using the satellite link it was to establish, as well as other unusual powers and privileges. But the monopoly and unusual powers had their price. For the first time, a communications system was seen as having implications for, and a role to play in, the US national interest and foreign policy. Governmental supervision of the new enterprise was required to ensure that the public interest objectives were met, and that was to be exercised partly through normal regulatory procedures and partly through presidential duties and powers of supervision. In addition, other agencies were required to continue their support in various ways.

Governmental supervision of the new enterprise was to be exercised chiefly through the President, through NASA, the FCC and through the State Department. Congress, of course, retains residual powers of legislation, has the right and duty of receiving reports made to it by the Corporation and other agencies, and has powers to hold Hearings.

The US President has considerable duties in relation to both the internal constitution of the Corporation as originally set up, and its business activities. Subject to the advice and consent of the Senate, he appointed its incorporators,[18] who were responsible for the initial stock offering. He also approved the Articles of Incorporation of the Corporation (s. 302).[19] He appointed one-fifth (that is, three) of the Corporation's first Board of Directors, again subject to the Senate's advice and consent, and these appointees serve for three years or until their successors have been appointed and qualified – a rolling programme instituted with the first presidential appointees, resulting in one appointee's term of office being completed every year (s. 303(a); Articles of Incorporation VIII, (s. 8.02(a)). As the other directors are elected annually, the presidential appointees could have had considerable opportunity to influence policy by reason of the continuity of thinking they represent, and this was remarked on in the early years; more recently the point has become occluded. Every year the President transmits a Report to Congress on the activities of the United States in pursuit of its national space programme, and may make recommendations for additional legislation that he considers necessary for the attainment of the objectives contained in the 1962 Act (s. 404(a)).

Again reflecting the 1961 Kennedy Statement, the US President was given extensive duties and powers in relation to the business activities of the Corporation under s. 210(a). He had (and to some extent still has) to aid the planning and development and foster the execution of a national programme for the expeditious establishment of the global satellite telecommunications system; provide for the continuous review of the development and operation of the system, including the Corporation's activities; coordinate the governmental agencies with telecommunications responsibilities to secure their compliance with the Act; supervise the relationships of the Corporation with foreign governments, relevant entities and international bodies to ensure that the Corporation's relationships were (and are) consistent with US national interest and foreign policy; ensure that timely arrangements were made for foreign participation in the establishment and use of the system; ensure the availability and use of the system for US government purposes except where a separate system was needed for unique governmental needs or (notably) 'otherwise required in the national interest'; and to help attain a proper use of the radio spectrum and the technical compatibility of the system with existing facilities at home and abroad. Such a list of powers and duties means that the President, if he so chose, could directly influence developments particularly at the early stages when the business was malleable.

The business depends upon foreign terminals and the technical efficiency of the satellite system. The presidential powers in relation to the foreign relationships and participation (s. 201(a)(4) and (5)) are extensive: such relationships were and are crucial to a valuable commercial network. Similarly, the presidential responsibilities in relation to the US national space programme and its interaction with the Corporation's business (s. 201(a)(1)) had effects on the development of the actual network, the other pillar of the enterprise.

The powers and duties of NASA were not as immediately effective as those of the President, but they did, and still do, carry weight. Under s. 210(b) NASA provided, at cost, the launch facilities for the satellite element of the system, as well as other services as required. These included advice on technical matters and cooperation with research, although again the Corporation paid for the help it received. It was noticeable, however, to me writing in 1965 that, in the enumeration of the responsibilities of NASA, NASA was to act as it deemed appropriate (s. 201(b)(2) and (3)). It was not to be the servant of the Corporation, and I undersand that with the setting-up of IN-TELSAT and the development of COMSAT's own laboratories and research programmes by various potential suppliers, NASA was not called on nearly as much as it might have been had the USA been able to establish the international system with a more limited foreign ownership and control.

On the other side of the governmental coin, the FCC was given wide regulatory authority (s. 201(c)). When examining these provisions in 1965 I thought that the powers there conferred were very wide indeed, and I ventured the statement that by their use and those of the President, the US government could exercise an extraordinary degree of control, greater than that exercisable over any private corporation in the history of the business enterprises in the USA.[20] Perhaps that was overstated, but the FCC powers and responsibilities are extensive and have been used not least in the last decade. In particular there have been several attempts to have the powers used by those who would wish to trim, if not overset, COMSAT's monopoly and reduce its position within the business of telecommunication services. This is not to say that the FCC has itself abused its powers, but that the various FCC hearings and inquiries which have involved COMSAT (notably the COMSAT Structure sequence of hearings as well as those into the ownership and operation of ground stations, and into the question of direct access to the INTELSAT space segment), have provided opportunity for other telecommunications entities to seek to have COMSAT's commercial position altered, and to have its tariffs and methods of doing business changed. These attempts have not been without result.

Naturally the FCC powers given by s. 210(c) include those normally available to the FCC in relation to any of the telecommunications enterprises which come within its supervisory jurisdiction by virtue of the Communications Act of 1934 (48 Stat. 1064; 47 USC 609). In the area of setting up and running the system these powers include that of ensuring effective competition in the procurement of equipment and services through the requirement of competitive bidding if appropriate, together with a duty to see that small business gets a share of the contracts (s. 201(c)(1)). The FCC approves the technical characteristics of satellites and the satellite terminal stations (s. 210(c)(6)). It is the agency required to authorize the construction and operation of terminal stations with a view to the public interest, convenience and necessity, and it has that role whether the construction is to be undertaken by COMSAT or another authorized carrier (s. 201(c)(7)). The FCC ensures that the satellite system is both compatible with other telecommunications networks and that it is interconnected with them (s. 210(c)(4)). Substantial additions to the satellite system require the FCC to consider them to be justified by public interest, convenience and necessity (s. 201(c)(9)), but the initiative for addition does not lie exclusively with the carriers. Of its own motion the FCC can require COMSAT and the other carriers to make additions to the system if it perceives public interest, convenience and necessity so to indicate (s. 201(c)(10)). Furthermore, at the request of the Secretary of State, and provided that it is technically feasible, the FCC may require COMSAT to establish a communications link with a particular foreign point (s. 210(c)(3)). Finally, as a major matter of the new creation, and necessary given the climate in which the undertaking was proposed, the FCC supervises the allocation of facilities in the new system to ensure the non-discriminatory use of, and access to, the system on just and reasonable terms and conditions to all present and future communications common carriers authorized to use the system (s. 201(c)(2)).[21]

The FCC also has power in regard to COMSAT's business operation. In particular, the FCC prescribes the accounting regulations and systems for the Corporation, and sets rate-making procedures intended to ensure that the rates for public services reflect the economies of the new facility (s. 210(c)(5)). In the 1978 Settlement Agreement, the FCC set a rate of return on investment for the Corporation of 11.48–12.48 per cent – the 1 per cent range being available if the enterprise is found efficient. The rate was a matter of contention and court action as well as FCC consideration, and was adopted in 1978 as a Settlement Agreement.[22] The FCC continues to monitor the matter of rates and is aided in this by other carriers, for complaints levelled at COMSAT by other parties to FCC proceedings often focus on rates and charges.[23] The question of tariffs was also integral to the restructuring of COMSAT described below. And in particular, the FCC itself has acted to require COMSAT to repay sums it is held to have overcharged customers in the period August 1984–December 1986 through a failure to keep to the 11.48–12.48 per cent band.[24]

As regards the formal constitution and structure of COMSAT, the FCC authorizes all stock issues except the initial offering, borrowings and assumptions of liabilities in order to make sure that such action is necessary and consistent with the purposes of the 1962 Act as well as with the overall public interest (s. 210(c)(8)).[25] Even the holding of COMSAT stock is not outside the FCC's control. By s. 304(f) on the application of an authorized US communications carrier, and after hearings on the matter, the FCC can compel any other authorized carrier owning shares in COMSAT to transfer to the applicant such number of shares as it determines. In making its decision, the FCC has to have in view the purposes of the Act and the overall public interest including the promotion of the widest possible distribution of stock amongst the authorized carriers. So at least says s. 304(f). It may be noted, however, that over the years the other authorized carriers have virtually and mostly voluntarily divested themselves of COMSAT shares[26] – a comment on their view of the Corporation's commercial potential and on the availability of other more attractive business within the US domestic telecommunications scene. Some have also gained permission for limited transborder services subject to coordination with INTELSAT.[27] Further, it seems that now some US companies wish to move into international space telecommunications bypassing both COMSAT and INTELSAT, a problem we will turn to in our consideration of INTELSAT.[28]

Finally, apart from such direct controls exercisable over the Corporation, under s. 404(c) the FCC has the duty to report to Congress annually, and at such other times as it may consider desirable, on anti-competitive practices, an evaluation of its own activities with a view to recommending any additional legislation it may consider necessary in the public interest, and 'an evaluation of the capital structure of the corporation so as to assure the Congress that such structure is consistent with the most efficient and economical operation of the corporation'. Special reports may also be instructed, as in the 1978 International Maritime Satellite Telecommunications Act.[29] It would be impossible for an outsider to express a useful view on the suitability of this last head of requirement in the light of COMSAT stockholding movements in the last 25 years.

The activities of COMSAT, particularly in its early years, were to have a clear and obvious bearing on the standing and reputation of the United States. Viewed from the standpoint of other countries, and despite specific disclaimers in s. 301 of the 1962 Act and in the Prospectus for the Corporation, [30] the prestige of the United States was at stake.[31] The various Congressional Hearings on the bills which led to the 1962 Act are clearly informed by this realization. Therefore, in addition to the presidential supervision mentioned above, COMSAT is required to notify the US Department of State whenever it enters into business negotiations with any international or foreign business entity (s. 402). The State Department may assist in the negotiations, but in any case advises the Corporation of relevant foreign policy considerations. It is not, however, clear to an outsider what this precisely means. COMSAT personnel tend to say that the Department of State advises, while the Department believes that it instructs.[32] The original arrangements between COMSAT and the US governmental agencies were fairly complex, and may still not now be entirely satisfactory because of the way in which their terms have been interpreted by the parties involved – as is shown by the divergent views of the participants just indicated.[33] The FCC has recommended amendment of the Act to clarify the position which, in its view, must be that in the last analysis the President as supreme authority within government should have the ability to instruct the Corporation on foreign policy matters.[34]

Nevertheless, there is also some FCC public documentation on the matter which indicates that perhaps no such amendment of the Communications Satellite is necessary. The internal US Procedures are on public file. An attempt by the Pan American Satellite Company to have COMSAT instructed to attempt to have INTELSAT business deferred so that US companies could act within the formulation of US policy on INTELSAT's new domestic services failed on the ground that these procedures had been complied with and provided a method by which public input into the internal US decision-making was possible.[35] On the other hand, the detail of the instruction which COMSAT may receive is fascinating. COMSAT disclosed part of one 'instruction letter' in an attempt to have part of the FCC decision on licensing of separate US international systems reconsidered. The question was the interpretation of the term 'long-term lease' which was the subject of an INTELSAT paper BG–66–20.[36] The letter states:

> The Signatory [COMSAT] shall not initiate any discussion of BG-66-20. Should a discussion prove unavoidable, the Signatory shall downplay the issue by merely noting the paper as a statement of opinion. Only if merited as a response to further discussion, the Signatory shall state that the Director General's interpretation of the joint letter from the Departments of State and Commerce to the FCC is without foundation. The Executive Branch is of the view that the comments are taken out of context.

The purpose of the letter was to clarify that the 'long-term lease or sale' aspect of the restrictions on separate system applies to occasional use television services. This issue is currently before the FCC.[37]

Such detail argues that COMSAT is indeed instructed when the appropriate part or parts of the US government so decides.[38]

By the system of reports and the wide powers which are given to the President, NASA, the State Department and the FCC, it was intended that the Corporation would operate in a manner consistent with the national interest and US foreign policy. But the broad powers of interference raise some questions as to the exact nature of the Corporation. The 1962 Act states by s. 301 that the Corporation is not an agency or establishment of the US government, but a corporation for profit, and it is best to consider it as a legal person albeit subject to controls and disabilities which do not affect the average company in the US legal environment. The reason for rejecting the idea that the entity be considered as *sui generis* is that its genus is quite apparent.[39] The Corporation is a company within the normal meaning of that word. It has a share capital, and was incorporated under the laws of the District of Columbia, USA, on 1 February 1963.[40] It has legal personality with wide powers of action,[41] is capable of holding property in its own name,[42] and of entering into contracts.[43] It is subject to the District Courts of the United States,[44] may sue and be sued, complain and defend in its corporate name,[45] and has perpetual succession by that name.[46] The entity therefore clearly falls within the classification 'corporation' in the commonly accepted jurisprudential meaning of that term.

However, the capital structure of the Corporation is unusual. While the non-voting shares provided for under the 1962 Act were of the standard kind, the common stock was divided for the purpose of the initial offering into two series.[47] Series I shares were available for public purchase;[48] Series II shares comprised 50 per cent of the voting stock and were issued to common carriers authorized to hold the stock once the FCC had ascertained that their ownership of the stock would be consistent with the public interest.[49] Further, under s. 304(b)(3) 50 per cent of all other stock issues were reserved for carriers and, while only carriers might hold more than 10 per cent of the stock, the total carrier holding was restricted to 50 per cent at any one time. Again while, as noted, no stockholder or syndicate other than a carrier might own more than 10 per cent of the outstanding issued stock of the Corporation,[50] an additional restriction was placed on foreign ownership, foreign participation being limited to a maximum of 20 per cent of the stock held by persons other than communications carriers (that is, 10 per cent of the whole).[51] In these ways it was intended to spread ownership of the new Corporation as widely as possible, and this was successfully achieved. However, subsequently by FCC initiatives, and as COMSAT became less attractive to corporate investors, the carriers stockholding diminshed to virtually zero.[52]

The Corporation's affairs are administered by a Board of Directors, fifteen in number, all of whom are citizens of the USA.[53] As already mentioned, there are three presidential appointees who serve for periods of three years each or until a successor is appointed, with a staggered rotation of period.[54] The other 12 directors are elected annually by the appropriate constituency, six to represent Series I shareholders and six to represent Series II shareholders.[55] The Corporation has a President and other officers who are named and appointed by the Board of Directors.[56] All the officers of the Corporation are appointed on a full-time basis and must be US citizens.[57]

In order to implement the purposes and policy of the 1962 Act, to achieve its objectives and carry on business, the Corporation was authorized to plan, construct, own and operate a communications satellite system and the necessary ground stations. It specifically was to provide communications channels for hire, and conduct all necessary research for the implementation of its purposes. Satellite launch facilities were to be purchased from the US government, but responsibility for technical specification remained with COMSAT.[58] The Articles of Incorporation list the powers more fully, and the additional powers there enumerated which are not stated in the Act itself are those of any normal US company.[59]

The Certificate of Incorporation was issued on 1 February 1963 and the first issue of stock in the Corporation was made to the public on 2 June 1964.[60] Five million shares at no par value were offered at a price of US$ 20 per share, with a maximum of 50 shares allowed to each customer.[61] This ensured the wide distribution of shares amongst the share-buying American public, as was required by s. 304(a) of the 1962 Act. Series II shares were also placed without difficulty.[62] The success of the offer was mainly due to the glamour of the enterprise and not to any guarantee of quick dividends, since uncharted waters lay ahead and there was little hope of the Corporation paying dividends before 1970.[63] The magazine *Newsweek* accurately characterized the prospectus as a 'litany of caveats'.[64] The Prospectus repeatedly stressed that the Corporation was attempting to do things that had not been tried before, that the system would run at a loss for at least its first few years, that the system was open to being intentionally jammed, and that the Corporation might be called upon to provide services which would be contrary to its business judgement to provide. The public, however, seemed willing to run these and all the other risks.[65]

There undoubtedly were risks. The system might have been uneconomic to provide. There was potential competition for many of the services through cable, and alternative global systems might have come into being, resulting in the market being unable to support a commercial system of the kind envisaged in the 1962 Act.[66] But things worked out. COMSAT was able to engage in its business, particularly through its being designated the US Signatory to the Operating Agreement for the International Telecommunications Satellite System, as described in the discussion of INTELSAT below, and also through being designated as the US Signatory in the International Maritime Satellite Organization (INMARSAT), again as described below. The development stages of COMSAT took some 30 months from the first shareholders' annual meeting on 17 September 1964. Fully commercial operations with conventional business accounting began on 1 May 1967.

Thereafter, much of COMSAT's history in the following 25 years is a matter of national and international business which, although fascinating, lies outside the scope of interest of this book. However, certain legal developments do require examination since they have international repercussions. I must warn, however, that what follows is not to be taken as a full statement of telecommunications and telecommunications law within the United States. For that, other sources must be looked to.

Naturally, for COMSAT, first the interim, and then the permanent, arrangements for INTELSAT were of major importance. As indicated in the discussion of INTELSAT itself, there were many reasons why these turned out as they did. Certainly the 1962 Communications Satellite Act followed the 1961 Kennedy Statement in making reference to foreign participation. However, it is not clear from the Act and the debates leading to it whether any particular form of foreign participation was intended. Was there a hope that the foreign participation could come through investment in the COMSAT Corporation? Certainly the Act speaks of it being the 'policy of the United States to establish, in conjunction and in cooperation with other countries' a commercial communications satellite system. (s. 102(a)). The 'United States participation in the global system' was to be 'in the form of a private corporation' (s. 102(c)). But other language in the Act is very clearly directed towards the establishment of an American enterprise with limited foreign participation. Foreign shareholding in the Corporation can amount to only 20 per cent of the issued Series 1 stock (that is, 10 per cent of the total stock),[67] which meant that, even arithmetically, foreign participation in COMSAT was insufficient to meet the requirements of other countries. There would be insufficient control over the policies adopted by the Corporation even if all the foreign participants acted in concert. Further, even if all they did decide to act unanimously, no person who is not a US citizen can sit on the Board or be appointed an officer of the Corporation.[68] Last, in many of the potential foreign participant countries, telecommunications was in state hands, and a state was unlikely to consider that it was adequate or expedient (to say nothing of dignified) for its participation in an international satellite telecommunications system to be through the holding of stock in a US company. It might therefore seem that the intention was to enter into an international agreement or set of agreements with foreign telecommunications entities and set up the system on the analogy of the cable agreement structures. In that case who would provide the hardware? Foreign governments were suspicious of the new Corporation's commercial intentions. There certainly is a view outside the USA (and to an extent within it) that Congress and COMSAT did orginally intend for the USA to be the provider of a system from which others would buy services. COMSAT would allow others to have their own ground stations, but the satellite system itself and its US connections and ground stations would be a US business. Even if this is not objectively true, it remains that that is how the matter appeared to those on the other side of the table (or Atlantic). It follows that the precise nature of the foreign participation in the enterprise had to be first fought for and then carefully negotiated.

The interim result was an arrangement which suited COMSAT quite well. The 1962 Act had mandated it to establish a satellite system as expeditiously as possible (s. 102(a)) and the interim arrangements allowed that. Following negotiations, an interim arrangement was worked out under which a satellite system could be begun and extended, the permanent arrangements to be negotiated later in the light of experience of the working of the interim arrangements.[69] This was to become INTELSAT, discussed later in Chapter 3.

As discussed elsewhere, in the interim arrangements a consortium of joint-venturers was established with an Interim Communications Satellite Committee, (the ICSC) as its major decision-making forum. States members of the Agreement designated each national telecommunications entity to act as its Signatory to the Special Agreement and its participant in the new enterprise. Costs, voting and ownership were assigned on a quota basis which was initially negotiated among the parties, and then was related to use of the system. COMSAT itself had 61 per cent under the original quota allocation.[70]

Ownership of the space segment had to be surrendered to the new entity, Interim INTELSAT, the parties owning the space segment in undivided shares proportionate to their share of first the costs and then utilization of the system, with a minimum of 0.05 per cent. Again, decision-making at the Board level was weighted in accordance with the share of costs. However, although it was always clear that COMSAT, as US Signatory of the Interim Operating Agreement, would have a major voice, 14 matters which went to the heart of the system to be established were reserved for decision by special majority. For such a majority the affirmative vote of the US Governor and of Governors voting at least a total of 12.5 votes was required.[71]

However, after 60 days' negotiation, the level of votes additional to the USA required to make up the special majority was reduced to 8.5 votes in decision on certain enumerated matters. These were decisions as to the type of space segment, the approval of budgets by major categories, the approval of contracts over US$500,000 for space segment equipment, and as to satellite launchings including decisions as to the launch source and elated contractual arrangements.[72]

Highly important for COMSAT were art. VIII of the Intergovernmental Agreement and art. 12 of the Special Agreement. Under the first of these COMSAT was to act as the manager in the design, development, construction, establishment and the day-to-day operation and maintenance of the space segment. Of course, COMSAT was required to act in accordance with the general policies laid down by ICSC which operated as a steering committee of the joint venturers engaged in the enterprise, but COMSAT's role was therefore secure and massively influential in the creation of the new global system. This was further spelled out in the Special Agreement, under art. 12 of which the Corporation prepared the annual programmes and budgets subject to the approval of the ICSC; made recommendations as to the type of space segment; carried out research, operated and maintained the space segment; advised the ICSC and its members; arranged for technicians nominated by Signatories to take part in design and equipment specification; and arranged for the free use of data and inventions obtained before the signing of the two Agreements.

However, there were storms ahead. In particular we have yet to deal with the renegotiation of the Agreements, the position regarding INMARSAT, the FCC inquiry following on INMARSAT, the Earth Station Ownership, the Direct Access and the Authorised User II and III decisions to which I will come shortly.

The negotiation of the permanent arrangements for INTELSAT caused major changes in COMSAT's position both within that organization and in its standing within the domestic US telecommunication business arena. COMSAT's very insistence on its control of decision-making in Interim INTELSAT and its position as manager of the system which allowed the establishment of the global coverage to be attained in the years following 1964 stored up trouble for it. The European nations were determined to preserve and develop their space industries, and wanted to garner what they could from procurement decisions, technology transfer and technology-sharing arrangements. Indeed, there is a view that they would have preferred a slower progress to the eventual satellite system so that they might gain more through its development. But COMSAT was determined that the new venture should go ahead speedily with itself as manager. The other major point which stuck in the throats of the foreign negotiators was that Interim INTELSAT was – again at the insistence of the US negotiators – a body without international or national legal personality. That meant that it had no legal being apart from its membership in the joint venture. The side-effect of this was that COMSAT as manager under the ICSC was responsible for the legal side of the enterprise, and that the satellite system, when it was set up, would be owned by the parties to the Agreement not by the organization itself. If the arrangements failed, the US participant would reap benefit.[73] The *quid pro quo* of foreign agreement to the COMSAT package turned out to be the assent of the United States to the renegotiation of the interim arrangements in terms of art. IX of the Intergovernmental Agreement. Permanent arrangements would be established by an international conference to be convened by the USA, with the intention that the definitive arrangements 'will be established at the earliest practicable date, with a view to their entry into force by 1st January 1970' (art. IX(c)).

While the interim arrangements were being set up other problems had emerged, notably as to the relationship between COMSAT and the US government. In April 1966, for example, the ICSC approved on the recommendation of COMSAT the placing of the INTELSAT III contract with TRW Inc., a US company. In accordance with s. 201(c)(1) of the 1962 Agreement, COMSAT had asked the FCC for its approval of the contract in February 1966. However, the FCC had not replied by time of the ICSC's decision in April 1966, and Hughes Aircraft, the builder of Early Bird (INTELSAT I) and INTELSAT II, objected in the FCC proceeding to the placing of the contract elsewhere. COMSAT had to return to the ICSC and say that the placement of the contract would have to wait until the FCC acted. Needless to say, the other members of Interim INTELSAT were outraged. Was their organization to wait on the deliberations, or be subject to the consent of a domestic governmental agency of one of their members?

Another problem was technology transfer. COMSAT sought to have Interim INTELSAT buy the best equipment available at lowest prices. This was commercial instinct at work. However, it meant in general that contracts would go to US companies. The European nations naturally wanted some rough distribution of contracts proportionate to investment in the system in order to feed their own struggling space industries and keep abreast with the state of the art in such matters.[74]

Suffice it to say that, in the renegotiation, COMSAT paid a price for its earlier stances and their associated problems. Whether a different solution would have been obtained had the emphasis not been quite so clearly on the commercial side of the satellite system is difficult to say. But during the 'interim' phase of the organization the original 14-member consortium had become an international body large by any standards, with 83 members, many of which were developing countries. The only major category of countries which were absent was the Communist bloc.

The satellite system was up and running. It clearly would continue, and the significance of the new facility had been brought home to governments round the world. However, the very increase in size and the extended membership of Interim INTELSAT, together with the interests of the developed and developing countries in procurement matters,[75] meant that the negotiations of the permanent arrangements had a political aspect which was considerably more to the fore than had been the case in the negotiation of the interim arrangements. Ministries of Foreign Affairs and of Technology as well as telecommunication entities and government departments were all involved.

In the eventual agreement on the permanent arrangements for INTELSAT COMSAT's role was diminished. In the first place, INTELSAT became itself a juridical entity with full legal capacity including that of concluding agreements with states and other international organizations; of entering into contracts, of acquiring, holding and disposing of property; and of being party to legal proceedings (Intergovernmental Agreement, art. IV(a)). Parties to the new agreements were required to take any necessary steps to make INTELSAT an effective legal person within their own jurisdictions (art. IV(b)). In this way, the problem of lack of legal being in most states, including the US, was elided.

Second, the new organization was given four organs: the Assembly of Parties, the Meeting of Signatories, the Board of Governors, and the Executive Organ (art. VI(a)). However, clearly it was to be effectively led by its Board of Governors, successors to many of the responsibilities of the ICSC.

Third, specific provision was made in art. VI(a) for an Executive Organ to be one of the four organs of the new organization. The duties of the Executive Organ and the terms of appointment and the duties of its head, which was to be the new post of Director-General, were laid out in art. XI. It followed that the Executive Organ was to take over management functions in due course, and in any event was to provide a source of advice to the Board separate from COMSAT, which was (eventually) to become merely one (albeit the largest) of the Signatories. Such changes would, however, take time. Article XII, therefore, provided for a transitional period in which INTELSAT was guided by a Secretary-General with an appropriate staff, that post vanishing with the assumption of office of the Director-General (art. XII(a), (b) and (c)).

Article XII(a)(ii) and (e) secured the position of COMSAT as contractor for management services in the transition period until the new

Executive Organ was fully operational. The responsibilities of COMSAT and guidelines for the Management Services Contract were laid down in Annex B to the Intergovernmental Agreement and, by art. XII(a)(i), the INTELSAT Board was required to arrange the contract in terms of art. XII(c). In addition, by art. XII(a)(iii) and (f), the Board of Governors was required as soon as possible, and in any event within one year of the Agreement entering into force, to initiate a study designed to 'provide the information necessary for the determination of the most efficient and effective permanent management arrangements' for the organization. By art. XII(g) the Board was required, within four years of the entry into force of the Agreement between Parties, to submit to the Assembly of Parties a comprehensive report including the results of the study, and to make recommendations as to the organizational structure of the Executive Organ. This was done.

Progress was swift. In 1973 the Fifth Meeting of the Board of Governors approved the structure of the Executive Organ (BG–5–3, approving BG–5–26 (later amended *inter alia* by BG–7–3, approving BG–7–31), thus laying out the nature of the arrangements which INTELSAT itself would adopt to deal with management matters. The Secretary-General of INTELSAT entered into the Management Services Contract with COMSAT on 1 August 1974. The permanent management arrangements were approved in October 1976 by the Second Assembly of Parties (AP–2–3; approving AP–2–9 which is also MS–5–5 and BG–22–50), to take effect from 11 February 1979. As it happened, progress in setting up the Executive Organ ran ahead of schedule and the Management Services Contract with COMSAT was amended to provide for its termination on 31 December 1978 (BG–34–3; c.f. BG–35–63). The new permanent management arrangement's control over management services was therefore effective from 1 January 1979. One immediate result was that contractors other than COMSAT were able to deal directly with the Executive Organ without having to go through COMSAT, or having their bids for contracts considered by COMSAT as part of its role as management contractor under which it advised the Board as to the placing of contracts. However, the links with COMSAT were not wholly severed: certain were retained, though put on a new contractual footing. Thus BG–35–55 of November 1978 contains a draft three-year laboratory services contract expiring on 31 December 1981, and BG–34–66 of September 1978 contains draft Technical Services Contracts of six and four years respectively, dealing with services in respect of the introduction into service and the procurement for the INTELSAT V series. Again, a Maintenance and Supply Agreement was also made for purposes set out in BG–35–79 under which COMSAT was responsible for receiving faulty equipment and repairing it, testing, storage, maintenance and replenishment of inventory, and obtaining non-inventory material. Similar matters to these three have been entered into subsequently. The point is, however, that COMSAT no longer had its protected status vis-à-vis INTELSAT but was open to competition in the placement of these contracts.

One other point has to be made in respect of the assumption of functions by the Executive Organ of INTELSAT. As part of this process, INTELSAT took over the actual operation of the satellites and, not surprisingly, many of the COMSAT personnel who had been employed in COMSAT's discharge of that function for INTELSAT moved over into the INTELSAT's employment. The same happened in other areas of the new Executive Organ's responsibilities. While this was good for INTELSAT, it was not to COMSAT's benefit.

At the time that the Corporation was losing a considerable degree of its favoured position within INTELSAT other difficulties were emerging for it. First, questions of COMSAT's structure were to be reviewed. Then, as what in retrospect can be seen as a major review by the FCC of the whole area of satellite telecommunications, inquiries and decisions (including in some cases modifications of prior policy) were made in relation to earth station ownership, direct access to the INTELSAT space segment, and in the categories of those authorized to use the satellite system. Not all of these developments were to the detriment of COMSAT, it being given, for example, authority to engage in business activities other than those connected with the provision of access to the INTELSAT system for other carriers: it now can provide service directly to non-carrier customers. But it has all meant a major change in the balance, and in some cases the very direction, of COMSAT's business efforts.

In the first place, COMSAT, foreseeing that its position within INTELSAT was to diminish, very sensibly sought to change the balance of its activities by developing other areas of expertise. Because it had gained much experience and expertise working within the nascent INTELSAT system, and through experimental work done by its laboratories and so on, it was therefore well placed to enter international competition – for example, in helping the design and establishment of foreign domestic satellite systems. Second, the same expertise might have made COMSAT well placed to involve itself in the internal US market. Third, there were also other international developments afoot, although these were unlikely to be of the scale or significance of the development of INTELSAT. In particular, COMSAT had been active in the US side of the setting up of INMARSAT, although, in the nascent INMARSAT as with early INTELSAT, it had not endeared itself to the other negotiators.

Trouble was also lurking for COMSAT at home, spurred by the very success COMSAT had achieved in the INMARSAT negotiations.

When the appropriate legislation was enacted by the US Congress, those who had earlier opposed the compromise which produced the Corporation, those who had over the years been its commercial competitors, and others who simply thought that the time had come for some further scrutiny of COMSAT took their opportunity. The International Maritime Satellite Telecommunications Act 1978, 1 November 1978, (Public Law 95–564; H.R 11209; 47 USC 751) added a new Title V to the Communications Satellite Act of 1962, making it competent for COMSAT to act as the US participant in INMARSAT. Section 505(a) of the new title required the FCC to 'conduct a study of the corporate structure and operating activities of the corporation, with a view to determining whether any changes are required to ensure the corporation is able to effectively fulfil its obligations and carry out its functions under' the 1962 Act and the Communications Act of 1934. The study was to be submitted to Congress within 18 months of the 1978 Act coming into force, was to be detailed and had to contain any recommendations the FCC considered necessary or appropriate for legislative or other action.

The FCC issued its Interim Report and Notice of Inquiry on the question of the COMSAT structure on 19 October 1979.[76] The Report, the 'COMSAT Study', was adopted by the Commission on 22 April 1980, and published on 1 May of that year.[77] It is a bulky work. Broadly, it concluded that, under the terms of the 1962 Act (its parent legislation) COMSAT could lawfully engage in activities which were not inconsistent with its statutory function. The Corporation was well placed technically and in other ways to develop new business applications of satellite technology which would be in the public interest and which would contribute to the development of space technology. A distinction was therefore made between COMSAT's 'jurisdictional activities' (that is, those connected with its participation as the US entity in the INTELSAT and INMARSAT organizations) and its 'non-jurisdictional' activities (that is, all other businesses which it might develop).

However, there were problems in allowing COMSAT simply to proceed on the basis of the corporate structure it had developed up to that time. Public policy questions might well arise through conflict of interest between COMSAT's jurisdictional and non-jurisdictional activities, through the competitive advantages in non-jurisdictional activities accruing by virtue of its jurisdictional expertise and reputation, and through the potentiality for market edge that might result from faulty allocation of costs (and therefore cross-subsidization) between the types of major activities. The FCC therefore proposed the separation of COMSAT's structure into two distinct elements, the one to handle its jurisdictional and the other its non-jurisdictional activities. Arms-length dealing between these, and refined accounting and rate-setting mechanisms were also recommended. In addition, it was thought necessary that governmental oversight of the Corporation should be increased to ensure that users were protected, that competition was secured and that conflicts of interest were avoided.

Within six months, COMSAT responded by a major restructuring of itself. Two main groupings were established – a parent organization was to deal with (but not be confined to) the jurisdictional activities, while non-jurisdictional activities were to be hived off into various subsidiaries and partnerships with other telecommunications entities.

The parent COMSAT was divided into a Headquarters Division and the World Systems Division. The Headquarters Division dealt with policy and provided support activities as required for other sections: it was not involved in day-to-day operations. The World Systems Division had functional responsibilities for COMSAT's involvement in INTELSAT and INMARSAT, including engineering and research and development in space applications and other telecommunications, working under contract to many other telecommunications entities. Within the World Systems Division, responsibilities were divided among International Communications Services (the Division's carrier segment), INTELSAT Technical Services (working under contract to INTELSAT and INMARSAT) and COMSAT Laboratories. Administrative groups dealt with legal matters, contracts and procurement, finance and personnel.

With regard to COMSAT's non-jurisdictional activities, the main subsidiary, COMSAT General Corporation, was reorganized, and new subsidiaries, COMSAT General Integrated Systems and COMSAT General Telesystems Inc., were created, COMSAT General Satellite Systems provided domestic US service for AT&T through the Comstar satellites, and a maritime system for the US Navy which was partially leased to INMARSAT. Systems Technology Services provided systems and applications engineering and systems operation services worldwide. Environmental Research and Technology Inc. was a COMSAT subsidiary providing environmental data and services.[78] The Satellite Television Corporation was to provide a subscription television service by satellite for domestic US use. In addition there were such cooperative enterprises as Satellite Business Systems, which was a joint venture with IBM and Aetna Life and Casualty, an insurance giant.

On 29 October 1980 the FCC issued its Notice of proposed Rule-Making, seeking comment on both the COMSAT Study and COMSAT's reorganization.[79] The First COMSAT Structure Order followed.[80] This more or less agreed the overall structure developments and affirmed that COMSAT could lawfully engage in activities other than its jurisdictional activities in respect of INTELSAT and

INMARSAT. However, further detail on tariff-setting and cost allocation was required.[81] When that data was forthcoming the First Structure Order was slightly amended in the Second Structure Order
which stipulated for certain cost allocation practices.[82] This was further reconsidered and some further modifications (including a tariff
rate of return set at 11.48 per cent) were set out later that year in the
Final Structure Order (Second Structure Order reconsideration).[83]
Most recently, an additional Report and Order dealt further with
accounting procedures intended to permit the FCC more accurately
to trace any infringement of its requirements, to allow the COMSAT
subsidiaries to be independent of each other and to ensure that cross-
subsidization would not occur.[84]

While these developments were taking place under the aegis of
s. 505 of the 1978 International Maritime Satellite Communications
Act, further steps were taken which were also triggered by the US
participation in INMARSAT. When the FCC was considering the interconnection of the INMARSAT facilities with other telecommunications
services, as it was required to do under s. 503(g) of the 1978 Act, it set
up initial operational arrangements.[85] Questions of the ownership of
earth stations – although raised by one of the parties to the proceeding (Western Union International, Inc.) – were left to one side, but
only for a period. The FCC later took the issue on board, as an
important matter, in 1982.

Earth station ownership is a matter over which the FCC has
jurisdiction under s. 201(2), (7) and (9)–(11) of the Communications
Satellite Act of 1962. In a 1966 Report and Order[86] the FCC, after due
inquiry, established a policy by which ownership of the earth stations
in the USA which were used for international communications
through the INTELSAT system, were owned by a consortium of carriers – COMSAT, AT&T, International Telephone and Telegraph
World Communications, Inc. (ITT), Hawaiian Telephone Company,
Radio Corporation of America (RCA) and Western Union International, Inc. (WUI). This consortium was established as the Earth
Station Ownership Committee (ESOC) by agreement between the
carriers, COMSAT having a 50 per cent holding.[87] In 1969 the ESOC
Agreement was amended, with the FCC's consent, to add Guam.[88] In
fact, the consortium worked as three separate consortia, different
allocations of ownership dealing with earth stations on the US mainland, on Hawaii and for the Guam segment. In each of these,
COMSAT retained a 50 per cent holding, with the other participants'
share varying from region to region.

In 1982 the FCC released a Notice of Inquiry on the question of
earth station ownership.[89] Many submissions were received. The
Notice of Proposed Rule-Making was issued on 20 April 1984,[90] and
the Earth Station Ownership Report and Order was released on 18

December 1984.[91] The stages of the Inquiry, Notice and Report and Order are lengthy and contain a review of law, policy, technology, economics and operational factors. The result was a modification of the 1966 Earth Station Ownership Policy permitting the ownership and operation of earth stations connecting with the INTELSAT service external to the carriers' consortium (ESOC). The internal relationships of ESOC has also been altered, although COMSAT, retains its 50 per cent holding through its World Systems Division. The FCC retains its supervisory and licensing role vis-à-vis the construction of earth stations independent of the ESOC stations but, clearly, competition with COMSAT has potentially been once more increased by the 1984 Earth Station Ownership Order.[92]

Naturally, such a step is closely associated with two other questions which the FCC has been considering and on which it also has adopted a final ruling (final for the present, that is). These are first, the questions of direct access to the INTELSAT satellites by the US communications carriers (that is, access without the requirement to go through or purchase service from COMSAT) and, second, the status of 'authorized user' of the system. As it happens, these three were to terminate in a single court decision.

In matters of US access to the INTELSAT system, COMSAT had operated as a carriers' carrier in terms of the FCC's original policy decision in 1966 under which the Commission designated the authorized users of the satellite system.[93] Access to the INTELSAT service for the US international telecommunications carriers passed through COMSAT which charged for its service. Typically, the result was that the INTELSAT charge for a unit was more than trebled. Accordingly, and with the onset of deregulation in telecommunications, this arrangement was put under scrutiny.[94] In response, various communications common carriers sought direct access to the INTELSAT space segment, arguing that the immediate result would be efficiency and a reduction in the charges made to end-users. On 30 March 1984 the FCC denied these requests, but proposed two other options as alternatives on which it sought opinion.[95] These options were, first, access to the space segment through a lease arrangement under which COMSAT would receive only a fee for provision of the facilities and, second, an IRU arrangement for a satellite channel.[96] In both instances COMSAT would have to split up its tariff and set separate rates for its various components, not all of which would therefore operate in every use of the system. Those to whom these ideas were attractive argued that competition, with all its attendant benefits, would therefore be enhanced; others, however, believed the opposite, direct access being a ruse under which certain carriers would shield themselves from the possibility that customers would go to COMSAT itself for transmission capacity rather than go through the carriers. Furthermore, any removal of customers from COMSAT would produce a necessary increase in the Corporation's own tariff for its customers, and the public interest would not thereby be served. COMSAT also argued that its role as access to the international system was statutory, was good policy, and made certain other points relating to internal US law (for example, on confiscation of its rights without 'due process'). Suffice it here to state that the FCC's eventual conclusion was that direct access by the US international carriers to the INTELSAT space segment would not be of significant benefit, and it terminated the proceeding.[97] The Report was later appealed to the courts, but I truncate this discussion here, because we must first outline the third strand in the skein, the matter of the 'authorized user' decisions, the FCC final decision in that matter being appealed simultaneously with the Direct Access decision.

As indicated above, the FCC has jurisdiction in terms of s. 305(a)(2) and (b)(4) of the 1962 Communications Satellite Act in the designation of communications carriers and foreign and domestic US entities as 'authorized users' of the international telecommunications satellite system which COMSAT was to initiate. When the system was established, various communications carriers sought such designation, and by its Authorized User I decision of 1966,[98] the FCC both designated certain telecommunications entities as authorized entities and limited the Corporation to a role as a communications common carrier's common carrier. In 1980 the FCC issued a Notice of Proposed Rule-Making.[99] Aeronautical Radio had asked for authority to offer a communications service for aircraft (cf. the maritime communications service), and raised questions as to it being given status as an authorized entity under the 1966 Order.[100] The FCC, observing that the telecommunications industry had undergone change since the 1966 Authorized User I decision and questioning how far its 1966 criteria remained valid, felt the time had come to review COMSAT's position and sought comment on that and its tentative review of the telecommunications industry.

In Authorized User II,[101] the FCC concluded that the 1962 Act gave the Commission broad powers to set the limits of COMSAT's operations; that it was the Commission in application of a policy directed to the public interest, and not the Act, which had limited COMSAT to acting as a carrier's carrier; that its 1966 Authorized User I decision was based on policy considerations; and that it remained free to alter that determination should it find that circumstances had changed. Furthermore, circumstances had changed, and the FCC accordingly decided to modify the 1966 decision by, first, permitting COMSAT to provide services direct to customers through a separate subsidiary and, second, by permitting non-carrier users to lease satellite transmission capacity from COMSAT's World Systems Division.

This was not a universally popular determination. Court action followed and on 13 January 1984, in *ITT World Communications, Inc. et al.* v *FCC* (1984) 725 F. 2d 732 (DC Cir. 1984), the US Court of Appeals for the District of Columbia Circuit vacated and remanded the Authorized User II decision. It was the view of the Court that the FCC had abused its discretion by attempting, by means of Authorized User II, effectively to restructure the telecommunications industry: it also required the FCC to consider the issues in the Direct Access and in the Earth Station Ownership proceedings, which were then before the Commission, prior to implementing the Authorized User II decision. The Court took the view that these three matters were inextricably intertwined, and, to achieve a proper decision, they had to be settled together and in harmony with one another. The Court was worried lest these other two areas of decision might affect competition between COMSAT and others in providing satellite services ('intramodal' competition) and competition between cable and satellite services ('intermodal' competition), but it did uphold the FCC's right to permit COMSAT to provide service direct to customers.

The FCC responded on 30 March 1984 by terminating the Direct Access proceeding, as described above. Again, as described above, on 20 April 1984 the Notice of Proposed Rule-Making on Earth Station Ownership was issued (the Report and Order in that proceeding being made final on 4 December 1984). One has the impression of the decks being cleared. On 30 April 1984 the FCC issued a Further Notice of Proposed Rule-Making in the Authorized User II proceeding,[102] and adopted its Second Report and Order in the matter (Authorized User III) on 19 December 1984.[103] In it, the FCC confirmed the change of policy contained in Authorized User II, permitting COMSAT to broaden the scope of its activities by providing service to the public. The FCC considered that extension of activities would broaden customer choice, would remove arbitrary restrictions on access to the communications market, would lead to a freer market with fewer economic distortions, and would exercise a downward pressure on telecommunications rates. In Authorized User III, the FCC sought to clarify what it had stated in Authorized User II and extensively discussed both the comments of parties and the Federal Court's decision.

Authorized User III was appealed to the courts along with the FCC Report and Order in the Direct Access proceeding (as to which, see above). While that appeal was pending, the Commission itself denied a petition for reconsideration of Authorized User III.[104] On 31 October 1986 the FCC proceedings in both Reports and Orders were upheld.[105] The Court dismissed argument that the FCC had failed to follow the Court's orders in the earlier case[106] in which the FCC had been held to have abused its discretion in its attempt in Authorized

User II to restructure the telecommunications industry. Nor were the FCC's decisions in either proceeding capricious or arbitrary. The third argument of the petitioners was that the FCC had failed to comply with its duty under s. 201(c)(2) of the 1962 Communications Satellite Act to ensure their equitable access to the international communications satellite system at just and reasonable rates. The Court considered this argument also unfounded. It thought that that argument was in reality one about COMSAT's rates, and held that its own proceedings were not the competent channel for such debate.

It cannot be said that matters are now quiet either for COMSAT or for other US participants in the satellite telecommunications business. Certainly the proceedings outlined above – the COMSAT restructuring, the Earth Station Ownership decision, Direct Access and Authorized User III – appear to settle COMSAT's position. It has been given an increased position within the telecommunications market which to some extent might compensate for the other ways in which its role has been diminished.

But there are other factors to be taken into account. COMSAT has had many other difficulties both in its position in the telecommunications market, and as to its rates and tariffs. It will also be affected by decisions as to its international position, and there is some evidence that it has lost the confidence of both the FCC and some national commenting bodies.

First, most of what has been said above dealt with COMSAT and the international telecommunications market. Development within the national US telecommunications market has also put strains on COMSAT. In particular, companies other than COMSAT and the telecommunications giant, AT&T (which had proposed a contractual arrangement between them), were given a prior access to the US domestic satellite market – an opportunity which they seized to the point of saturating the market.[107] Further, there were changes through the development of computer telecommunications traffic and through the split-up of AT&T and its resultant change from a defensive monopoly supplier of terrestrial telecommunications links to a more active market-seeker. COMSAT was involved in the initial stages of the computer development which in fact triggered the second – the end of the Bell System monopoly. It was earlier indicated, in discussing the development of COMSAT, that as part of its attempts to expand its activities COMSAT, together with IBM and Aetna Life Insurance, formed Satellite Business Services. COMSAT did not remain a partner for long. IBM continued to be interested in the use of satellites for intercomputer telecommunications traffic. The FCC was required to regulate such developments and, through decisions on what it classified as 'basic' and 'enhanced' services, provided opportunity for enterprises other than AT&T to enter the telecommunications market through satellites.[108] That 'incursion' into 'its' domain provoked AT&T first to attempt to defend its monopoly through the use of Congress, but brought counter-argument and weakened its position in a long-running debate as to the wisdom of permitting the Bell System to exercise its virtual monopoly on long-distance communications. The result was the break-up of the Bell System with the divestiture of AT&T of the local Bell System companies from 1984 under a settlement agreed in 1982 between it and the Department of Justice and approved by Judge Greene of the Federal District Court for the District of Columbia.[109]

During, and as a result of, these developments COMSAT encountered severe competition in its attempts to diversify its activities, not least from those companies which, in the 1970s, were permitted to engage in the creation and operation of US domestic satellite systems and it appears not always to have made the best decisions.[110] In particular, COMSAT's attempt to establish an internal US service using the Comstar series of satellites has run into difficulty, and that produced problems for the company in the mid-1980s. In 1986, therefore, COMSAT and the Contel Corporation agreed the merger of their two companies and duly applied to the FCC for approval of such a step.[111] GTE Spacenet Corporation sought a belated extension of time to file comments on the application, but was unsuccessful.[112] As it happened, however, difficulties were encountered in negotiating the merger, and by mutual agreement the proposal was dropped.[113] This may have been unfortunate. The businesses of the two corporations could well eventually have been brought into a useful harmony, and certainly such a link to internal US telecommunications would have benefited COMSAT. Steps continue to be taken, however, short of merger and it will be intriguing to see how matters develop, particularly since the Japanese International Satellite Joint Users Organization (JISO) *inter alia* thought it desirable to ask the FCC to place conditions on its approval of the most recent transfers between the two corporations.[114] The FCC considered the imposition of such conditions to be outwith its jurisdiction.

Second, COMSAT has run into severe trouble in connection with the rates and tariffs it charges its customers. The question of its rates came before the FCC in 1965 when the initial INTELSAT system was being established.[115] Thirteen years and many proceedings and court actions later[116] in 1978, an acceptable rate of return on investment was set at 11.48–12.48 per cent by a Settlement Agreement.[117] Notwithstanding, when reviewed in proceedings begun in 1985 COMSAT was adjudged to have exceeded its permitted level of return in various ways and was ordered to repay, in a manner to be negotiated, some US$ 39 million accrued through excess revenues for the period 1983–86 – a sum which had been originally tentatively set at US$ 61.7

million. The matter of tariffs,which shows COMSAT in deep waters, was settled finally (?) in December 1988.[118]

Third, it seems that COMSAT has antagonized the FCC by the way in which it has conducted itself during some of the various FCC proceedings. This can be seen documented in the proceeding just discussed. One cannot but regret that the FCC found it necessary to comment, for example, that the record in that proceeding is 'pockmarked with instances of COMSAT's failure to respond' (April 6 Order, para 82), and that:

> . . . the Commission's effort to implement its statutory mandate to ensure just and reasonable rates continues to be frustrated, as in the past, by COMSAT's unresponsiveness. COMSAT has responded to several fundamental issues in the Designation Order by resubmitting altered and deficient information already provided with the tariff transmittals, and has failed to provide the supporting studies and workpapers required by the Order (April 6 Order, para 3).

The FCC ends its Conclusion thus:

> 'We again emphasize that COMSAT's failure to fully satisfy these and subsequent requirements [of the Order], in light of its past history, will occasion a thorough review of administrative actions available to enforce compliance (April 6 Order, para 84).

This kind of language, and allegations from the appointed regulatory authority having jurisdiction over COMSAT, does not augur well for the company's standing and hence for its future role in US international telecommunications arrangements.

Fourth, while COMSAT has been affected by the domestic changes in the US telecommunications industry which are represented and reflected by the Inquiries and Reports and Orders dealt with above, we must also mention a US change which impinges directly on satellite usage for international telecommunications. By its International Circuit Distribution Report and Order of 24 March 1988[119] the FCC has abandoned the requirement of what was called 'balanced loading – that is, that international telecommunications be split between satellites and cables, according to FCC guidelines. Instead, the FCC will rely on an agreement between COMSAT and AT&T indicative of the latter's intentions as to satellite use, and will keep a watching brief. Even so, it is now possible that satellite traffic will decline if it fails to meet the requirements of open competition, and that will affect COMSAT.

Fifth, COMSAT will be further significantly affected by developments in the so-called 'separate systems' proceedings.[120] COMSAT is still umbilically linked to INTELSAT. If the USA persists in its drive to force the development of US competition to INTELSAT through licensing separate satellite systems to the extent that INTELSAT is damaged, COMSAT will also necessarily be damaged. Indeed, COMSAT will suffer through any reduction in the profitability of INTELSAT or INMARSAT (although the latter organization is significantly smaller and hence less important in this context). But in addition, if INTELSAT is wilfully damaged by the USA it may be that other clients of COMSAT services might respond by reducing their business with COMSAT because it is a US national. Be that as it may, competition through such separate systems (or through the new fibre optic international cables) will affect both COMSAT and INTELSAT. It suffices here, however, to mark that COMSAT's profits and future will also suffer by such developments.

However, before we leave COMSAT, certain positive things can be said. First, without it, the international structures for satellite telecomunications would be quite different. COMSAT's place and role in the creation of the INTELSAT organization were fundamental. Its role in the development of the INTELSAT system was also highly important. Decisions taken when it was first the manager and operator, and then under the management services contract in the permanent INTELSAT structure, form the foundation of what has been achieved by the organization. Again, COMSAT remains the major participant in the present-day INTELSAT, with voting strength to match. Its share has decreased over the years, but unless the Communist countries seek entry to INTELSAT (which seems unlikely), COMSAT's share will not reduce much further. The USA is the major user of the international satellite system, and hence COMSAT's share will reflect that usage. That said, there is the implicit threat in the entry of separate systems into the market, particularly if much US traffic is diverted to such systems. That possible development must cause concern to objective commentators. We can but hope that wise decisions are taken as to the introduction of separate systems. INTELSAT was created to provide the international participation in the single global system envisaged in the Kennedy Statement of July 1961 and in the Communications Satellite Act of 1962. COMSAT is the US participant in that organization and must not be crippled in its international role through the vagaries of an isolationist and self-regarding US telecommunications policy. Pessimists might say that that has already happened.

With minor variation, the same points may be made about COMSAT's place in the development and functioning of INMARSAT. The difficulties indicated below which may be emerging as to COMSAT's place in the provision of an air mobile service under the aegis of INMARSAT seem strange to an outsider, although one can see and understand the pressure for an independent or an alternative additional service-provider within the domestic US air mobile market. Again, we may but hope that decisions are taken which will not cripple COMSAT's international role.

The OTA Report, *International Cooperation and Competition in Civilian Space Activities*, 1985,[121] notes three options for consideration for the future of COMSAT. The first is to continue with the present regime, COMSAT remaining the USA's sole owner of its INTELSAT investment and having the monopoly of access to INTELSAT with, however, a somewhat increased Congressional supervision of the FCC's role in rate-setting and in the separation of regulated from unregulated activities. The second option is to require the rapid transformation of COMSAT into an ordinary carrier and remove its privileged position vis-à-vis INTELSAT. Third, the OTA identifies the possibility of reducing COMSAT into merely a sole conduit to INTELSAT. The report discusses these briefly, states that options two and three would require further elaboration, but then recognizes that each of these two, while providing some solution to problems of internal US competition, would themselves cause other problems in the other areas. It is to be hoped that this warning is heeded.

Little comfort can, however, be taken from FCC decisions in 1987 in relation to INMARSAT. In January 1987 the FCC refused authorization to COMSAT to participate in the proposed aeronautical service to be instituted in accordance with the revision of the INMARSAT Convention and Operating Agreement.[122] In part, this decision was arrived at because the prospective customer for the service which COMSAT proposed said it did not wish it in the form proposed; but more importantly, the FCC accepted argument that the proposed service was *ultra vires* of COMSAT's powers. Neither the Maritime Satellite Act of 1978 nor the 1962 Communications Satellite Act contemplated COMSAT's providing such services. On the contrary, COMSAT's monopoly in relation to mobile services (conferred by Congress through the 1978 Maritime Satellite Act) was restricted to the maritime services of INMARSAT. Any monopoly for aeronautical services would have to be similarly constituted, if that was the will of Congress: it was not to be created by FCC fiat.

Competition in the provision of aeronautical services was different from that in the maritime field, and separate consideration would have to be given as to how such services should be provided for the USA within the context of the INMARSAT developments. The FCC announced that it would return to the question in the immediate future, and in March 1987 did so. In 'In the Matter of Provision of Aeronautical Services via the INMARSAT System,'[123] the FCC gave Notice of Proposed Rule-making, asking for comment and counterproposals on a scheme to provide aeronautical services on a competitive basis, holding that a competitive basis would best serve the (US?)

public interest. Competition could be provided through separate aeronautical satellite services, and also through a variation of the provision of INMARSAT services within the USA. US carriers could be authorized to provide services through the INMARSAT system, and would interact with the organization itself through a consortium of such users, the aeronautical consortium (AEROCON).[124] Changes both to the INMARSAT constitution and the relevant US legislation would be required. This proposal is considered in the discussion of INMARSAT itself. Here it suffices to note that COMSAT's position is being eroded. Certainly COMSAT would continue to have a role and would continue to provide services, but its position as the sole channel between the USA and the international organization of telecommunications would be altered. Indeed, the FCC discussion contemplates the possible substitution of two US Signatories to the INMARSAT Operating Agreement in order to meet internal US requirements. If that happens, it will result in a further, undesirable, weakening of COMSAT as the US participant in international satellite telecommunications.

Once more, one wonders whether the public discussion of these options is helpful to US participation in the organizations. The USA seems prone to act first, and negotiate later, to talk domestically and then expect others to fall into line.[125] It is sad to see such a confusion between 'leadership' (which is alleged) and cooperative initiative (which others would welcome). Leadership in international affairs does not consist in either an insistence on or a naive expectation of compliance by others. General De Gaulle (to name an historic example) shows that such tactics do not work for long.

On the other hand, we may note thankfully that in autumn 1987 the FCC did dismiss an application to establish an entirely separate global aeronautical communications satellite system.[126] However, the dismissal was on the grounds that the proposal was not consistent with FCC rulings on spectrum allocation[127] and because Arinc had not demonstrated its financial qualification to provide the services. The dismissal is without prejudice to a future application by Arinc. The possibility of a separate international organization for aeronautical communications, or of some revised form of INMARSAT with COMSAT not the US Signatory for air-mobile purposes, remains. At its worst, then, COMSAT might disappear through legislative action. Short of that, COMSAT might still vanish were it to find itself no longer the sole US channel to any international telecommunications system. In that case, the Corporation would appear to be redundant, for it was designed, and still is tailored, to be that sole US channel. Were COMSAT ever remodelled to be one channel among many to the international organizations, if its US competitors were able selectively to remove the more lucrative parts of its business, COMSAT would probably find it difficult to meet its financial obligations in the organizations. In such a case, COMSAT would have to be replaced under the INTELSAT and the INMARSAT Agreements.[128] What one would hope is more likely, however, is that COMSAT will be permitted to retain its major role within the provision of the US link to international telecommunications, though subject to close scrutiny by the FCC and its 'competitors' and those who buy service from it for further retail. In the 1980s that scrutiny has proved effective as is shown in the 1985/7 FCC proceeding which resulted in COMSAT having to repay excess charges.[129] That supervisory mechanism having worked once, it will be easier to work it again. Maybe there is something to be said for allowing competitors to police each other's activities.

NOTES

1 See the useful statement, *Communication Satellites: Technical, Economic and International Developments*, a Staff Report prepared for the Use of the Committee on Aeronautical and Space Sciences, US Senate, 87th Cong. 2nd Sess., 25 February 1962.

2 See, 'Who should own and operate the system?', ibid. ch. IV, pp. 41–62.

3 Some would argue that the provisions of the Space Treaties bind only their members. Others consider that the principles of such treaties are now part of customary international law, and hence binding on all states. See Chapter 1, 'General international law' and 'Space law'.

4 Cf. Chapter 6, section 'The European Launcher Development Organization', p. 248.

5 Report. *Communications Satellite Act of 1962, and Minority View on HR 11040*, and relative *Hearings* before the Committee on Foreign Relations of the US Senate, 87th Congress, 2nd Session. A table of the estimated costs of the Telstar launchings is given at p. 264 of the *Hearings*.

6 See the Narrative Summary and Documentary Annex to *Documents on International Aspects of the Exploration and Use of Outer Space*, Report prepared for the Committee on Aeronautical and Space Sciences, US Senate. 88th Congress, 1st Session. Doc. no. 18, 9 May 1963. These developments are outlined above, Chapter 1, 'Advent: Communications and the Space Age'.

7 Cf. the reaction reported by the horror writer, Stephen King, in his *Danse Macabre* (London: Futura, 1982) pp. 17 and 21–2.

8 See *Satellite Communications (Military-Civil Roles and Relationships)*, Report prepared by the Military Operations Subcommittee of the Committee on Government Operations, US HR, 1964, 88th Cong. 2nd. Sess. and relative Hearings, *Satellite Communications – 1964*, March–August 1964, Parts I and II.

9 The Council was to be abolished as from 1 July 1973: Reorganisation Plan No. 1 of 1973, 38 Federal Register 9579, 18 April 1973.

10 Thus, a year later, NASA stated its purpose in its communications satellite programmes as being 'to assist in the early establishment of operational communications satellite systems, and to support the continuing development and expansion of such systems through a program of research, development, and flight testing of techniques and concepts designed to insure (sic) the realisation

of the full capabilities of communications satellites': *NASA Budget Estimates, Fiscal Year 1964*, vol. II, p. 7.

11 See *Satellite Communications: (Military–Civil Roles and Relationships)*, op. cit., p. 13, which also cites testimony given to the Subcommittee by a NASA witness: See its Hearings, *Satellite Communications – 1964*, op. cit., March–May 1964, Part I, p. 230.

12 Department of State *Bulletin*, 16 January 1961, p. 77; *Documents on International Aspects of the Exploration and Use of Outer Space, 1954–1962*, Staff Report prepared for the use of the Committee on Aeronautical and Space Sciences, US Senate, 1963, 88th Cong. 1st Sess., Doc. no. 18, at p. 186.

13 *Public Papers of the Presidents: John F. Kennedy* (Washington DC: US Government Printing Office, 1961) pp. 403–5; *Documents on . . . Outer Space, 1954–62*, op. cit., pp. 202–4 at p. 203.

14 Excerpted in *Documents on . . . Outer Space, 1954–1962*, op. cit., pp. 207–8 at p. 208, with omission of government responsibility elements; available in full in *Public Papers of the Presidents: John F. Kennedy* op. cit, pp. 529–32; *Space Satellite Communications, Review of the Report of the Ad Hoc Carrier Committee*, Part 2 of Hearings before the Subcommittee on Monopoly of the Select Committee on Small Business, US Senate, 87th Cong. 1st Sess., pp. 733–4; in *Satellite Communications – 1964*, Hearings before a Subcommittee of the Committee on Government Operations, US HR, 1964, 8th Cong. 2d Sess., Part 1 at p. 590–1; A. Chayes, T. Ehrlich and A.F. Lowenfeld, *International Legal Process*, vol. 1 (Boston, Mass.: Little Brown 1968) pp. 636–7.

15 See 'In the Matter of an Inquiry into the Administrative and Regulatory Problems Relating to the Authorization of Commercially Operable Space Communications Systems, FCC Docket No. 14024; 1. Notice of Inquiry, adopted 29 March 1961 and released 3 April 1961; 2. First Report, adopted and released 24 May 1961; 3. Memorandum Opinion and Order, released 25 July 1961; 4. Supplemental Notice of Inquiry, adopted 21 July 1961 and released 25 July 1961, which set up an Ad Hoc Carrier Committee to consider how a joint-venture could be established; 5. Petition of General Telephone and Electronics Corporation and others for reconsideration of No. 4; 6. Statement by the FCC before the Subcommittee on Communications of the Committee on Commerce, US Senate, 1 August 1961; 7. Report of the Ad Hoc Carrier Committee, 12 October 1961. Apart from the FCC files, these are printed as follows. In *Communications Satellites, Part 1*, Hearings before the Committee on Interstate and Foreign Commerce, US HR, 1961, 87 Cong. 1st Sess.: 1. at p. 10; 2. at p. 12; 3. at p. 16; 4. at p. 19. In *Space Satellite Communications*, Hearings before the Subcommittee on Monopoly of the Select Committee on Small Business, US Senate, 1961, 87th Cong. 1st Sess. Part 1: 1. at p. 492; 2. at p. 493; 3. at p. 496; 4. at p. 499; 5. at p. 184; 6. at p. 485. No. 7 is available in *Space Satellite Communications: Review of the Report of the Ad Hoc Carrier Committee*, (the confusingly retitled volume which is Part 2 of the Hearings immediately previously cited), at p. 669.

16 See, generally, *Progress Report from FCC–1965*, Hearings before the Subcommittee on Communications of the Committee on Commerce, US Senate, 1965, 89th Cong. 1st Sess.

17 Were the copyright in the Statement to be enforced, or enforceable, a case could be made out for its infringement.

18 See: 1. *Communications Satellite Incorporators*, Hearings before the Committee on Commerce, US Senate, 88th Cong. 1st Sess., 11 March 1963; 2. *Nomination of Incorporators*, Hearings before the Committee on Aeronautical and Space Sciences, US Senate, 88th Cong. 1st Sess., 19 March 1963.

19 The Articles of Incorporation of the Communications Satellite Corporation are printed, (1963) II ILM 395.

20   F. Lyall, 'Law and Space Telecommunications', unpublished LL.M. thesis, Institute of Air and Space Law, McGill University, (1965) at p. 85.

21   The 'Authorized User' concept and the relevant FCC proceedings are discussed below.

22   'In the Matter of Communications Satellite Corporation Investigation into Charges, Practices, Classifications, Rates and Regulations', FCC Docket No. 16070, Memorandum Opinion and Order, adopted 9 May 1978, released 23 May 1978, Release No. FCC 78–312, (1978) 68 FCC 2d 941; the COMSAT Rate Case.

23   Thus, in a dispute over direct access to the INTELSAT space segment for other US international carriers, much argument turned upon the matter of COMSAT's rates: In the Matter of Regulatory Policies Concerning Direct Access to INTELSAT Space Segment for the US International Service Carriers, 97 FCC 2d 296 (1984). Recent FCC action on Direct Access is discussed below.

24   'In the Matter of Communications Satellite Corporation', CC Docket Nos. 80–634 and 85–268, Memorandum Opinion and Order, adopted 3 April 1987, released 6 April 1987, (1987) 2 FCC Rcd 3706: limited reconsideration, Memorandum Opinion and Order, adopted 10 December 1987, released 28 January 1988, Release No. FCC 87-388, 37238, (1988) 64 Rad. Reg. 2d (P & F) 524. See n.118 below.

25   Cf. 'In the Matter of the 1985–1987 Consolidated Capitalization Plan of the Communications Satellite Corporation (COMSAT)', Memorandum Opinion and Order, Slip Opinion, Release No. FCC 86–15, released 21 January 1986, adopted 7 January 1986.

26   Domestic Communications Satellite Facilities, (1972) 38 FCC 2d, 665 at pp. 679–80.

27   For example, 'In re the Application of American Satellite Company', et al., Order, Authorization and Certificate, Slip Opinion, adopted 20 September, 1985, released 30 September, 1985. Coordination with INTELSAT is dealt with in relation to INTELSAT, Chapter 4, p. 173.

28   See Chapter 4, section 'US Separate Systems', p. 182.

29   S. 505 of the 1978 Act triggered the COMSAT structure proceedings discussed below.

30   Prospectus of the Communications Satellite Corporation, 1. The preliminary form of the Prospectus is printed (1964) III ILM 571. The final form was available from stockbrokers handling the first stock issue, and is printed in Satellite Communications – 1964 (Part 1), op cit. pp. 597–657.

31   Cf. G.A. Almond, 'Public Opinion and the Development of Space Technology, 1957–60' in J.M. Goldsen (ed.), Outer Space in World Politics, (New York and London: Frederick Praeger, 1961) p. 71.

32   Although based on personal observation, this point has other authority. See paras. 439–40 of the Final Report and Order in the COMSAT Study, 1980: In the Matter of COMSAT Study – Implementation of Section 505 of the International Maritime Satellite Telecommunications Act, CC Docket No 79–266, 77 FCC 2d 564, released 1 May 1980.

33   See: 'Procedures for US Government Instruction of the Communications Satellite Corporation in its Role as the US Representative to the Interim Communications Satellite Committee', printed as Appendix F to the Final Report and Order in the COMSAT Study. The Interim Communications Satellite Committee was the equivalent of the Board of Governors of INTELSAT in the days of the interim arrangements for INTELSAT: see below and Chapter 3.

34   See the FCC Final Report on the COMSAT Study just cited, para. 442.

35   See 'In the Matter of Communications Satellite Corporation; Participation in INTELSAT's Planned Domestic Services, Memorandum Opinion and Order', File No I–S–P–88–001, adopted 25 November 1987, released 30 November 1987.

36   See Chapter 4, section 'US Separate Systems' at n. 173, p. 187.
37   'Establishment of Satellite Systems Providing International Communications'; [The Separate Systems Order, Further Reconsideration] Memorandum Opinion and Order, adopted 23 October 1986, released 5 November 1986, Release No. FCC 86–471, (1986) 1 FCC Rcd 439 at para 23.
38   Cf. S.A. Levy, 'Private Diplomacy and Public Business: Public Supervision of the Communications Satellite Corporation' *Chicago Law Review*, vol. 45, 1978, pp. 418–49. I have also heard allegations that in some early discussions the COMSAT delegation had an open line to the State Department during Board Meetings.
39   Contra, S.D. Estep, 'Some International Aspects of Communications Satellite Systems', *NW. University Law Review*, vol. 58, 1963, p. 237.
40   Prospectus, p. 3. The Articles of Incorporation of the Communications Satellite Corporation, printed, *Nomination of Incorporators*, Hearing before the Committee on Aeronautical and Space Sciences, US Senate, 19 March 1963, 88th Cong. 1st Sess., pp 43–51. The volume also contains the Byelaws of the Communications Satellite Corporation (as adopted 4 February 1963) at pp. 51–63, and a section-by-section annotation of the Articles as a 'legislative history' of their drafting at pp. 112–23.
41   See generally Act, s. 305 and Articles of Incorporation, art. III, s. 3.01 and 3.02.
42   Act, s. 305(a)(1) and (3), s. 305(b)(2): Articles, art. III s. 3.02(b)(1), (3), (5), (13), (14) and (15).
43   Act, s. 305(b)(4): Articles, art. III s. 3.02(4), (6), (7).
44   Act, s. 403(a).
45   Articles, art. III, s. 3.02(11).
46   Articles, art. III, s. 3.02(10).
47   Articles, art. V, s. 5.03(a).
48   Articles, art. V, s. 5.03(b).
49   Act, s. 304(b)(1), (2): Articles, art. V, s. 5.03(b).
50   Act, s. 304(b)(3): Articles art. V, s. 5.02(c).
51   Act, s. 304(d): Articles, art. V, s. 5.02(d).
52   'Domestic Communications Satellite Facilities' (1972) 38 FCC 2d 665 at pp. 679–80. Final Report and Order, In the Matter of COMSAT Study – Implementation of Section 505 of the International Maritime Satellite Telecommunications Act, (1979) 74 FCC 2d 564, released 1 May 1980, adopted 22 April 1980, para. 54 and n. 17.
53   Act, s. 303(a): Articles. art. VIII, s. 8.02.
54   Act, s. 303(a): Articles. art. VIII, s. 8.02.
55   Act, s. 303(a): Articles. art. V, s. 5.04(b), art. VIII, s. 8.02.
56   Act, s. 303(b).
57   Act, s. 303(b): Articles, art. VIII, s. 8.10.
58   Act, s. 305.
59   Articles, art. III.
60   Prospectus, p. 7: 'Offering of Common Stock'.
61   See terms of Purchase Contract, Prospectus, p. 51. Over 130,000 individuals bought shares.
62   The original offering taken up by the carriers authorized to hold shares is listed, Prospectus, p. 7.
63   Cf. Statement of Leo D. Welch, Chairman, Communications Satellite Corporation, *1964 NASA Authorisation, Part 4*, Hearings before the Subcommitee on Applications and Tracking and Data Acquisition of the Committee on Science and Astronautics, US HR, 88th Cong. 1st. Sess., 1963, at p. 3297. In fact, the first quarterly dividend was paid in December 1970, at a rate of 12.5 cents per share. This quarterly rate rose to 14 cents in 1972, 17 cents in 1973, 20 cents in January

1974 with a leap to 25 cents that June and 35 cents in October 1977. The dividend in 1978 was US$ 2.00; 1979, US$ 2.225; 1980, US$ 2.30. In 1983 it had dropped partly through the reorganization dealt with below, and partly through business losses to US$ 1.175. In 1984 and 1985 it was US$ 1.20 per share on a net loss in the latter year because of discontinuing some operations, and also the establishment of a reserve against projected losses in the direct broadcasting satellite business (if I have understood the figures). Although these figures do have some impressionistic meaning, it should be noted that they are not strictly comparable over the years through inflation, the issuance of further stock offerings and repurchases.

64 *Newsweek*, 18 May 1964, p. 87.
65 Cf. Prospectus, p. 4 'The Venture and its Risks', and p. 9 'Satellite Communications'.
66 There was the possibility that if the commercial venture failed, the military authorities might have taken over the system for their purposes. See *Satellite Communications – 1964, Parts 1 and 2*, op. cit.
67 Act, s. 304(d): Articles, art. V, s. 5.02(d)).
68 Act, s. 303(b): Articles, art. V, s. 5.04(b) and art. VIII, s. 8.02 and 8.10.
69 'Agreement Establishing Interim Arrangements for a Global Commercial Communications Satellite System, and relative Special Agreement', Cmnd. 2436, 1964.
70 Annex to the Special Agreement.
71 Intergovernmental Agreement, art. V(c).
72 Intergovernmental Agreement, art. V(d).
73 See R.R. Colino, *The INTELSAT Definitive Arrangements: Ushering in a New Era in Satellite Telecommunications*, (Geneva: European Broadcasting Union, 1972) pp. 3–5; J.T. Kildow, *INTELSAT: Policy-Maker's Dilemma* (Lexington, Mass.: Lexington Books, 1973) pp. 3–18, 43–58.
74 Kildow, *INTELSAT*, op. cit., pp. 52–3.
75 The developing countries wanted a good but inexpensive system. The USA wanted a state-of-the-art system so far as compatible with price, which would have favoured US contractors. The other developed countries wanted their 'share' of procurement contracts and know-how. See the discussion of the negotiations below, Chapter 3, 'Interim INTELSAT: the negotiations' p. 79.
76 'In the Matter of Implementation of Section 505 of the International Maritime Satellite Telecommunications Act', 74 FCC 2d 59, October 19, 1979 released; adopted 18 October 1979.
77 'In the Matter of COMSAT Study – Implementation of Section 505 of the International Maritime Satellite Telecommunications Act, 77 FCC 2d 564, 1 May 1980 released; adopted 22 April 1980.
78 Environmental Research and Technology Inc. was sold to Resource Engineering, Inc. of Houston, Texas in 1985: see *COMSAT: The Magazine of the Communications Satellite Corporation*, no. 17, 1985, at p. 3. The same page intimated the sale of another subsidiary, Compact Software, Inc. to the Communications Consulting Corporation.
79 'In the Matter of Changes in the Corporate Structure and Operations of the Communications Satellite Corporation', 81 FCC 2d 287 (1980), CC Docket No. 80–634.
80 'In the Matter of Changes in the Corporation Structure and Operations of the Communications Satellite Corporation', First Memorandum Opinion and Order, 90 FCC 2d 1159 (1982), reconsideration denied 93 FCC 2d 701 (1983), appeal sub nom. *RCA Global Communications, Inc. et al.* v. *FCC*.
81 It may also be noted that, in a separate proceeding, COMSAT was permitted, again through a fully separated subsidiary, to enter the ship earth station

manufacturing market: 'In the Matter of Changes in the Corporate Structure and Operations of the Communications Satellite Corporation (Ship Earth Stations)', 90 FCC 2d 488, CC Docket No. 80–634, released 18 May 1982, adopted 13 May 1982.

82   'In the Matter of Changes in the Corporate Structure and Operations of the Communications Satellite Corporation', Second Memorandum Opinion and Order, 97 FCC 2d 145; released 20 April, 1984; adopted 30 March 1984.

83   'In the Matter of Changes in the Corporate Structure and Operations of the Communications Satellite Corporation', 99 FCC 2d 1040 (1984). As to the 11.48 per cent, see above and below. The FCC has found COMSAT to have accrued excess revenue and has ordered repayment amounting to US$ 31 million.

84   'In the Matter of Changes in the Corporate Structure and Operations of the Communications Satellite Corporation', Slip Opinion, CC Docket 80–634, FCC 85–178, adopted 11 April, 1985, released 19 April, 1985. This Report and Order is referred to as the Third Report and Order in the COMSAT Structure proceeding.

85   'In the Matter of of Implementation of Requirements of the International Maritime Satellite Telecommunications Act', CC Docket No. 79–35, Memorandum Opinion and Order, adopted 26 April 1979, released 30 April 1979, Release No. FCC 79–248, (1979) 71 FCC 2d 1069.

86   'In the Matter of Amendment of Part 25 of the Commission's Rules and Regulations with respect to Ownership and Operation of Initial Earth Stations in the United States for Use in Connection with the Proposed Global Commercial Communications Satellite System', 5 FCC 2d 812 (1966), Docket no. 15375, Release No. FCC 66–1133, the 'Second Report and Order'. The First Report in the matter, setting Interim Policy, was adopted on 12 May, and released on 13 May, 1965, Release No. FCC 65–401, (1965) 38 FCC 1104. Under it, COMSAT was authorized to construct, own and operate the first three initial earth stations – one in Hawaii and one in each of north-east and north-west US mainland. That First Report was reconsidered and policy slightly modified by Order adopted 23 February and issued 25 February 1966, Release No. FCC 66–176. Both the First Report and its amendment are printed in *Progress Report on Space Communications*, Hearings before the Subcommittee on Communications, Committee on Commerce, US Senate, 1966, 89th Cong. 2nd Sess., at pp. 31–52.

87   See Second Report and Order, paras 18–31. COMSAT was required to transfer its ownership of the three initial earth stations dealt with in the First Report and Order to the ESOC, Second Report para. 29.

88   In 'Re Radio Corporation of America Global Communications, Inc.', (1969) 18 FCC 2d 1037.

89   'In the Matter of Modification of Policy on Ownership and Operation of US Earth Stations that Operate with the INTELSAT Global Communications Satellite System, Notice of Inquiry', (1982) 90 FCC 2d 1458, CC Docket No. 82–540, released 17 August 1982.

90   'Modification of Policy on Ownership and Operation of US Earth Stations that Operate with the INTELSAT Global Communication Satellite System, Notice of Proposed Rule-Making', (1984) 97 FCC 2d 444.

91   'In the Matter of Modification of Policy on Ownership and Operation of US Earth Stations that Operate with the INTELSAT Global Communications Satellite System, Report and Order', adopted 4 December 1984, released 18 December 1984, (1984) 100 FCC 2d 250.

92   In 1985, partly in response to the Earth Station Ownership decision, COMSAT sold to AT&T, effective 1 January 1988, its 50 per cent share in three US international earth stations, and announced the closure of two others. COMSAT is to focus its international earth station business on urban gateway earth stations

that provide specialized services. See *COMSAT: the Communications Satellite Corporation Magazine*, no. 17, 1985, at p. 2.

93   Authorized Entities and Users, (1966) 4 FCC 2d 421, reconsidered (1967) 6 FCC 2d 593: see below.

94   'In the Matter of Regulatory Policies Concerning Direct Access to INTELSAT Space Segment for the US International Service Carriers, Notice of Inquiry', (1982) 90 FCC 2d 1446, CC Docket No. 52–548.

95   'Regulatory Policies Concerning Direct Access to INTELSAT Space Segment for the US International Service Carriers', (1984) 97 FCC 2d 296.

96   The IRU is the 'Indefeasible Right of User' concept used in cable agreements under which it is possible to purchase rights of use but not of management or control in a cable. There is usually a cost related to those of construction and maintenance over the life of the cable.

97   'In the Matter of Regulatory Policies Concerning Direct Access to INTELSAT Space Segment for the US International Service Carriers, Report and Order', (1984), adopted 30 March 1984.

98   'In the Matter of Authorized Entities and Authorized Users under the Communications Satellite Act of 1962', (1966) 4 FCC 2d 421, Docket No. 16058, release no. FCC 66-677, adopted 20 July 1966, (also printed, *Progress Report on Space Communications*, Hearings before the Subcommittee on Communications, Committee on Commerce, US Senate, 1966, 89th Cong. 2nd Sess., at pp. 20 31); reconsideration granted in part (the modification was minor), (1967) 6 FCC 2d 592, release no. FCC 67-164, adopted 1 February 1967.

99   'In the Matter of Aeronautical Radio, Inc. Petition for Declaratory Ruling that it is an Authorized User of the International Telecommunications Facilities Provided by the Communications Satellite Corporation under the Communications Satellite Act of 1962, CC Docket No. 80–170, Notice of Proposed Rule-Making, adopted 22 April 1980, released 6 May 1980, (1980) 77 FCC 2d 535.

100 At its next similar attempt, the FCC dismissed without prejudice an application by Aeronautical Radio Inc. to establish a global aircraft communications satellite system on grounds of failure to conform to the FCC's rules for spectrum use and failure to show financial standing for the service proposed: 'In the Matter of Aeronautical Radio Inc. (Arinc); For authority to construct an aviation satellite system (AvSat)', adopted 9 September 1987, released 10 September 1987, Release No. FCC 87–289, 2 FCC Rcd 5990, (see below).

101 'In the Matter of Proposed Modification of the Commission's Authorized User Policy Concerning Access to the International Satellite Services of the Communications Satellite Corporation', CC Docket No. 80–170, Report and Order, adopted 5 August 1982, released 19 August 1982, Release No. FCC 82–357, (1982) 90 FCC 2d 1394.

102 'In the Matter of Proposed Modification of the Commission's Authorized User Policy Concerning Access to the International Satellite Services of the Communications Satellite Corporation, Further Notice of Proposed Rule-Making', (1984) 98 FCC 2d 158.

103 'In the Matter of Proposed Modification of the Commission's Authorized User Policy Concerning Access to the International Satellite Services of the Communications Satellite Corporation, Second Report and Order, adopted 19 December 1984, released 11 January 1985, Release No. FCC 84–633, (1984) 100 FCC 2d 177.

104 'In the Matter of Proposed Modification of the Commission's Authorized User Policy Concerning Access to the International Satellite Services of the Communications Satellite Corporation', CC Docket 80–170, Slip Opinion, adopted 19 June 1985, released 28 June 1985, Release No. FCC 85–323.

105 *Western Union International, Inc., et al.* v. *FCC*, (1986) 804 F. 2d 1280 (DC Cir.).

106 *ITT World Communications, Inc. et al.* v. *FCC* (1984) 725 F. 2d 732 (DC Cir.).

107 'In the Matter of Establishment of Domestic Communications-satellite Facilities by Non-governmental Entities', Docket no. 16495, Second Report and Order 16 June 1972, (1972) 35 FCC 2d 844; Memorandum Opinion and Order adopted 21 December 1972, released 22 December 1972, Release No. FCC 72–1198 (1972) 38 FCC 2d 665.

108 'In the Matter of Amendments to Sec. 64.702 of the Commission's Rules and Regulations (Second Computer Inquiry)', (1980) 84 FCC 2d 50. The distinction between basic and enhanced services was sought to be introduced into INTEL-SAT thought by the USA in relation to the 'separate systems' question. See Chapter 4, section 'US Separate Systems', p. 182.

109 *United States* v. *American Telephone and Telegraph Company*, Modification of Final Judgement, Civil action nos. 74–1698 and 82–0192, (1982) 552 Fed. Supp 131, aff'd sub nom. *Maryland* v. *United States* (1983) 460 US 1001. See J. Hill, *Deregulating Telecoms: Competition and Control in the United States, Japan and Britain* (London: Frances Pinter, 1986) pp. 50–77; and P. Temin, *The Fall of the Bell System: A Study in Prices and Politics* (Cambridge: Cambridge UP, 1988).

110 It is perhaps symptomatic of their commercial reservations that other carriers have divested themselves of the share of COMSAT stock they were entitled in terms of the Communications Satellite Act of 1962: see 'Domestic Communications Satellite Facilities (1972)', 38 FCC 2d 665 at pp. 679–80.

111 Public Notice, released 18 November 1986, (1986) 1 FCC Rcd 721.

112 'In the Matter of Application of Communications Satellite Corporation and Contel Corporation for Approval of Merger of the Two Companies', Order, adopted 15 December 1986, released 16 December 1986, File No. ENF 87–05, (1987) 2 FCC Rcd 76.

113 'In the Matter of Application for Consent to Merger of Contel Corporation and Communications Satellite Corporation', Order, adopted 22 July 1987, released 30 July 1987, File No. ENF–87–05, (1987) 2 FCC Rcd 4462.

114 'In the Matter of Communications Satellite Corporation, COMSAT International Communications, Inc., COMSAT Technology Products, Inc. and American Satellite Company d/b/a/CONTEL ASC: Application for Authorizations to Transfer Control of and to Re-issue Commission Authorizations; COMSAT International Communications, Inc. and COMSAT Earth Stations, Inc.; Application to Assign Commission Authorizations from COMSAT International Communications, Inc., to COMSAT Earth Stations, Inc.', Memorandum Opinion and Order. File Nos. ENF–87–18 and ENF–87–20, adopted 13 November 1987, released 16 November 1987; (1987) 2 FCC Rcd 7202.

115 'In the Matter of Communications Satellite Corp.', Docket no. 16070, Memorandum Opinion and Order, 23 June 1965, modified 28 July 1965 (1965) 1 FCC 2d 533.

116 In 1975 the FCC noted that the record in the proceeding was 'large even by regulatory proceedings – aproximately 20,000 pages; Proposed Finding, Conclusion, Replies and Briefs of nearly 1700 pages were also filed: 'In the Matter of Communications Satellite Corporation Investigation into Charges . . . (1975) 56 FCC 2d 1101 at 1107.

117 'In the Matter of Communications Satellite Corporation Investigation into Charges, Practices, Classifications, Rates and Regulations', FCC Docket No. 16070, Memorandum Opinion and Order, adopted 9 May 1978, released 23 May 1978, Release No. FCC 78–312, (1978) 68 FCC 2d 941; the COMSAT Rate Case.

118 'In the Matter of Communications Satellite Corporation', CC Docket Nos. 80–634 and 85–268, Designation Order Mimeo No. 0672 released 8 November 1985, with Erratum Mimeo No. 1702 released 27 December 1985; Memorandum

Opinion and Order, adopted 3 April 1987, released 6 April 1987, (1987) 2 FCC Rcd 3706 (the April 6 Order): limited reconsideration, Memorandum Opinion and Order, adopted 10 December 1987, released 28 January 1988, Release No. FCC 87–388, 37238, (1988) 64 Rad. Reg. 2d (P & F) 524. See for the conclusion of the Saga: 'In the Matter of Communications Satellite Corporation; Tariff No. 101 Implementing Second Memorandum Opinion and Order in CC Docket No. 80–634, Tariff FCC Nos. 101 and 14 Implementing Reconsideration Order in CC Docket No. 82–540, Tariff No. 103 Establishing Separate, Unbundled Space Segment Offering' CC Docket No. 85–268, Phase II, adopted 5 December 1988, released 8 December 1988, FCC Release No. 88–396 37558; (1988) 3 FCC Rcd 7164.

119 'In the Matter of Policy for the Distribution of United States International Carrier Circuits Among Available Facilities During the Post-1988 Period', CC Docket No. 87–67, Notice of Proposed Rule-Making (1987) 2 FCC Rcd 2109; Report and Order adopted 24 March 1988, released 14 April 1988, FCC Release No. 88–122.

120 Chapter 4, section 'US Separate Systems', p. 182.

121 OTA, *International Cooperation and Competition in Civilian Space Activities*, US Congress, Office of Technology Assessment, OTA–ISC–239, (Washington, DC: US Government Printing Office, July 1985).

122 In the Matters of COMMUNICATIONS SATELLITE CORPORATION: Application for Authority to Participate in an INMARSAT Program to Provide for Aeronautical Bandwidth on Second Generation Satellites; Application for Authority to Establish and Operate Leased Channel Facilities via the INMARSAT System for use in the Provision of Aeronautical Services, Action': Memorandum Opinion and Order, adopted 7 January 1987, released 12 January 1987, Release No. FCC 87–22, (1986) 1 FCC Rcd 390.
On INMARSAT see Chapter 5 below.

123 CC Docket No. 87–75, Release No. 87–106; adopted 26 March 1987, released 30 March 1987.

124 The consortium would operate like the INTELSAT or INMARSAT Boards, with the voting weight of members of the consortium being linked to usage of the INTELSAT system: Notice, para. 18. Membership of AEROCON would not be open to owners or users of other satellite systems: Notice, n. 15.

125 Cf. Chapter 4, section 'US Separate Systems', p. 182.

126 'In the Matter of Aeronautical Radio Inc. (Arinc)'; For authority to construct an aviation satellite system (AvSat); adopted 9 September 1987, released 10 September 1987, Release No. FCC 87–289, 2 FCC Rcd 5990.

127 The FCC has responsibilities in regard to spectrum use, exercised in its Spectrum Allocation Order, (1986) 61 RR 2d 165, 207.

128 As explained in the appropriate chapters, Parties to the Agreements must replace defaulting Signatories which they have designated.

129 'In the Matter of Communications Satellite Corporation', CC Docket Nos. 80–634 and 85–268, Designation Order Mimeo No. 0672 released 8 November 1985, with Erratum Mimeo No. 1702 released 27 December 1985; Memorandum Opinion and Order, adopted 3 April 1987, released 6 April 1987, (1987) 2 FCC Rcd 3706 (the April 6 Order): limited reconsideration, Memorandum Opinion and Order, adopted 10 December 1987, released 28 January 1988, Release No. FCC 87–388, 37238, (1988) 64 Rad. Reg. 2d (P & F) 524. See n.118.

3 The International Telecommunications Satellite Organization: Its formation and structure

INTRODUCTION

The International Telecommunications Satellite Organization, INTELSAT, is the dominant organization in the area of public service satellite telecommunications. It was the first international organization created to serve the needs of public telecommunications by satellite, has the greatest traffic carried by any such facilities, operates worldwide, has the largest number of members – 115 at April 1988 – and contains within its constituent agreements the obligation, binding on all its members, that they shall coordinate their other satellite telecommunication activities with those of INTELSAT.

Needless to say, such an organization was not easily born. There was much negotiation on political, economic and technical matters before the structure we have today was arrived at. Those birth pangs, and the route by which INTELSAT arrived at its current form, have provided models for most of the other satellite organizations which have subsequently come into being. In particular, such organizations as INMARSAT and EUTELSAT have found it advantageous to exist in an interim form before definitive arrangements were agreed for them.

INTERIM INTELSAT: THE NEGOTIATIONS

The US Communications Satellite Act of 1962 was, as noted above, a compromise between those who supported a private and those who supported a public enterprise. Irrespective of that debate, to an outsider there does seem to have been an expectation in the USA that the United States would provide a satellite telecommunications system, and other countries would purchase that service from COMSAT. It seems that at first the USA had hoped that their global network would be established by means of bilateral agreements between the USA and individual foreign countries.[1] Such a system would have had as its model the various bilateral and limited multilateral agreements which regulated the worldwide cable network of the time. That system could have been 'benevolent', or it could have permitted a strictly 'commercial' approach, but whichever might have been adopted, the US dominance among its potential partners, both in satellite technology and the ability to launch the necessary satellites, would have given it a whip hand. In such a context it is difficult to envisage that purely commercial considerations would long have been excluded, and that before long, as the US interest became dominant, the commercial might well have displaced other considerations from the enterprise. The attempts by US companies in the 1980s to introduce competing separate systems into international public telecommunications services are but a delayed manifestation of that potential.[2]

Whatever the truth of the climate of expectation within the USA, other countries feared how things might develop. For example, the UK, which was a leader in international communications in the early 1960s, had much at stake. It had a major investment in cables, and *inter alia* was part-owner of the then operational northern transatlantic cables. The development of a satellite system might threaten that UK cable investment, and future plans might also be jeopardized. New cable technology was about to be introduced which would have met the projected need for cable capacity for the next few years, and transistorized cable systems were to be introduced in the near future. These planned networks might be threatened by the satellite system. Even if the new cable systems were in position first, although that would render the satellite system less commercially viable, it was unlikely that satellites would be abandoned for such reasons. Further, and on the level of international politics, the UK cable network naturally channelled cable communications through London, a feature which helped to maintain UK links round the world. A satellite system could talk as easily to Washington as to London, and London might therefore find itself bypassed.

The development of a new satellite system would pose further problems for the UK. An easily perceived danger was that the UK would be excluded from participation in the new system as anything other than a client-buying service from a foreign provider. Such fears were articulated, for example, in the British House of Commons on 26 February 1964, by Roy Mason, MP, in a supplementary to an Oral Question about the progress of the negotiations to Reginald Bevins, the then Postmaster General:

... there is now a growing feeling that, according to the trend of present talks, we shall finally end by starving the transatlantic cable of telegraphic communications from America and assisting COMSAT to get off the ground, and that Britain will merely end by renting a line from the Americans.[3]

Other countries balanced such questions differently. For some European countries, an advantage was seen in the ability to talk to the USA and elsewhere by satellite from their own ground stations, without having to go through cable links through states bordering on the Atlantic. The existing investment of these countries in the cable systems was minimal, since they only paid for their use of cable circuits (except in the case of IRU circuits). For such countries, the opportunity to invest in the creation and exploitation of the new medium, and to gain through technological 'spin-offs' from its development for domestic industry were considerations more important than the preservation of the previous international tele-communications systems. However, the initial US attitude made uncertain whether satisfactory opportunities for investment and return on that investment would be given to non-US investors. Nor was it clear that satisfactory arrangements would be made for the embryonic space industries outside the USA to share in the development work which the new system would require.

The option of buying into the Communications Satellite Corporation as a commercial undertaking was not available on a satisfactory basis. By s. 304(d) of the constituent Act and art. V, s. 5102(d) of the Articles, foreign participation in COMSAT was limited to 10 per cent of the overall common stock (20 per cent of Series I stock). The chances of one foreign government or telecommunications entity being able to purchase the entire 10 per cent were passing small. A single foreign stock-owner would therefore not be a significant voice within COMSAT. Nor would the 'foreign voice' have been effective even if all the foreign stockholders had acted in concert. It was suggested in some quarters that such a bloc could be significant, but that seems unlikely. To have influence the 'foreign bloc' would have required the cooperation or the goodwill of a large number of holders of Series 2 stock, who at that time were all US common carriers. In addition, even were they to decide to act unanimously, foreigners would be left in an odd position, for a person who is not a US citizen cannot be a member of the Board of COMSAT or be appointed as an official of the Corporation.[4] Again, were the foreign holder of COMSAT stock to be a governmental telecommunications agency,[5] there would be the peculiarity of an arm of a foreign government working within a corporation subject to US legislation and other governmental controls.[6]

The practical alternative to participation in COMSAT itself was some sort of international organization or arrangement. However, other possibilities had to be explored, in part out of good investment practice, and in part so as to secure a bargaining position from which a satisfactory eventual agreement might be reached. If you have few cards to play, it sometimes helps to invent some.

International telecommunication requires cooperation at both ends of any link. The USA realized that bilateral arrangements would have to be made for satellite links, just as such arrangements have to be made for the operation of a cable system.[7] A network of bilaterals could have allowed in their terms for individual countries or their telecommunications agencies to negotiate a share of the business with a US principal. On the other hand it was clear that if they engaged in isolated and individual negotiations, non-US countries were likely cumulatively to end up with a worse deal than if they negotiated jointly. To improve their bargaining position they had to act in concert and had to make it plain that separate bilateral negotia-tion was not in prospect. If to that could be added the prospect of they themselves starting a rival satellite system, so much the better. In theory, therefore, there were two possibilities: first, that a US sys-tem might, by coordinated boycott, have no international traffic and therefore be restricted to internal US usage only; and second, that international systems rival to a COMSAT-established system might emerge.

So much for theory. When international discussions began, the possibilities were found not merely to be theoretical. It was indeed clear that various groupings of countries were interested in the pro-vision of international telecommunications services. Competitive international satellite services might well have developed had there not been agreement on a global system, each fighting for their own share of the telecommunications market and directly affecting the viability of the US global system.

Of the potential groupings, two were clear and significant – Europe and the British Commonwealth. Various conferences were held to consider suggestions for action within the Commonwealth, of which the most important for our purposes was held in spring 1962. At that conference the position of the British government was made plain. The view was that a single global system was preferable, which would obviously require sharing in the US project. The conference agreed and emphasized the need for cooperation.[8]

Talks were held in Washington in October 1962 between Canada, the UK and the USA, and some progress was made. More steps were taken at a special meeting of the Telecommunications Committee of the Conference of European Postal and Telecommunications Admin-istrations (CEPT) in December 1962,[9] and a European Conference on

Satellite Communications was set up in May 1963. (It is significant that Australia was also associated with this Conference.) At its first meeting, the European Conference agreed that the European views should be harmonized, and that discussions should be entered into with the USA.

As noted, the USA had intended to negotiate bilateral agreements to give the new facility access into other countries' telecommunications systems. The united European front in favour of a group approach and a multilateral agreement caused this idea to be abandoned, although not without the Europeans having to go quite far in proof of their determination. At a meeting on 26 November 1963, the European Conference proposed the setting-up of a counterpart to the COMSAT, which would have been financed by shared capital contributions from the members of the Conference.[10] This consortium approach was a distinct threat to the USA, requiring her to abandon her prior hopes of retaining a monopoly of ownership and control in the global satellite system. In fairness (or by way of defence) I also have to say that this European proposal, though meant to be implemented if that proved necessary, seems all along to have been primarily intended for use as a bargaining weapon and was willingly abandoned when that was possible. The UK's attitude was typical of the non-US view, and can be illustrated by statements by Reginald Bevins, the Postmaster General, in the exchange in February 1964 from which I have already quoted. He stated:

> The Government's objective is the creation of a world system for satellite communications in partnership with the United States, the Commonwealth and European and other countries, on conditions which will give countries investing in the system a share in its design and character, and in its management and control, and an opportunity to provide some of the equipment.
> ... [T]he Government's view is that the only way of preventing an American monopoly in this sphere is to join a partnership with the United States and other countries and so secure the right to influence the course of events.[11]

Such a statement makes it plain that the threat of a rival enterprise was made only to gain a satisfactory deal with the negotiators from the United States. The statement also makes clear the importance to countries other than the USA of questions of ownership, control, design and supply. They neither wished to finance a US monopoly, nor see the considerable industrial and technological benefits going solely to the US economy and military capacity. Such considerations were sufficiently important for them to require the USA to set aside hopes of establishing a single monolithic commercial entity in the field of satellite telecommunications.

On the other hand, it would be unfair to give the impression that the USA had never considered a partial foreign ownership of the satellite link until it was forced upon her.[12] There were elements within the USA which considered that an international organization was the best way to solve such matters. However, as in accordance with the American ethic (The 'American Way'), telecommunications in the USA was (and is) traditionally a matter of private enterprise and profit for those engaged in the business of providing the service: the US approach to the matter was that of the businessman seeking not to give away any more than was necessary. That must be stated clearly. There is a tendency nowadays to say that the USA had all along intended the creation of an organization such as INTELSAT (which is even presented as entirely a US creation), and that the USA was actuated entirely by benevolence: there is a correlative tendency to react angrily to any suggestion that the history was different. The evidence is to the contrary: the initial position taken by the USA was not unmixedly altruistic.

In the face of the unified European position, the United States had to give way.[13] Any commercially viable satellite system had to gain a share of the heavy communications traffic across the north Atlantic, and no European country was willing to enter into a bilateral arrangement with the USA. Accordingly, the European concept of a consortium had to be explored. This was done at a series of meetings, the most important of which were held in February and April 1964. At the first of these, the establishment of the consortium was agreed; and at the latter the position of COMSAT as the day-to-day manager of the interim system was clarified. It was further agreed that it was more important to get a system up and running than to iron out the constitution of the necessary organization on a permanent basis – definitive arrangements could be negotiated once experience had been gained in the problems of creating, financing and running a global system. COMSAT's announcement that it was to proceed with the launch of the synchronous satellite 'Early Bird' in spring 1965 concentrated minds; thus, after a number of meetings, two agreements were initialled and were opened for signature at Washington for a period of six months from 19 August 1964. These were the constituent documents of Interim INTELSAT.

INTERIM INTELSAT: THE AGREEMENTS

Two linked Agreements formed the constitution of Interim Intelsat.[14] The necessity for there being two agreements, both at the interim stage and in the definitive arrangements is explained by the different constitutional provision for the control and operation of telecommunication services in different countries. In countries such as France and the Netherlands telecommunications are under direct government control and there is no special entity through which the government acts. In many countries, including most of the British Commonwealth, a special government department exists for the purpose.[15] In states such as the USA and Japan, the provision and financing of telecommunication facilities are under private corporations.[16] It was therefore necessary that two agreements be arrived at to encompass the provisional constitutional arrangements for the new satellite system. One was an Intergovernmental Agreement dealing largely with organizational matters, and the other, a Special Agreement between the telecommunications entities of the several states, dealt with the commercial operation of the system itself. Each state member of the Intergovernmental Agreement, could either itself sign, or it could designate a national telecommunications entity to sign the Special Agreement between the operational telecommunications entities (Intergovernmental Agreement, art. II(a)).

The consortium established by the Interim Arrangements was not a new juridical entity. There was a form of intergovernmental organization in that the agreements did set up machinery for the establishment and running of the global network. However, the principal body under the Interim Arrangements, the Interim Communications Satellite Committee (ICSC), was not otherwise recognized by law or treaty, being little more than a steering committee of a number of joint-venturers, existing for the purpose of the convenient administration of the affairs of the consortium as a whole. It could not sue or be sued; it did not own property; it was a creature of contract between the parties to the agreements, but in law nothing more. Indeed, it was a jurisprudential freak.

The duration of the Interim Arrangements was one of the main topics of discussion in the negotiations which led to their adoption. The Europeans did not wish them to continue too long, and desired a more permanent constitutional structure to be adopted as soon as the practical problems of the establishment and operation of the system had been solved. In addition, the Interim Arrangements were very favourable to the USA, giving it a substantial measure of control over policy as well as a veto on all decisions – something else which other participants in the venture did not wish to see prolonged. On the other hand, with COMSAT already in existence and ready to go about its business, the USA wanted to get something arranged quickly. In view of that general desire to get started and the legal constraint of s. 102(a) of the 1962 Communications Satellite Act which specifically required COMSAT to establish a global system as expeditiously as possible, the US negotiators were willing to concede the early revision of the Agreements in order to gain the early creation of the milieu within which COMSAT might function. What was therefore agreed was that within one year of the initial global system becoming operational, and at latest by 1 January 1969, the Interim Communications Satellite Committee was to make recommendations on the definitive arrangements for the running of the enterprise and an international conference would be held to discuss the matter. (Interim Intergovernmental Agreement, art. IX)). That timetable slipped somewhat, but after negotiations the definitive agreements, replacing the Interim Arrangements, came into force in 1973.

Broadly speaking, the aims of the two sides in the discussions were attained in the Interim Arrangements. The USA in general, and COMSAT in particular, was able to proceed with some security to the creation of the global satellite system. Conversely, the Arrangements allowed nations and their telecommunications entities to have a say in the nature of the system, and a share in the advantages which it was to bring. Countries had a voice in the policy of the body which was to create the satellite system. Reasons of state required that they had some say on such questions as the type of ground stations to be used, in the type of system, its design and its operation. They wanted to participate in the setting of rates, tolls, and the extent to which they (and other states) might participate in the system through the dedication of transponders and the like. The problem in the negotiations was how to accommodate all these interests.

The solution to the majority of these difficulties, which was adopted both in the Interim and in the later Definitive Arrangements, was a quota system. It had been recognized at a very early stage in the negotiations that to call a major international conference on telecommunications was not the way to tackle the setting-up of a global system. While there was clearly a place for the developed nations, if all countries had been invited to such a conference and had all had an equal voice in the discussions (as would, for example, have been the case if the UN had organised such a conference), it would have been unlikely that any agreement at all would have been arrived at. So a selection was made on the simple basis that the countries to participate in the discussion would between them represent some 90 per cent of the telephone traffic in the world.[17] The invited participants were the USA, Western Europe, Australia, Canada and Japan. But even within that group, it was clear that some countries were more important than others.

The whole enterprise was to be costly and, conceivably, very risky. It was agreed that the various participants in the enterprise should share these costs and risks, having regard to their estimated telecommunications traffic. Proportions were agreed and ranged from 61 per cent (applicable to the USA) to a minimum of 0.05 per cent (applicable to the Vatican City State). On a global basis, the division was that the USA had its 61 per cent, Europe 30.5 per cent and the rest of the world (Australia, Canada and Japan) 8.5 per cent. The actual shares were: Australia 2.75 per cent; Austria 0.2 per cent; Belgium 1.1 per cent; Canada 3.75 per cent; Denmark 0.4 per cent; France 6.1 per cent; West Germany 6.1 per cent; Ireland 0.35 per cent; Italy 2.2 per cent; Japan 2.0 per cent; the Netherlands 1.0 per cent; Norway 0.4 per cent; Portugal 0.4 per cent; Spain 1.1 per cent; Sweden 0.7 per cent; Switzerland 2.0 per cent; UK 8.4 per cent; USA 61.0 per cent and Vatican City 0.5 per cent.

The quota based on share of cost was carried over into the matter of ownership of the space segment of the global system. Countries were unwilling that the USA, although bearing the largest share of the cost, should itself own the space segment. Therefore the Interim Arrangements specified that each Signatory to the Special Agreement owned the space segment in proportion to its share of the cost of establshing that segment. In Scots Law we would call this a pro-indiviso share of the subjects of ownership: 'The space segment shall be owned in undivided shares by the signatories to the Special Agreement in proportion to the costs of the design, development, construction and establishment of the space segment.' (Intergovernmental Agreement, art. III.)

The quota was also used as a solution to the problems of decision-making among such a large group. Clearly it was right that each country should have the same proportionate weight in the decision-making process as it bore a share of the costs of the enterprise: 'He who pays the piper calls the tune.' However, even though there were only 19 prospective Signatories to the Special Agreement, that was too large a body to take decision, particularly when some of the 19 had a very small proportionate shareholding. A policy-making body with such disparities within it, and of that number, was undesirable. The solution was the Interim Communications Satellite Committee (the ICSC), composed of one representative from each of the Signatories to the Special Agreement whose quota was not less than 1.5 per cent, and one representative from any two or more signatories whose combined quotas totalled not less than 1.5 per cent and who agreed so to be represented.

However, even a body so constituted could be considered not to be entirely satisfactory were it to have operated simply on the basis of a vote weighted in proportion to quota. The initial US quota was 61 per cent, and, as will be explained, there was an agreement as to the quota to be allocated to those acceding to the arrangements which meant that the US share would never fall below 50.63 per cent (Intergovernmental Agreement, art. XII(c)). The USA would always therefore have a simple voting majority in the ICSC. This matter caused extreme difficulty, requiring careful balancing so that COMSAT should not be too fettered since it was bearing the largest share of the costs, and yet ensuring that other participants should not lose their identity or influence. The problem was solved only on the last day of the final conference in Washington. By art. V(c) of the Intergovernmental Agreement, the Interim Committee was required to try to act unanimously, but could take decisions by majority vote. However, in 14 enumerated matters, any decision had to have the concurrence of representatives whose total votes exceeded that of the USA by not less than 12.5 votes. These matters were:

1   the choice of the type or types of space segment to be established;
2   the establishment of general standards for the approval of earth stations for access to the space segment;
3   the approval of budgets by major categories;
4   the adjustment of accounts;
5   the establishment of rates for satellite users;
6   decisions on additional capital costs between $200 and $300 million;
7   the approval of contracts over $500,000 in value for equipment for the space segment;
8   the approval of matters relating to satellite launchings, the launch source and contracting arrangements for launchings;
9   the approval of quotas for new Signatories;
10   the determination of financial conditions for the accession by new Signatories after the initial period for signature of the Agreements;
11   decisions relating to the withdrawal of a member;
12   the recommendation of proposed amendments to the special agreement;
13   the adoption of rules of procedure of the Interim Committee and advisory subcommittees;
14   the approval of costs and fees paid to COMSAT as manager of the system.

In short, the USA itself could not railroad through decisions on these matters by recourse to its otherwise overwhelming voting strength. On the basis of the initial quotas, in the ordinary case of dissension on the principal features of the initial system, the agreement of the UK (8.4 per cent) and that of either France (6.1 per cent) or West Germany (6.1 per cent) with the USA would suffice to allow a decision to be taken. Three was thus then the minimum number of countries which the USA would have had to persuade on these 14 matters. However, in an additional four categories (namely numbers 1, 3, 7 and 8 above which all related directly to the establishment of the satellite system), if matters were not decided within 60 days on the USA plus 12.5 per cent rule, the additional votes required for their passing was to be reduced to 8.5 per cent (Intergovernmental Agreement, art. V(d) and (e)), which was (no doubt quite by chance) the same as was required for a meeting of the Interim Committee to be quorate (Intergovernmental Agreement, art. V(b)). Again on the basis of the original quotas, this meant that mathematically the establishment of the basic satellite system could be decided by the vote of the USA (61 per cent) and the UK (whose initial share was 8.4 per cent) plus one other, other than the Vatican (0.05 per cent). Steps were thus taken to ensure that acrimonious disputes should not delay the establishment of the space network for more than two months, and in these tortuous ways some balance was obtained between the interests of COMSAT and those of other participants. It seems to have been successful. So far as I am aware, it was never necessary to make use of the 12.5 per cent rule, let alone the reduced 8.5 per cent voting margin provision.

The European negotiators had two other preoccupations which had to be satisfied by the Interim Agreements. These were the interrelated matters of equipment procurement and technological 'spin-off' or 'fall-out'. It was obvious that the creation of the global system would require considerable sums of money, and also that much technical expertise would be gained by those who did the work. Other countries were unwilling to finance projects which might only increase the already great superiority of the USA. On the other hand, it was also obvious that the USA was the leader in space technology, and that the others were not very well advanced in such matters. One object, therefore, of the initial Agreements was to ensure that, while the best equipment and service were provided at the best price for the system, where there was no clear margin in favour of one bidder, contracts for research, development and supply would be distributed among the companies of the Signatory states in approximate proportion to their quotas (Intergovernmental Agreement, art. X). That principle was to apply not only to main contracts, but also to subcontracts, and it is notable that such a practice has been carried forward into the definitive Agreements. Further, except where the ICSC otherwise determined, it was to secure contractually that all inventions, technical data and information acquired in work on the space segment were to be available for use by each Signatory on fair and reasonable terms (Special Agreement, art. 10(g)).

For the United States in general, and COMSAT in particular, the most important clauses in the Agreements were art. VIII of the Intergovernmental Agreement and arts. 10 and 12 of the Special Agreement. The first of these made COMSAT the manager in the system, and the others spelled out in more detail its powers and responsibilities in relation to budget, research, proposals on a wide variety of matters, contracts and its duties of reporting to the Interim Committee. These provisions assured COMSAT of its role in space telecommunications, for the immediate future at least.

The Interim Agreement and the Special Agreement were opened for signature at Washington on 20 August 1964 for a period of six months. Austria alone of the 19 initial negotiators had not signed by 5 October 1965, the date of the first meeting of the Interim Committee. But it was never the intention of the original participants in the negotiations that they alone should take part in the projected tele-communications system. The Agreements were not between a closed list of countries. As implied above, accession to the Agreements was thereafter competent, and there were special rules for the assigning of a quota to each new Signatory to the Special Agreement, and a pro rata reduction of that of others. That reduction was, however, subject to the limitation that the combined initial quotas of all Signatories other than those listed in the Annex to the Special Agreement (the original negotiators) should not exceed 17 per cent (Intergovern-mental Agreement, art. XII(c)). The mathematical result of this appar-ently curious provision was that the US quota would never reduce below 50.63 per cent, with a consequent effect on voting and decision-making which produced the other usual rules we have dealt with above.

Withdrawal from membership of interim INTELSAT was also com-petent, with rules for the readjustment of quotas and settling of accounts (Intergovernmental Agreement, art. XI). But the trend was always upwards. Thus, in my McGill University thesis of April 1965, I note that at March 1965 – that is within the six months of the opening of the Interim Agreements for signature – the membership of Interim INTELSAT was 45, the US quota being the 56 per cent.[18]

DEFINITIVE INTELSAT: THE NEGOTIATIONS

By art. IX of the Interim Agreement the ICSC was required to report to all states party to that Agreement its recommendations as to the definitive arrangements for the international global satellite telecom-munications system. The timetable called for that report to be submit-ted within one year of the initial global system becoming operational, and in any event for it to be presented not later than January 1969. This was done, and the definitive arrangements were worked out by plenipotentiary conference convened by the USA in terms of art. IX(c). To this conference were invited all the then INTELSAT mem-bers (of 68 eligible, 67 attended), as well as governments not mem-bers of INTELSAT but which were members of the International

Telecommunication Union (29 sent observers). The UN and the ITU were also invited.

There were many competing, and occasionally conflicting, points of view on what should be the structure and the remit of the new organization. In the main it was clear that there would be an international organization – the decision on Interim INTELSAT had dealt with that question. However, within that broad framework there were important decisions to be made. Was the new organization to be an international person, or an association of joint-venturers? How was it to relate to its members, and to users of the INTELSAT global system? How was it to be organized?

Particular mention may be made of COMSAT's role in the successor organization, and the changes made in the new INTELSAT were to have important effects on COMSAT's own existence.[19] Again, the relationship between the global organization and other systems had to be considered. All these matters were explored in confidential meetings and discussions.[20] As far as relevant (and in some cases as far as known), the objectives of the various participant and the extent to which they were attained are indicated later in relation to specific matters within the Agreements. Broadly, it may be said, however, that the eventual package was not as favourable to COMSAT as it might have wished. A number say that was because of COMSAT's behaviour both in the prior organization and in other telecommunications endeavours, some adding that there was also a distrust of the relationship between COMSAT and the US State and other Departments and a disquiet about US space diplomacy in general.[21]

The Plenipotentiary Conference met from 24 February–21 March 1969 at Washington, resumed for five weeks from 16 February to 20 March 1970, continued through the agency of an Intersessional Working Group (IWG), and the final resumed Plenipotentiary Conference was held from 14 April–21 May 1971. During this period membership of INTELSAT had risen from 68 to 79. Seventy-eight members participated in the final Conference, and the texts of the Agreement between governments and the Operating Agreement between the telecommunications entities involved were adopted on 21 May 1971 by an affirmative vote of 73 with none against, while France, the Malagasy Republic, Mexico and Monaco abstained from voting.

The Agreements were opened for signature on 20 August 1971, seven years to the day after the Interim Arrangements had opened for signature. The definitive INTELSAT are therefore contained in the Agreement relating to the International Telecommunications Satellite Organization (INTELSAT),[22] and the relative Operating Agreement with Annex.[23] Their entry into force was slightly complicated. By art. XIX the Intergovernmental Agreement might be signed by governments in a period terminating by either their entry into force or the elapse of nine months, whichever should be the earlier. Signature might be by any state member of Interim INTELSAT or any other state member of the ITU (art. XIX(a)). In terms of art. XX, however, the entry into force of the Agreement was tied to the Interim Arrangements and membership of Interim INTELSAT. The Agreement entered into force 60 days after it had been signed, provided that all appropriate domestic constitutional processes (for example, ratification, acceptance, approval or accession) had been completed by two-thirds of Interim INTELSAT's membership, representing at least two-thirds of the investment quotas under the Special Agreement, and provided that in the case of each state Signatory, the appropriate telecommunications agency had signed the Operating Agreement.

Within the period from the conclusion of the definitive Agreements to their inception, one other member joined Interim INTELSAT. A minimum of 54 out of 80 members was thus required for the entry into force of the Agreements. In fact, on 14 December 1972 (somewhat less than half the time period allowed) the requirements were met, and the 60-day period began to run. The quota represented by these 54 was 83.93 per cent – well above the two-thirds quota minimum which had been set. The Definitive Arrangements therefore entered into force on 12 February 1973. In terms of art. XX(e) of the Agreement and art. XV of the Interim Agreement and art. 23(a) of the Operating Agreement and art. 16 of the Special Agreement, that entry into force superseded and terminated the earlier Arrangements.

THE DEFINITIVE ARRANGEMENTS

The Agreement and Operating Agreement relating to INTELSAT, signed at Washington on 20 August 1971, came into force on 12 February 1973.[24] As in the case of the Interim Arrangements and for the same reasons, two interlinked Agreements are involved. The Agreement establishing INTELSAT is an agreement between governments, having the status of an international treaty. The Operating Agreement, which is equally fundamental to the working of the organization, is an agreement between telecommunications entities. Each Party to the Agreement must either itself sign the Operating Agreement, thereby taking upon itself the rights and duties of a Signatory for the purposes of the Agreement, or it may designate a public or private telecommunications entity to sign the Operating Agreement (art. II (b)). This provision therefore accommodates the situation where telecommunications within a state are a governmental function, the responsibilities of a public body, or that of a private operating company. It also permits the government to act, even in the case where ordinary telecommunications are not a governmental function.

The new organization is an international organization. It is not a consortium. Unlike its interim predecessor, INTELSAT has legal or juridical personality (art. IV). It is not in law the consortium or association of joint-venturers as was Interim INTELSAT. INTELSAT has 'the full capacity necessary for the exercise of its functions and the achievment of its purposes', including specifically the capacities to conclude agreements with states or other international organizations, to contract, to acquire and dispose of property and to be party to legal proceedings (art. IV(a)). That provision might itself be enough, given the international law on the implied powers of international organizations. However, as an added safeguard, article IV(b) of the Agreement requires each Party to it to 'take such action as is necessary within its jurisdiction for the purpose of making effective in terms of its own law the provisions' of art. IV.[25] Further to that end art. XV(c) required the Parties to the Agreement other than the USA[26] to conclude 'as soon as possible' a protocol to regularize the position of INTELSAT as an international organization within their jurisdiction, and to ensure that in every state its privileges etc were uniform. Accordingly a Protocol on INTELSAT Privileges, Exemptions and Immunities was opened for signature on 19 May 1978 and entered into force on 9 October 1980.[27] INTELSAT's headquarters are in Washington, DC, and the appropriate arrangements normal in such cases are contained in the Headquarters Agreement between it and the USA.[28]

The INTELSAT Agreements are 'open' agreements in the sense that their membership is open to all or virtually all states. Membership of INTELSAT was open to any state signing the Agreement during the period in which it was opened for signature, and which was either a party to the Interim Agreement, or was a state member of the ITU (art. XIX(a)). Since that period, membership has been open to any state meeting these qualifications, and which accedes to the Agreement (art. XIX(b)). However, the Agreements are available only as a package and would-be members cannot pick and choose to which elements of them they will consent. The Agreement specifically excludes any reservation being made to it (art. XIX(d)). This is important. Under the General Law of Treaties, reservations to treaties are competent if not excluded by the treaty itself. Broadly, a reservation is an announcement made by a state or government on joining an international treaty that it does not consider itself bound by specified parts of the treaty, and the effect is that, as between the parties, the treaty is modified for its application to that state in the way it has indicated.[29] Although that seems a reasonable way to do business –

allowing the generalities of agreement to stand in treaty form while permitting some 'local' variation as to detail, if accepted by the other parties to the treaty – in modern times, reservations have bedevilled many multi-lateral agreements. It was a wise decision not to subject INTELSAT to that problem. An international organization with operational capacities and responsibilities cannot function efficiently if it is governed by provisions of its constituent documents which vary from member to member and which differ as to the rights and obligations between members and between members and the organization.

Amendment of the Agreement is competent. Any Party may submit a proposed amendment to the Executive Organ, which promptly distributes it promptly to all Parties and Signatories (art. XVII(a)). The Assembly of Parties considers the matter either at its next ordinary meeting, or at an extraordinary meeting provided that the proposal has been distributed at least 90 days prior to the meeting. The Assembly takes into account any views and recommendations expressed by the Meeting of Signatories or the Board of Governors (art. XVII(b)), but in taking decisions on such matters the Assembly has no special rules as to quorum or voting (art. XVII(c)). However, such a matter being clearly a matter of substance, it requires a two-thirds affirmative vote of Parties present and voting (in terms of art VII(f)). The Assembly may modify any proposal before it, and may also take decisions on any further amendment which has not gone through the formal requirements of circulation, but which is consequent upon a proposed amendment, whether or not it has modified that proposal.

The approval of the Assembly of Parties is not, however, itself sufficient to bring about the amendment of the Agreement. The Agreement is an international treaty, and it follows that the normal processes for the amendment of an international treaty have to be gone through. The adopted amendment is therefore subject to the approval, acceptance or ratification by Parties, depending upon what constitutional process each Party adopts in such matters. A successful amendment enters into force for all Parties to the Agreement, whether or not they have approved it, provided, of course, that they are not so incensed as to withdraw from the organization (art. XVII(e)). The amendment comes into force ninety days after the Depositary (the USA, as according to art. XX) notifies all Parties that the amendment has received a stipulated level of approval by Parties. The amendment, however, cannot enter into force earlier than eight or later than 18 months from its approval by the Assembly (art. XVII(f)).

There are two possible levels which an amendment can reach, which will allow it to enter into force. First, an amendment may be approved by two-thirds of the Parties, provided that cumulatively they, or their designated Signatories of the Operating Agreement hold at least two-thirds of the total investment share of the organization (art. XVII(d)(i)). Alternatively, if the level of approval reaches 85 per cent of the total state membership of INTELSAT at the date of the approval of the amendment by the Assembly of Parties, the amendment will come into force, irrespective of the total investment share held by the approving states or their Signatories (art. XVII(d)(ii)).

Mutatis mutandis, the rules and procedures for the amendment of the Operating Agreement, which are contained within its art. 22, are the same as for the amendment of the Agreement. The only changes are that amendments may be proposed by any Signatory, by the Assembly of Parties or by the Board of Governors (art. 22(a)), and that it is the Meeting of Signatories which considers proposals (art. 22(b) and (c)). The Meeting has powers to modify proposals and deal with other consequential amendments (art. 22(e)). Amendments to the Operating Agreement require similar levels of approval by Signatories as are stipulated for in the case of Parties' approval of amendments to the Agreement (art. 22(d)). Like amendments to the Agreement, amendments to the Operating Agreement cannot enter into force later than 18 months from their approval by the Meeting of Signatories (art. 22(f)).

It may seem strange that amendments to the Agreement can come into effect only between eight and 18 months from their approval by the Assembly of Parties, and that a similar 18-month limit operates in the case of amendments to the Operating Agreement. If we deduct the 90-day notification period stipulated for by art. XVII(e) and art. 22(f), that does not leave very long in terms of international action. There are two issues, the low and the high limit. First, the 8-month low limit for amendments to the Agreement may be wise, since the normal cycle for the Assembly is two years. The effect is to concentrate the matter: unduly swift (and perhaps badly thought-out) action is prevented, yet constitutional change is possible. That low limit constraint seems less necessary in the case of the Operating Agreement, where the Signatories will pay attention in any case, but the Agreement itself may not be given a high priority by a government. Eight months is a fair balance between these two approaches. Second, the imposition of the 18-month time-limit for approval of amendments to either Agreement is certainly sensible. It means that the Organization will not have a variety of amendments lying around, any of which might come into force unexpectedly and even when many have forgotten about them, as can happen in the case of some multilateral agreements which have no such provision. The Parties and Signatories, and equally importantly the Board of Governors, will be aware of amendments which are out for approval and will take these into account. A terminus is, however, welcome as a constraint upon both Parties and Signatories, discouraging them from being dilatory, and also has the advantage of 'killing off' amendments after a reasonable period of consideration by members. In that connection also, we should note once again that, when it comes into force, a successful amendment comes into force for all (art. XVII(e); art. 22(e)). This is useful. Other international agreements have become patchworks, as different parties to them have accepted different amendments. In an organization like INTELSAT, this needs to be avoided. There must be a single constitution known to and binding all parties. Only thus can the purposes of the organization be efficiently achieved.

CONSTITUTIONAL STRUCTURE

Introduction

INTELSAT's constitutional structure as established by the Agreements, is quadripartite and took effect in its final form after a five-year transitional period. The provisions for the transitional arrangements, contained in art. XII and Annex D of the Agreement and the Annex to the Operating Agreement, were duly complied with and need not here be traversed. The current and final structure is in accordance with art. VI of the Agreement.

INTELSAT has four organs: the Assembly of Parties; the Meeting of Signatories; the Board of Governors; and an Executive Organ, operating under a Director-General. The Executive Organ and the Director-General are responsible to the Board of Governors. Although, by art. VII(a), the Assembly of Parties is designated as the principal organ of INTELSAT, art. VI(b) provides that no organ is to interfere with the discharge of any function attributed to another organ except to the extent provided for by the Agreements. This is neither an anarchists' charter, nor a source of paralysis. Art. VI(c) requires that the Assembly of Parties, the Meeting of Signatories and the Board of Governors shall each 'take note of and give due and proper consideration to any resolution, recommendation or view made or expressed by another of these' three organs acting within the sphere of its own competence. It seems that so far, the good sense of the members of the different organs has not overstrained the meaning of the words. 'Taking note' and 'due and proper consideration' mean what they say and no more. No organ is bound to do what another resolves, recommends or expresses being its view: yet due weight is given to all relevant and intrinsically cogent contributions by other organs.

Before discussing the structure, a comment may usefully put matters into perspective. Its justification will become apparent during the discussions, but it is as well to isolate it here, since it is to draw attention to one of the organization's major strengths and a reason why that organization has been so successful during a period when many other international organizations seem to have been floundering. The comment is this: the effect of the structuring of INTELSAT and the various checks and balances within it, has been to vest power in the Board of Governors – a Board composed of persons who are primarily skilled in telecommunications matters. The organization has therefore been able to work well towards the attainment of its aims – the provision of telecommunications facilities. Thus, at first sight the structure is hierarchical. The Assembly of Parties is designated as the principal organ (art. VII(a)), and is the plenary meeting of the governmental participants in the organization. Below it, as it were, is the Meeting of Signatories which is attended by the telecommunications entities and authorities which are the Parties to the Operating Agreement. The Board of Governors is a much smaller body and is representative of Signatories and groups of Signatories.

Nevertheless, this hierarchy is if anything inverted from its apparent form. The composition, respective powers and responsibilities, and the frequencies with which each body meets are the true tests of importance. Seen in that light, politics and national considerations are pushed to the periphery (although are not entirely absent) from the decision-making within INTELSAT. The body is run by engineers and telecommunications administrators: the result is the efficient service which we now have.

The working languages of the several organs and of INTELSAT itself are English, French and Spanish (art. XXI(a)), and documentation is produced for the three deliberative organs in the three languages – the suffixes 'E', 'F' and 'S' as necessary appearing in the document numbering.[30] All documentation is available to Parties and Signatories, and copies may be requested by them of any document (art. XXI(b)). They also may make documentation and information available to others. Access to documentation for third parties was relatively easy until the 1980s when INTELSAT found itself subject to potential competition from the US 'separate systems' applicants. At that stage an increased degree of confidentiality was found necessary – which explains some gaps in the documentation and discusion below. The US Freedom of Information Act is, of course, not irrelevant.

The Assembly of Parties

The Assembly of Parties is the plenary meeting of all the states which are party to the Intergovernmental Agreement, and, as already noted, it is the principal organ of INTELSAT (art. VII(a)). As is to be expected in a gathering composed of the representatives of sovereign states, each member of the Assembly has one vote (art. VII(f)). The quorum for the Assembly is a majority of the parties (art. VII(f)). Decisions on matters of substance are taken on the affirmative vote of at least two-thirds of the members present and voting. Decisions on procedural matters are taken by the simple majority of Parties present and voting, who also decide whether a matter is substantive or procedural. The particular Rules of Procedure for the Assembly of Parties were adopted in accordance with art. VII(g) and are published separately.[31]

In principle, the Assembly of Parties meets every two years, although it may decide otherwise from meeting to meeting (art. VII(d)). In addition to its ordinary meetings, it may meet in extra-ordinary session, convened either at the request of the Board of Governors or of one or more Parties, supported by at least one-third of the Parties including the requesting Party. Such requests are addressed to the Director-General. Any request for an extraordinary meeting must state the purpose for which it is sought and, if the request meets the condition of support, the Director-General must arrange for the meeting to be held as soon as possible (art. VII(e)). Each party is responsible for its own costs of representation at an Assembly, but the expense of running the meeting is an administrative cost of the organization itself (art. VII(h)).

The function of the Assembly of Parties is to give consideration to 'those aspects of INTELSAT which are primarily of interest to the parties as sovereign states. It [has] power to give consideration to general policy and long term objectives . . . consistent with the prin-ciples, purposes and scope of activities of INTELSAT' as provided for by the Agreement. Its power in these matters is, however, only that of making recommendations to the other organs, and, in terms of the day-to-day functioning of INTELSAT, the role of the Assembly is minimal. Thus, the Assembly can transmit its views or make recommendations to any other organ of INTELSAT on any matter of general policy or long-term objective (art. VII(c)(i)). It considers and expresses views on reports presented to it by the Meeting of Signa-tories or the Board of Governors on the implementation of general policy and on the activities and long-term programme of INTELSAT (art. VII(c)(vi)). It makes recommendations as to the establishment, acquisition or utilization of space segment facilities other than those of INTELSAT itself (art. VII)c)(vii)). Most importantly, and almost amounting to a power of decision, the Assembly expresses, in the form of recommendation, its findings about the establishment, acquisition or use of space segment facilities separate from that of the INTELSAT system (art. VII(c)(vii)). The effect of a negative recommendation has not yet been tested in practice, but overshadows developments and proposals in the USA in the 1980s.[32]

The Assembly does have certain powers of decision and determination in other matters which could be of importance in appropriate circumstances. Thus, it decides on proposals for the amendment of the Agreement, and may propose and can express views and make recommendations on the amendment of the Operating Agreement (art. VII(c)(iii)). Its other powers of decision include the determination of measures to be taken to prevent INTELSAT's activities from conflicting with any general multilateral convention which is consistent with the INTELSAT Agreements and which is adhered to be at least two-thirds of the parties (art. VII(c)(ii)). An example of such might be the ITU Convention or measures adopted by the ITU in regard to, say, the use of the geostationary orbital position. The Assembly also has power to authorize, either generally or specifically, the utilization of the INTELSAT space segment, or the provision of satellites and facilities separate from that space segment for specialized telecommunications services (art. VII(c)(iv)). It takes decisions in connection with what art. VII(c)(vi) calls 'the withdrawal of a party from INTELSAT', but which, in fact, deals with the expulsion of a member from the organization (art. VII(c)(vi)). It decides on formal relationships between INTELSAT and states, whether Parties or not, and between INTELSAT and other international organizations (art. VII(c)(ix)). It considers complaints (art. VII(c)(x)), selects the legal experts for the settlement of disputes by arbitration through the mechanisms of Annex C to the Agreement (art. VII(c)(xi)) (see below). It confirms the appointment of the Director-General by the Board of Governors (art. VII(c)(xii)), and adopted the organizational structure for the Executive Organ (art. VII(c)(xiii)). Finally, art. VII(c)(xiv) gives the Assembly power to exercise any other powers given to the Assembly elsewhere in the Agreement – a tidy but not altogether helpful clause.

In addition to these powers of decision, the Assembly of Parties reviews the general rules established by the Meeting of Signatories for the approval of earth stations, the allotment of space segment capacity, and the rates of charge for the utilization of the INTELSAT space segment (art. VII(c)(v)). Such a review has been provided for to ensure that, in accordance with art. VII(b)(v), the rules on such matters are non-discriminatory – a matter of great concern to certain of the states involved.

Important though the powers of decision could be, in fact they rarely come into play. The major part of the Assembly's functions is ordinarily the expression of views or the making of recommendations. Since it is an intergovernmental body, and politicians are the usual delegates from each country, the views and recommendations expressed by the Assembly are not usually concerned with technical aspects of the telecommunications service. Nevertheless, the political input is useful, for there must be an element of political consideration in the development and functioning of any international organization. That said, however, to an outside observer, the main usefulness of the Assembly of Parties seems to lie in allowing professional politicians scope for their interests without allowing that to impinge too directly on the organization's real business. It may be a straw in the wind, but the minutes of the Assemblies are interesting in their format. Of all the minutes of INTELSAT organs, only the minutes of the Assembly are ascribed to individual speakers, and are virtual transcripts of the speeches made by the delegates.

The Meeting of Signatories

The Meeting of Signatories is composed of representatives of all Signatories to the Operating Agreement meeting in plenary session. As Professor Bin Cheng points out,[33] in many ways the Meeting of Signatories resembles the shareholders' meeting of a limited liability company, for the financial interest in INTELSAT lies with the telecommunications entities which are the Signatories to the Operating Agreement. However, as Professor Cheng also points out, unlike the shareholders' meeting, each representative at the Meeting has one vote (art. VIII(e)): shareholders at a company meeting have the voting strength represented by their individual shareholding (together with any proxies held). As we shall see, the scale of financial interest is of crucial importance in INTELSAT only in the affairs of the Board of governors.

The quorum for the Meeting of Signatories consists of the representatives of a majority of Signatories (art. VIII(e)). Its Rules of Procedure are a matter for the Meeting itself, and detailed Rules have been adopted in terms of art. VIII(f).[34] In matters of procedural and substantive decision, the Meeting of Signatories follows the same pattern as that indicated for the Assembly of Parties. Matters of substance require a two-thirds majority present and voting; procedural matters are settled by simple majority; whether a matter is one of substance is determined by simple majority (art. VIII(e)).

Ordinarily the Meeting of Signatories is held annually (art. VIII(c)). Extraordinary meetings can be called through a procedure similar to that required for an extraordinary Assembly. They are held as soon as possible, and the agenda for such a meeting is restricted to the purpose indicated in the request which led to its being called (art. VIII(d)). As with the Assembly, each party is responsible for its own costs of representation at a Meeting of Signatories, but, for accounting purposes, the expense of running the meeting is an ordinary administrative cost of the organization itself (art. VIII(g)).

The function of the Meeting of Signatories has been accurately characterized by Professor Cheng as relating 'essentially to matters of high policy',[35] on account of the particular interests and competence of the individual members of the Meeting, comprised as it is of representatives of telecommunications administrations and entities. Not for such are the governmental and political responsibilities which are the fodder of the Assembly of Parties. Equally, matters of the day-to-day running of the system, and recommendations and planning as to future developments are, in practice, matters for the Board of Governors acting jointly with the planning arm of the executive organ. The Meeting takes an overview, basing much of its consideration on material transmitted to it in the form of working papers and recommendations by the Board, which (along with the Executive) must, by definition, possess the most relevant information, and which is accustomed to dealing with the material and material of the global telecommunications system.

Art. VIII(b) lists various functions and powers as belonging to the Meeting of Signatories. Like the functions and powers of the Assembly, these may be divided between powers involving decision-making and powers to make recommendations and give views. It is noticeable among the listing of powers how often the Meeting of Signatories is instructed to take into account the views of the Board of Governors, entirely apart from the overriding duty to give due and proper consideration to the views of the Board and the Assembly which is imposed both generally by art. VI(c), and specifically by art. VIII(a)).

The Meeting of Signatories expresses views to the Board of Governors on the Annual Report and annual financial statements which the Board submits to it (art. VIII(b)(i)). It also expresses views and makes recommendations on any proposed amendments to the Agreement and Operating Agreement (art. VIII(b)(ii). It also considers and gives its views on complaints by Signatories submitted to it directly through the Board of Governors and complaints submitted through the Board of Governors by users of the INTELSAT space segment which are not Signatories to the Operating Agreement (art. VII(b)(vii)). Finally, prior to the adoption of the permanent management arrangements, it expressed views on the proposals for the Executive Organ, which were suggested to it by the Board of Governors (art. VIII(b)(x)).

The Meeting of Signatories has decision-making powers in relation to many financial and operational matters; understandably since the Signatories bear the costs of the organization. It decides on recommendations made by the Board for the raising of INTELSAT's financial ceiling (art. VIII(b)(iv)). Based also on recommendations by the Board, it established the general rules concerning the approval of earth stations for access to the INTELSAT space segment, the allotment of space segment capacity, and the rate of charge for utilization of the space segment (art. VIII(b)(v)). The Meeting also takes certain decisions in the process of the expulsion of a Signatory (art. VIII(b)(vi)). Its approval is required for the provision of certain domestic telecommunications services by INTELSAT (art. VII(b)(ix)).[36] It also partakes of a general catch-all provision in art. VII tautologously allowing it to exercise any powers given to it elsewhere in the Agreement or Operating Agreement (art. VII)(b)(xii)). Finally, and of considerable effect in the functioning of the Board of Governors, the Meeting of Signatories has the regular function of annually determining the minimum investment share which entitles a Signatory or group of Signatories to be represented on the Board of Governors (art. VIII(b)(xi)).

The Board of Governors

The Board of Governors is not misnamed, being for all practical purposes, the governing body of INTELSAT. The number of Governors forming the Board can fluctuate, but the Agreement specifies, through a fairly complicated set of interlinking provisions, that, exclusive of up to five 'geographic' Governors (discussed below), the Board consists of approximately 20 members, rising in appropriate circumstances to as many as 22 (art. IX(b)(iii)–(iv)). Thus the Board in 1988 consists of 27 members, the 22 plus the five.

As we have already seen, it is one of the functions of the Meeting of Signatories annually to determine what is known as the 'minimum investment share' for the purpose of establishing the membership of the Board of Governors (art. VIII(b)(xi)). Possession of that minimum entitles a Signatory, or a group of Signatories which have agreed so to be represented, to a seat on the Board of Governors. The 'investment share' of each Signatory was first determined as at the date of entry into force of the Operating Agreement, and for purposes of the Board is redetermined on 1 March of each year (art IX(h) of the Agreement and art. 6(c)(ii) of the Operating Agreement). The investment share attributable to a Signatory is equal to its percentage of all use of the INTELSAT space segment by all Signatories (art. 6(a)) as amended by agreement.[37]

The calculation and setting of the minimum investment share required for representation on the Board of Governors is dealt with by art. IX(b) of the Agreement. The initial minimum investment share was that equalling the investment share of the investment share of the Signatory thirteenth in descending order of size of the initial investment shares of all the Signatories (art. IX(b)(i)). Thereafter, the Meeting of Signatories has determined the minimum investment share 'guided by the desirability of the number of Governors being approximately twenty' (art. IX(b)(ii)). This figure of 20, however, excludes the special category of up to five Governors whose function is to represent a regional or geographic area (art. IX(b)(ii) and (v)), to which we will come shortly.

There are three kinds of Governor. The classification depends upon whether Governors:

1   represent a single Signatory;
2   represent a group of Signatories whose cumulative investment shares are more than the minimum investment share currently set for the purpose of membership of the Board;
3   represent groups with an investment share less than the current minimum, but which are represented on the basis of geographic area.

Each Signatory having an investment share of more than the minimum set annually by the Meeting of Signatories is entitled to be represented by one Governor whom it appoints (art. IX(a)(i) with art. IX(c)). In addition, one Governor represents each group of two or more Signatories which do not have their own individual governors, and whose combined investment shares is not less than the minimum (art. IX(a)(ii) with art. IX(c)). Normally, but not necessarily, these groups have some sort of regional commonality.

The remaining kind of Governor represents at least five Signatories which are not otherwise represented on the Board of Governors, and which lie within one of the five regional groups defined for the purposes of representation within the ITU. These regions are: the Americas; Western Europe; Eastern Europe and Northern Asia; Africa; and Asia and Australia. A Governor representing such a group holds office regardless of the total investment share which is attributable to the members of the group, but the number of governors holding office under the category cannot exceed two for any of the ITU regions, nor may there be more than five governors in all holding office under the category (art. IX(a)(iii)). In addition, it should be repeated that these governors do not count for the purpose of the determination of the minimum investment share by the Meeting of Signatories.

A Signatory or group of Signatories fulfilling the requirement for representation under any of these three types of representation is entitled to be represented on the Board of Governors (art. IX(c)). A single Signatory nominates its own representative and, obviously, groups negotiate who shall be their representative within the group.

As indicated, where the representation is of a 'regional' nature, special rules may come into play. A group seeking representation through a 'regional governor' rather than on the basis of minimum investment shares must seek such representation in writing. The request becomes effective immediately upon its receipt by the Executive Organ, provided that the maximum number of such 'regional governors' has not been met. Where there already are two such regional governors for a particular region or there are already five regional governors in all, the request for representation by the new group has to be submitted to the next ordinary meeting of the Meeting of Signatories (art. IX(c)). It is the duty of the Meeting of Signatories annually to determine which regional group shall be represented on the Board of Governors. If the limit of five 'regional governors' prescribed in art. IX(a)(ii) is potentially to be exceeded, the Meeting of Signatories must first select that group from each region which has the highest investment share. If the number of groups so selected is less than the limit of five regional governors in all, the number is made up through the selection of the group having the next highest investment share until in the case of a region there are two groups selected, or the total limit of five has been reached (art. IX(d)). Although such provision sounds complicated, the device does ensure that the matter of regional or geographic distribution takes precedence in the determination of the regional governors. To permit regional representation to be given merely on the basis of size of groups might militate against a proper geographic distribution. Two groups from two regions, each with relatively high investment shares (though less than the minimum), could otherwise mean that the regional representation on the Board of Governors would be limited to three regions. Telecommunications statistics indicate that size of group alone might even produce the representation of only two regions on the Board under the regional concept. As it is, the needs of each of the five regions must first be satisfied when there is a conflict between a large number of groups seeking such representation.

Continuity on the Board is ensured through a Governor remaining the representative of his Signatory or group of Signatories until the next determination of the minimum investment shares (art. IX(e)). This means that changes in investment share following adjustment during the tenure of the Governor is disregarded. Only if a Signatory withdraws from either the organization itself or a group, with the result that total investment share of the remainder of the group no longer entitles it to representation does the Governor lose his position on the Board (art. IX(e)).

There are two ways in which a Board maybe quorate – by the level of voting participation of those present or by simple numbers. First, a quorum for any meeting of the Board of Governors consists of a majority of the Board having at least two-thirds of the total voting participation of all Signatories and groups of Signatories represented on the Board (art. IX(i)). Thus, the absence of a number of Governors holding small voting shares is unimportant, and the Board is not thereby paralysed. Second, a quorum may be made numerically if not more than three Governors are absent, irrespective of the voting participation which they represent (art. IX(i)). This reduces the impact of the absence of up to three Governors representing major investment shareholders.

The Agreement specifically provides that the Board of Governors must endeavour to take decisions unanimously (art. IX(j)), and it is remarkable how well that instruction has been observed. Reading the Minutes of the Board, the unanimity of decision is remarkable, although, given the views which are noted in the minute of the discussion, on occasion one is led to wonder what the precise balance of opinion might have been had the matter gone to a vote. As it is, the secretarial and minuting arrangements of the Board of Governors must be considered to be models of their kind, and quite exemplary. They are highly stylized, and this may assist in the agreement on decisions which they record.

The form of the minutes states that the Board undertook consideration of a particular item of its agenda, and what were the background papers on the matter. It then narrates that the Board 'took note that' and 'considering that' 'decided'. The 'taking note' and 'considering that' statements are succinct summaries of points raised in the discussion. However, no points are ever attributed to particular Governors. Of course, on occasion, one might deduce which Governor was likely to have advocated a particular point, but such guesses can sometimes be wildly wrong.

The stylized minuting, and the decision-making by consensus gives a Governor considerable freedom. He can go back to the Signatory or Signatories which appointed him and show that the points of concern to them were made, although the eventual decision was to the contrary. The Governor's own responsibility within the argumentation is masked, as is his role in the attainment of the consensus. This is useful, for it removes him from the danger of being considered a delegate of his Signatory and makes him more truly a representative. Thus, he is allowed to consider rather the best interests of the INTELSAT system as a whole.

The stylizing and summarizing also saves a researcher from wading through the verbiage encountered in the minutes of some other organizations where it sometimes appears to be the intention of a speaker to utter at length (often confusedly) the views of his own state on matters which may or may not be relevant to the matter in hand, or to engage in an ego-trip, or both. That advantage apart, I gather that the Minutes of INTELSAT Board are not too remote from the material they chronicle. Discussions within the Board are very businesslike, points being stated succinctly partly because the Board is composed of specialists who do not need extended presentation of points, and partly because there is much discussion between Governors prior to Board meetings.

That said, it is, of course, possible for the Board to take decisions by voting and, in that, the investment shares of the Signatory or Signatories which the individual Governors represent are of crucial importance.

The first point to be noted is that, unlike the case of the deliberations of the Interim Communications Satellite Committee, no Governor of INTELSAT may cast more than 40 per cent of the total voting participation of all Signatories represented on the Board. If the voting participation of any Governor would exceed that limit, the excess is, for voting purposes, distributed equally among all the other Governors on the Board (art. IX(g)(iv)). Such equal distribution to the other Governors has an interesting effect. Proportionately it is the smallest investment holders who are increased most, thus augmenting their voice within the body as a whole.

The result of this provision is that the Governor representing the Communications Satellite Corporation (whom for convenience we can call the US Governor) has a maximum voting weight of 40 per cent within the Board of Governors.[38] This is radically different from the first days of Interim INTELSAT, when the US representation was slightly in excess of 61 per cent. It is also different from the provisions of Interim INTELSAT under which the US participation would never have become less than 50.63 per cent. Needless to say, such a reduction in the US voice was not achieved without lengthy discussion, not to say argument, and as will be seen, there are other provisions which do ensure that, on many matters, the USA retains an effective veto. Nonetheless, and despite that qualification, it must be recognized that the USA has accepted a constitutional provision under which the American interest is a minority interest, even though it is likely always to remain the largest single interest within the organization.

Apart from such questions of redistribution, the position is that each Governor has a voting participation equal to whatever part of the investment share of the Signatory or Signatories he represents is derived from the use of the INTELSAT space segment for ordinary services provided by the organization. These are:

1   international public telecommunications;
2   domestic public telecommunications services between areas of a
    state which are separated by areas of the high seas or by the juris-
    diction of another state,
3   domestic telecommunications services in difficult areas which are
    not already linked by ground systems, and for which approval
    has been given by the Meeting of Signatories (art. IX(f)).

In the case where a Signatory has been permitted an investment share
smaller than that which would ordinarily have been allocated to it,
that reduction affects its potential voting participation. A Signatory
which has been granted a greater investment share has its voting
potential correspondingly increased. However, there is one exception. Where a Signatory has been allocated the absolute minimum
investment share of 0.05 per cent, and it forms part of a group for the
purpose of representation on the Board of Governors, its voting participation is taken as that 0.05 per cent, irrespective of its actual utilization of the INTELSAT space segment (art. IX(g)(iii)). In short, the
minimum voting participation within the Board of Governors held
by any Signatory is always 0.05 per cent. The maximum is 40 per
cent.

It follows that the total voting participation within the Board of
Governors is not necessarily equal to the total of the investment share
of all Signatories. It is quite competent for a Signatory not to be
represented on the Board of Governors. It may not have the minimum investment share required for representation, or it may decline
to be a member of a group having representation, either on the basis
of its total investment share or as a regional group. Again, an investment share may exceed the share attributable to utilization of the
space segment for the indicated public purposes. This can, and does,
produce some very efficient work by the Conference Secretariat section of the Executive Organ. Prior to each meeting they calculate the
various total voting potentials of Governors, and the total vote
required for various purposes in terms of the Agreement, the Operating Agreement and the Rules of Procedure of the Board, so that
decisions which require special majorities are, if a vote is required,
dealt with with the expertise and skill which the Board has come to
expect from its Secretariat. The potential votes required can alter as
new information becomes available, or as, say, a Signatory forfeits its
voting share through a default in paying dues.

Substantive matters are decided by the Board by an affirmative
vote complying with one of two requirements as to majority. These
requirements are similar to those for the quorum of the Board. First,
such a majority may consist of a majority of the Board having at least
two-thirds of the total voting participation of all Signatories and

groups of Signatories represented on the Board. In the calculation of the voting participation for this purpose, any distribution of the 'excess over 40 per cent', of any Signatory's share is taken into account. Second, and alternatively, the majority needed for a decision on a matter of substance is the total number of Governors making up the Board for the time less three, regardless of the voting participation which they represent. The alternative, based on numbers only, prevents the Board being paralyzed by the opposition of less than three Governors representing the largest investment shares (art. IX(j)(i)).

The effect of these provisions is that, at present, calculations show that numerically the US Governor plus three other Governors from those representing the five next largest investment shares (whether or not these represent single Signatories), can take decisions on matters of substance. At the same time, the altenative formula of total number minus three allows the possibility that such decisions could be taken without the approval of the US Governor and two others (even the next two in terms of voting participation). These, then, are the two extreme positions which can be envisaged; decisions by the US Governor plus three with large participation, and decisions contrary to the will of the US Governor and two of the largest participants. It is good that, as a matter of fact, neither extreme has so far been approached in the working of the Board.

Procedural questions are settled far more easily. A simple majority of Governors present and voting will determine a procedural matter, unless that matter has other rules prescribed for it in the Rules of Procedure of the Board of Governors (art. IX(j)(ii)). The decision whether a matter is one of procedure or of substance is in the first place decided by the Chairman of the Board, but he may be overruled by a two-thirds vote of Governors present and voting, each having one vote on the matter (art. IX(k)). However, matters are usually dealt with in terms of the Rules of Procedure which were adopted by the Board in terms of art. IX(m).[39]

The Board of Governors ordinarily meets four times per year (art. IX(n)), but it is competent for it to meet as often as necessary for the performance of its functions (art. IX(n)). The quarterly meeting has been found, however, to be quite sufficient under normal circumstances, the three-month interval allowing time for the decisions of the previous meeting to be carried forward, reports to be drafted, and so on.

The functions of the Board of Governors are set out mainly in Article X of the Intergovernmental Agreement. Twenty-seven separate heads are given, fleshing out the generalized responsibilities of the Board given by art. X(a) for 'the design, development, construction, establishment, operation and maintenance of the

INTELSAT space segment', and for carrying out other INTELSAT activities which have been appropriately authorized. The enumerated headings include the adoption of policies, plans and programmes; procurement procedures; financial decisions; decisions regarding the intellectual property rights in necessary inventions; approval of earth stations; the establishment of conditions for access to the space segment; the appointment and removal of the Director-General; dealing with the staffing of the Executive Organ; arranging contracts; dealing with the withdrawal of a Signatory; and generally expressing views and making recommendations on all necessary matters both to the Meeting of Signatories and the Assembly of Parties.

In most of these issues the Board of Governors took up matters which had already been the province of the Interim Telecommunications Satellite Committee (ICSC). However, as was only to be expected, the Board of Governors has pressed ahead, and is now well into the execution of its own plans and decisions, as opposed to implementing what the ICSC had already decided. The decade which has passed since the adoption and coming into force of the Permanent Arrangements has seen the Board build on the foundations created by its predecessor to the extent that these foundations are now no longer visible, save in minor respects.

The Director General and the Executive Organ

The Executive Organ of INTELSAT is headed by the Director General,[40] who is the chief executive of INTELSAT and its legal representative. He is directly responsible to the Board of Governors for the performance of all management functions, and acts in accordance with the policies and directives of the Board.[41] The Director is appointed by the Board of Governors, subject to confirmation of the Assembly of Parties. He may be removed from office 'for cause' by the Board acting on its own authority (art. XI(b)(i)–(iii)).[42] Given the importance of the Director General within the organization, it is not surprising to find that art. XI(b)(iv) of the Intergovernmental Agreement states that the 'paramount consideration' in his appointment 'shall be the necessity of ensuring the highest standards of integrity, competency and efficiency'. Notwithstanding, the article goes on to require the appointed individual to 'refrain from any action incompatible with [his] responsibilities to INTELSAT'.

Under the Director General is the Executive Organ. Until the Executive Organ was brought into being, INTELSAT's affairs were dealt with by a Secretary General whose position, together with his office, had existed transitionally between the coming into force of the

Agreements and bringing into force of the Permanent Arrangements in 1978. The Secretary Generalship had more or less the equivalent powers and responsibilities as that of the Director General, although during the transition period a considerable amount of INTELSAT management was contracted to COMSAT under a Management Services Contract (art. XII(e) and Annex B). Art. XI(a) of the Inter-governmental Agreement required that the structure of the Organ be implemented not later than six years after the entry into force of the Agreements, the Director-General under the Board being required to take all necessary steps to this end (art. XII(j)). Naturally, it was required that the management arrangements should be consistent with the basic purposes of INTELSAT, its international character, and its obligation to 'provide on a commercial basis telecommunications facilities of high quality and reliability' (art. XI(c)(i)). It was also required that, as part of the permanent management arrangements, the Director-General should arrange for the contracting out of certain technical and operational functions to other entities (such contracts being negotiated, executed and administered by the Director-General himself (art. XI(c)(ii))). Arts. XII(a), (f) (g) and (h) set out a timetable for studies and progress towards the permanent management arrangements. In accordance therewith, first a Special Committee and then a Working Group was set up within the Board of Governors which prepared three reports (BG–19–32, BG–20–26 and BG–21–48). The Board of Governors then submitted its report on 'Permanent Management Arrangements' (BG–22–50) to the Second Assembly of Parties meeting in Washington on 26 July 1976 (AP–2–9) where the proposed arrangements were duly approved (AP–2–4), and also to the Fifth (Extraordinary) Meeting of Signatories in Nairobi on 23–24 September 1976 (MS–5–5), which also approved (MS–5–4). Thereafter the Board dealt with the technical matters of the 'Office of the Director General' (BG–28–33) at its meeting in June 1977 (BG–28–3).

The Director-General assumed office in accordance with art. XII(i) on 31 December 1976. Although the former Secretary-General, Santiago Astrain, became the first Director General (cf. art. XII(c)), the change involved more than a mere change of title. The Management Service Contract with COMSAT continued for another two full years, being terminated as from 31 December 1978; the document 'Termination of the Management Services Contract' (BG–35–63) contained the amendments to the contract which gave its end-date and also obliged COMSAT to close in an orderly way those of its services which were to be assumed by the new Executive Organ. In that connection, it appears that when the permanent management arrangements were brought into being, many of the personnel who had been employed by COMSAT on the Management Services Contract were taken over into the employment of INTELSAT itself. It is generally agreed that these have made a smooth and interesting transition from being the employees of a commercial US company to becoming the employees of an international organization, in which (it is said) the responsibilities, goals and interests may be somewhat different.

From 1 January 1979 the management of INTELSAT has been wholly under the jurisdiction of the Director General and the Executive Organ, he being empowered to delegate such powers and responsibilities as he thinks necessary to subordinates within the Executive Organ (art. XI(d)(ii)). The members of the Executive Organ are appointed subject to the same 'paramount considerations' as the Director-General himself – that is, what is sought is 'the highest standard of integrity, competency and efficiency' (art. XI(b)(iv)). The Board of Governors has the responsibility of determining the number, status, and terms and conditions of appointment of all posts in the Executive Organ, but in this it acts on the recommendation of the Director General (art. X(a)(xx)).[43] The Board's approval is necessary for the appointment of senior officers reporting directly to the Director General (art. X(a)(xxi)). Appointment to less senior positions are the responsibility of the Director General himself.

RIGHTS AND OBLIGATIONS OF MEMBERS

The ordinary rights and obligations of members are indicated throughout the Intergovernmental Agreement, but art. XIV is so headed. In fact, this article is very significant, but to title it 'Rights and Obligations of Members' is rather misleading. Paragraph (a) is redundant since it obliges the Parties and Signatories to exercise their rights and carry out their obligations consistently with, and for the furtherance of, the principles in the Preamble to the Agreement. The General Law of Treaties holds that a Party shall not frustrate an international agreement to which it is a party. Nonetheless, such a statement may be of practical diplomatic use in either municipal or international disagreements, it being possible to point to the obligation and to the signature of the Party or Signatory to the Agreements.

More importantly, paragraph (b) of art. XIV gives all Parties and Signatories the right to attend and participate in meetings and conferences at which thay are entitled to representation under the Intergovernmental Agreement or Operating Agreement. Further, they are entitled to representation at any other meeting called or held by INTELSAT in accordance with the terms of that meeting. This paragraph may also seem tautologous, but the right is filled out with obligations on countries which are to host INTELSAT meetings. The Executive Organ is charged with the duty of ensuring that all Parties and Signatories entitled to be present at a particular INTELSAT meeting are given access to the host country for the duration of the meeting (art. XIV(b)). This is carried through by the requirement that the agreement between INTELSAT and the host country shall include provision for such general access of representatives. There is therefore no possibility that political differences could lead to the deficient representation of INTELSAT members at a meeting because the host country refuses to admit certain of the representatives. This is but one of the ways in which the INTELSAT Agreements have been drafted to ensure that politics extraneous to the creation and maintenance of the satellite communciations system is kept, as far as possible, remote from the organization's business. The political considerations which affect so many other aspects of international cooperation (and non-cooperation) are thus minimized. The intention of the Parties, and perhaps more cogently, of the Signatories and the persons skilled in the technical and administrative aspects of telecommunications whom they have put in office, is to establish and maintain a global system. The possibility of a host country interfering with representation for purely political reasons has been skilfully excluded. Where such difficulties might arise, the prior negotiation by the Executive Organ allows them to be resolved or side-stepped by re-locating the meeting.

The remainder of art. XIV contains perhaps the most crucial legal provisions in the Agreement, affectng as it does the whole development of international satellite communications. Their effects are fundamental to most other developments of space telecommunications facilities throughout the world because of the obligations it places on Parties, on Signatories and through the jurisdictional control exercised by Parties, on persons or companies within the jurisdiction of Parties. Any telecommunications system established by a Party, by a Signatory or by any other entity or person under the jurisdiction of a Party subsequent to the coming into force of the Agreement for a Party may interact with the INTELSAT system at a technical level, and, if the telecommunications traffic is of the public international service it may affect the INTELSAT system through taking traffic which it might otherwise carry. There are therefore two elements to the obligations imposed by art. XIV – that directed towards the technical compatibility of systems with the INTELSAT system, and that directed towards the economic protection of the INTELSAT business. We will discuss these more fully below, but the requirements may be indicated here. The provisions are tightly drawn, but may be summarized.

First, technical compatibility: this can arise with either a domestic or an international system, and in relation to public or to specialized services. When a Party or a Signatory, or a person or company within the jurisdiction of a Party intends individually or jointly to establish, acquire or utilize space segment facilities separate from the INTELSAT system for a domestic public telecommunications system the appropriate Party or Signatory involved has to consult the Board of Governors 'prior to the establishment, acquisition or utilization' of the necessary facilities, giving the Board all relevant information on the matter (art. XIV.(c)). In the case of international public telecommunications services (art. XIV(d)), and of specialized telecommunications purposes whether domestic or international (art. XIV(e)) information is provided to, and the consultation is with, the Assembly of Parties through the Board. In the appropriate case, the Board of Governors (for domestic public telecommunications services) and the Assembly of Parties (for all other cases) then, within six months (including by way of a special Assembly if necessary) (art. XIV (f)), makes findings by way of recommendation as to the 'technical compatibility of such facilities and their operations with the use of radio frequency spectrum and orbital space by the existing or planned INTELSAT space segment'.

Second, economic questions. In the case where the proposed space segment facilities are for international public telecommunications services, the Assembly is additionally provided through the Board with all relevant information to allow it also to consider, first, whether the proposal, if carried into effect, would cause 'significant economic harm to the global system of INTELSAT', and, second, assurances that the development will 'not prejudice the establishment of direct telecommunications links through the INTELSAT space segment among all INTELSAT members' (art. XIV(d)).

The single exception to what therefore amounts to a duty on Parties and Signatories to clear with INTELSAT all their other non-INTELSAT space telecommunications developments is in the case of a space telecommunication facility established for national security purposes (art. XIV(g)). That exception apart, it is clear that by the requirement of findings of technical compatibility and lack of significant economic harm (where that also is needed), INTELSAT is given the major position of authority in relation to all later developments which its Parties, Signatories or persons or companies under their jurisdiction may wish to undertake. The cynical might think that it would be possible for a Party or Signatory to go through the motions of consultation, and then proceed at will in defiance of whatever the Board of Governors or Assembly of Parties might express in their 'findings' which are recommendations, not decisions. However, art. XIV(a) requires Parties and Signatories to 'exercise their rights and meet their obligations under [the] Agreement in a manner fully consistent with and in furtherance of the principles stated in the Preamble and other provisions of [the] Agreement'. The Preamble clearly expresses the will and wish of the Parties *inter alia* to achieve 'a single global commercial telecommunications satellite system as part of an improved global telecommunications network'. It also states the determination of the Parties to provide 'the most efficient and economic facility possible consistent with the best and most equitable use of the radio frequency spectrum and orbital space'. Such language is incompatible with the cynical approach indicated. But, if such high expressions of determination and aspirations are thought too starry-eyed, there is art. XVIII and the provisions of Annex C about the settlement of disputes, and ultimately there is the provision in art. XVI(v)(i) by which a Party may be deemed to have withdrawn from the organization. This last is a polite way of providing for the expulsion of a Party, and which could be applied to the case of the defiant, the cunning or the obtuse, and which will be considered more fully below when we deal with withdrawal from INTELSAT.

MEMBERSHIP, WITHDRAWAL, SUSPENSION AND EXPULSION

Membership

So far we have assumed the notion of membership of INTELSAT. Before dealing with withdrawal it is logical first to outline the criteria by which Parties and Signatories become members of the organization.

Membership of INTELSAT was open to any state signing the Agreement during the period in which it was opened for signature, and which was either a Party to the Interim Agreement, or was a state member of the ITU (art. XIX(a)). Since that period, membership has been open to any state meeting these qualifications, and which accedes to the Agreement once any procedure required by its internal law – such as ratification – has been gone through; and membership dates from the day an instrument of accession is deposited with the Depositary for the Agreements, the US Government (art. XIX(c) with art. XX(b)). However, the Intergovernmental Agreement cannot enter into force for any state until either its government, or a telecommunications entity designated by it for the purpose, has signed the Operating Agreement (art. XX(d)). The Operating Agreement enters into force for a Signatory on the date that the Agreement enters into force for its designating Party (art. 23(a)).

Withdrawal

It is competent for a Party or a Signatory voluntarily to withdraw from INTELSAT. The withdrawal of a Party necessarily involves the simultaneous withdrawal of the Signatory which it has designated to represent it for the purposes of the Operating Agreement, or of the Party in its capacity as Signatory where this is the case (art. XVI)d)). Where a Party does withdraw, the Agreement and the Operating Agreement cease to be applicable to the Party and to the Signatory on the same date (art. XVI(d)).

The procedure is that a withdrawing Party must give written notice to the US Government which, by art. XXII is the Depository for the Agreement (art. XVI(a)(i)).[44] In the case of the withdrawal of a Signatory to the Operating Agreement, the decision to withdraw is communicated in writing to the Executive Organ of INTELSAT by the relevant Party (art. XVI(a)(i)).[45] A Party's notification of the withdrawal of its Signatory also constitutes the Party's acceptance of the Signatory's decision to withdraw (art. XVI(a)(i)). In the cases of both the withdrawal of the Party and the withdrawal of a Signatory, the withdrawal becomes effective either three months after the date of the receipt of the appropriate notice of withdrawal, or, if the notification of the withdrawal so stipulates, on the date of the next annual determination of investment shares by the Assembly of Parties following the expiry of that three month period (art. XVI(a)(ii), referring to art. 6 (c)(ii) of the Operating Agreement). However, when a notice of withdrawal is received, it has one immediate effect. The Party and its Signatory, or the Signatory alone as the case may be, immediately cease to have any rights of representation and any voting rights within the organization (art. XVI(g)).

When a Signatory withdraws from INTELSAT, the Party which designated it must either assume the capacities and responsibilities of Signatory for its interest, or it must designate a new Signatory to take over from the date of the withdrawal of its predecessor (art XVI(e)).[46] A Party which does not comply with this requirement must itself withdraw from the organization (art. XVI(e)).

As may be implied from the provisions as to the withdrawal of a Signatory, it is competent for a Party to substitute itself as Signatory in place of the one which it had formerly designated. This is done by giving written notice to the Depository, and there is no three month minimum period between the notice and its coming into effect.[47] Upon its assumption of the outstanding obligations of the previous Signatory, and signature of the Operating Agreement, the Party enters into its role as a Signatory, and the Operating Agreement ceases to be in force for the previously designated Signatory (art. XVI(f)).

Although membership of the ITU was one of the criteria for signature of the Agreement (art. XIX(a)) and is a criterion for accession to it (art. XIX(c)), no Party or its designated Signatory can be required to withdraw from INTELSAT as a direct result of a change in the Party's status with regard to the ITU (art. XVI(n)). Thus if a member is expelled from the ITU (as has happened) that does not *ipso facto* affect the member's relationship with INTELSAT.

Suspension and expulsion

Although the language of the Agreements is carefully couched, the provisions relating to withdrawal also contain mechanisms under which a Party or Signatory may be suspended, and even expelled, from the organization.

Suspension is a halfway house towards expulsion, and can be applied in the case of a Signatory to the Operating Agreement. The suspension of the rights of a Signatory is automatic in the case where it has failed to pay its share of capital requirement due under art. 4(a) of the Operating Agreement, as determined by the Board of Governors within three months of the payment falling due (art. XVI(c)).[48] The rights lost are those of representation within INTELSAT and of voting. If the payment is not made within a further three months, or if the Party designating that Signatory has neither substituted another Signatory, nor itself taken on the dutes of the defaulting Signatory, the Board of Governors considers the matter. After taking into account any representations made to it by the Signatory or its designating Party, the Board may then recommend to the Meeting of Signatories that the Signatory be deemed to have withdrawn from INTELSAT. If the Meeting, after having heard any representations from the Signatory, so decides, the Signatory is deemed to have withdrawn and the Agreement and Operating Agreement cease to be applicable to the Signatory from the date of that decision (art. XVI(c)). In the period between the suspension and the decision, however, the Signatory continues to have all the obligations and liabilities of that status (art. XVI(h)). If the Meeting of Signatories decides that the Signatory should not be deemed to have withdrawn, the suspension is immediately lifted and the Signatory resumes its voting and representation rights.[49]

In the case of obligations other than that of capital contribution, since suspension is not automatic there would seem to be a greater degree of discretion available. However, that is merely appearance. The mechanism would appear to operate as inexorably in cases of breach of INTELSAT obligations as in the capital contribution case. In fact, most examples have been financially related – as, for example, where a Signatory has failed to pay its utilization charges or amounts due under investment share increases arising from utilization – but the language is more indefinite referring only to obligations under the Agreements. Where a Signatory fails to comply with an obligation under either the Agreement or the Operating Agreement, the Board of Governors may resolve to take note of the failure. That resolution is communicated to the Signatory, and if the matter is not remedied within three months the Board considers the matter further. The Signatory and its designating Party may make representations. The Board may then suspend the Signatory's rights and recommend to the Meeting of Signatories that the Signatory be deemed to have withdrawn. Again, there is opportunity for representations, after which the Meeting may approve the recommendations; and the deemed withdrawal has immediate effect (art. XVI(b)(ii)). As before, in the period between the suspension and the decision, the Signatory continues to have all the obligations and liabilities of a Signatory (art. XVI(h)).

Where the problem lies with a Party, it is for the Assembly of Parties to act. The Assembly may act either on its own initiative, or act on notice received that a Party is failing to meet its obligations under the Agreement. In practice one assumes that such a notice would emanate from the Board, but conceivably it could also come from the Director-General or any other Party to the Agreement, for there is no prescribed channel through which such notice may come to the Assembly. The procedure thereafter is for the Assembly to hear any representations from the Party. If it finds that the failure to meet obligations has occurred, the Assembly may decide that the Party be deemed to have withdrawn from INTELSAT. The Agreement ceases to be in force for the Party from the date of that decision (art. XVI(b)(i)).[50] The only problem with this procedure would be that the Assembly of Parties meets every two years, which is too long a period for such matters to lie unresolved. To that end, art. XVI(b)(i) specifically provides that an extraordinary meeting of the Assembly of Parties may be called to deal with such a matter.

Financial and other effects of withdrawal

As indicated above, when a Party gives notice of its withdrawal from the organization, or when it transmits notice of the withdrawal of its designated Signatory, the Party or Signatory ceases to have any rights of voting or representation within the organs of INTELSAT immediately upon the receipt of that notification (art. XVI(g)). As a corollary, it also incurs no obligations or liability in respect of decisions taken after the receipt of the notice of withdrawal (art. XVI(g)). But in the case of a Signatory, unless the Board of Governors determines otherwise, the Signatory remains liable for its share of capital contributions, so far as these have not been paid and it remains liable for commitments entered into before the notice of withdrawal, and continues to be liable in law also for acts or omissions of INTELSAT prior to receipt by INTELSAT of the notice of withdrawal (art. XVI(g)), referring also to art. 21(d) of the Operating Agreement).

In the case of the substitution of a Signatory, or its replacement by its Party, there are no financial effects of interest to INTELSAT. The 'new' Signatory takes on the responsibilities of its predecessor immediately on the effective date of signature, and there will be no interval between the former Signatory laying down its burden and the new one taking it up (art. XVI(f) and (e)).

In the case of expulsion (properly, 'deemed withdrawal') a Party or a Signatory, as the case may be, incurs no obligation or liability in respect of INTELSAT business decisions taken after the date of the decision by the Assembly of Parties or Meeting of Signatories (art. XVI(k) and (j)). However, in the case of the Signatory, or where the Party is itself Signatory, unless the Board of Governors decides otherwise, liability remains for meeting the appropriate share of capital contributions for commitments entered into before the decision, and for liabilities for acts and omissions of INTELSAT in the same period (art. XVI(j) and (k)).

As noted above, the rights of a Signatory are suspended automatically by operation of art. XVI(c) in the case of its failure to pay sums due from it by way of capital contributions under art. 4(a) of the Operating Agreement within three months of the sums becoming due. If, at the end of the procedures thereafter, the Meeting of Signatories decides that the Signatory should not be deemed to have withdrawn, the rights of the Signatory remain suspended until payment of the amounts due is made (art. XVI(i)). In the case of suspensions by the Board of Governors under art. XVI(b)(ii) (the non-automatic suspension for failure to comply with any obligation under either Agreement, including grounds other than those relating to capital requirement and dealt with by art. XVI(c)), a subsequent decision by the Meeting of Signatories that the Signatory shall not be deemed to have withdrawn restores the Signatory's rights without reference to other financial questions.

When a Signatory does leave INTELSAT, its financial rights and obligations do not terminate immediately upon its withdrawal. Certainly, as noted, there are no new obligations laid on it, but the process of disentanglement can take time. Under art. 21 of the Operating Agreement, the Board of Governors must, within three months of the effective date of the withdrawal, notify the Signatory of its reckoning of the financial status of the Signatory within INTELSAT as at the effective date of the withdrawal. It will also propose terms of settlement of that account.

The notification of financial status contains a statement of all sums due by INTELSAT to the Signatory in terms of a valuation of its investment share of INTELSAT's assets (the precise calculation being carried out in terms of art. 7(b) of the Operating Agreement). Notice is also given of sums due from the Signatory in terms of any outstanding share of capital contribution, and share of liabilities incurred by the organization prior to notice of the withdrawal being received, together with any other sums due to INTELSAT (for example, utilization charges) (art. 21(b)). The Board of Governors decides over what period repayment shall be made to INTELSAT, with some regard to the rate at which other Signatories receive repayment of their capital contributions, thus ensuring that the other Signatories remain within the organization are not prejudiced (art. 21(c)). The Board also settles the rate of interest to be paid on outstanding amounts (art. 21(c)).[51]

Such financial arrangements are not, however, absolutely rigorous. It is possible for the Board to relieve a withdrawing Signatory, in whole or in part, of the sums due from it by way of capital contribution and incurred contractual liabilities (art. 21(d)). However, the Board cannot relieve the withdrawing Signatory of either non-contractual obligations arising from INTELSAT acts or omissions occurring prior to receipt of the notice of withdrawal, or its effective date as the case may be (art. 21(e)). Nor can the Board deprive the Signatory of any rights acquired by it in its capacity as Signatory, which would otherwise continue after the effective date of its withdrawal, and for which it has not been compensated under art. 21 of the Operating Agreement (art. 21(e)(ii)).

INVESTMENT SHARES

Investment shares: utilization

The provisions of the Operating Agreement as to the calculation of the investment share held by each Signatory are highly important. There is a clear financial effect, as the investment share is the measure of a Signatory's obligation to contribute to capital requirements (art. 4(a)). It is also the basis of representation on the Board of Governors, those Signatories having more than the minimum investment share set by the Meeting of Signatories being entitled to separate representation on the Board. In the case of the smaller members, their investment share may form part of a group entitled to group membership on the Board. Finally, the investment share, which weights the Signatory's vote on the Board of Governors, also directly afffects the relative voice which that Signatory has within the counsels of the organization. The provisions for determining and redetermining the investment shares are therefore crucial for the working of INTELSAT itself.

Broadly, and subject to specific modification in certain instances, the basic position is that each Signatory has an investment share in INTELSAT equal to the percentage which its direct utilization of the space segment of the global system bears to the total utilization of that segment by all Signatories (art. 6(a)).

The relating of the investment share to utilization is an interesting departure from the situation which obtained under the Interim Arrangements. Under the previous Agreements, Signatories were requried to bear a proportion of the costs of the planned system as a percentage of the total cost. These proportions were split by agreement in terms of the economic importance of the state concerned, its traffic-generating potential and the technological skills of the Parties to the Interim Agreement. It was that quota which was taken through into matters of voting and representation. It is a mark of the success of the Interim Arrangements that it was possible in 1973 to make utilization of the space segment the basis for the determination of investment shares. These shares serve the same function as the former quotas, but have the advantage of being realistically based upon proven interest in the system. Of course, the former method of dealing with the matter was effective, and, indeed, a viable global satellite telecommunications system was brought into being under it. It was that system's strength and stability that permitted its financial base and governmental system to be related to utilization in 1973.

The question therefore arises: why change from the Interim Arrangements? The short answer is that by so doing the Parties and Signatories more closely related their investment liabilities, and hence their returns on those investments, to the actual operation and their own use of the system. They are thereby brought more closely into the Arrangements, and are less likely to see the system as simply a service which they buy. If they wish simply to buy service, they could stay outside it and make appropriate arrangements with INTELSAT (as, for example, Communist countries do). Investment as the basis of representation, voting weight, financial liabilities for the establishment, working and development of the system, together with returns on contributions, ensures that Parties and Signatories see the organization as 'theirs', and not as something external, whether servant or taskmaster. It also means that an international organization has truly developed, forming a cooperative endeavour. Although operating commercially, INTELSAT is not a commercial organization bent on maximizing profit. On the other hand, neither is it an international aid organization, although it has successful programmes directed towards the needs of Third World countries. The investment share system helps in maintaining that balance between two extremes.

Investment shares are determined for each Signatory and ordinarily become effective for INTELSAT purposes on the first day of March each year (art. 6(c)(ii)). Special provisions were made for the initial introduction of the system, but these have not lapsed (art. 6(c)(i) and (ii)). Investment shares are also redetermined effective as at:

1 the date of entering into force of the Operating Agreement for a new Signatory (art. 6(c)(iii));
2 the effective date of withdrawal of a Signatory from the Organization (art. 6(c)(iv)); and
3 the date on which a Signatory which has newly become liable to pay charges for the use of the space segment through its own earth station requests that such a determination be made, provided that that request is made not less than 90 days from the date on which those charges become payable (art. 6(c)(v)).

Implicit in the last of these is the point that a Signatory might, by reason of its non-utilization of the earth segment through its own ground station, have a nil utilization figure according to INTELSAT records. In that case its proportionate investment share would be 0 per cent, with consequent effects on representation and voting. That would clearly be unsatisfactory, as would also the related possibility that an investment share computed simply on the basis of utilization might become ludicrously small. To cope with these potential problems, art. 6(h) sets a minimum investment share to be held by any Signatory as 0.05 per cent of the total investment shares – the same figure as was used in the Interim Arrangements.

On the other hand, the annual revision of the investment shares, and the possibility of a further revision within that period, coupled with the use of utilization as the yardstick, could mean that a Signatory might hold an investment share larger than it desires. An investment share carries with it a commitment to contribute to the financial workings of INTELSAT, entirely apart from any question of paying for the use of the system. A Signatory might therefore find itself bound to pay more by way of capital contribution than it wishes to. Such an automatic burden was considered by the negotiators of the definitive arrangements to be a deterrent to Signatories in their potential use of the global system. Accordingly art. 6(d) of the Operating Agreement provides a way round the difficulty. At each redetermination of investment shares, a Signatory whose share would otherwise be increased may request that it be allocated an investment not less than its existing investment share, and, in appropriate cases, not less than the quota which it held under the Interim Arrangements (art. 6(d)(i)). The procedure is for the Signatory to deposit its request with INTELSAT, indicating the investment share which the Signatory wishes to have allocated to it.[52] The request is then given effect to the extent that other Signatories are willing to accept increases in the investment share which would otherwise fall to them.

Such a reallocation of an increase in an investment share requires that Signatories notify INTELSAT of their willingness to accept an increase in their investment shares in order to permit such a redistribution of 'unwanted increases', indicating any limit there may be to that willingness (art. 6(d)(ii)).[53] If the total amount by which Signatories are willing to accept increases in their investment shares is greater than the total amount of requested reductions, the two figures are offset. The increases are distributed among the willing Signatories in proportion to their investment shares immediately prior to the revision of investment shares which has given rise to the problem (art. 6(d)(ii)). If the requested reductions exceed the total indicated acceptances of increase, increases are allocated among the accepting Signatories in accordance with their stipulations. The total allocated increase is then deducted from the total requested reductions, and the remainder is allocated as increases among the Signatories requesting reduction in the proportion which each requested reduction bears to the total requested reductions (art. 6(b)(iii)). All these alterations in investment shares by request and consent continue in effect until the effective date of the next annual determination of investment shares – that is to the first day of March following (art. 6(b)(iv)).

Where an investment share is to be determined on either the acceptance of a new Signatory, or a Signatory first becoming liable for utilization of the space segment through its own ground station, or through a reallocation in terms of requests for reductions, or consequent upon the withdrawal of a Signatory, the investment shares of all other Signatories are readjusted in the proportion that their immediately previously held investment shares bear to each other (art. 6(f)). However, in the case of the readjustment of investment shares consequent upon the withdrawal of a Signatory, art. 6(f) goes on specifically to provide that a Signatory holding the minimum 0.05 per cent investment share through the requirement of holding that minimum in terms of art. 6(h), shall not have its share increased in such a readjustment. Presumably this means that a Signatory whose share is arithmetically 0.05 per cent by reason of its utilization would be subject to an increase, but not a decrease, on such readjustment. There does not seem to have been such a case so far.

Consequential financial adjustments

As it is possible (and usual) that at each determination of investment shares, the shares attributable to Signatories changes, it is necessary that there shall thereafter be a financial adjustment between the Signatories. Were this not done, a Signatory could find that it has gained or lost in terms of its share of INTELSAT assets through the changes in utilization patterns over which it has no control. The adjustment takes place in terms of art. 7 of the Operating Agreement, which also contains provision for the valuation of the assets and liabilities involved. The arrangements regarding adjustments needed in the transfer from the interim to the definitive arrangements (art. 7(a)(i) and 7(b)(ii)(A)) having lapsed, it fails to consider only those provisions presently governing the matter.

The amounts of each adjustment are calculated when each redetermination of investment shares occurs, and involve simply the difference, if any, between the new and the immediately previous investment share held by a Signatory (art. 7(a)(ii)). For these purposes art. 7(b) establishes a fairly cumbersome, but clear, method of valuing the assets and liabilities involved. The original costs of all INTELSAT assets, including capitalized returns and capitalized expenses, are taken as recorded in the organization's accounts as at the date of the adjustment. From that is deducted the accumulated amortization, again recorded in the INTELSAT accounts as at the date of the adjustment, and loans and other accounts payable by INTELSAT as at that date (art. 7(b)(i)). That net or base figure is then adjusted by adding or deducting a further amount for the deficiency or excess in the payment by INTELSAT of compensation for use of capital from the entry into force of the Operating Agreement (12 February 1973) to the effective date of the valuation relative to the cumulative amount due under the Operating Agreement, at whatever rates of compensation for use of capital have been set by the Board of Governors. This calculation is done on a monthly basis, and relates to the net or base figure which is being adjusted (art. 7(b)(ii)(B)). From this final figure, the amounts due to and from Signatories whose investment shares have respectively decreased or increased can be easily established, being the difference between the old and new shares (art. 7(a)(ii), above).

Payments to and from Signatories are made by a date designated by the Board of Governors, and the rate of interest on unpaid sums thereafter is likewise set by the Board (art. 7(c)). This rate is to be the same as the rate set for unpaid capital contributions under art. 4(d)(art. 7(c)). That apart, a persistent failure to comply with obligations of adjustment comes into the category of general failure which the Board may take up, leading in the ultimate to the possibility that the defaulter may be deemed to have withdrawn from the organ-ization.[54]

Finally, it has already been indicated that one importance of the investment share is that it is the measure of the voting strength available to a Signatory within the Board of Governors. The question may therefore arise whether a Signatory could 'buy up' others' investment shares in order to increase its voting strength. Although this possibility exists, with a consequent increase in liabilities for capital contributions and so on, its effect is blunted by the provision of art. IX(g)(iv) which limits an individual Governor to casting up to 40 per cent of the total voting participation of all Governors on the Board, and the equal distribution of any excess above that figure among the other Governors.[55]

NOTES

1 Testimony of W.G Carter, Satellite Communications – 1964, Hearings before a Subcommittee of the Committee on Government Operations, House of Representatives, 88th Cong., 2nd Sess., 1964, Part II at p. 660.

2 Cf. A. Chayes, T. Ehrlich and A.F. Lowenfeld, *International Legal Process: Materials for an Introductory Course*, vol. 1 (Boston, Mass.: Little, Brown and Co., 1968) vol. 1, 'Problem IX. An International Operating Agency: The Communications Satellite Corporation and the International Consortium for Satellite Communications', pp. 632–704, where the trend of questions put to student users is oriented to the US interests in the matter.

3 1963–4 690 *HC Deb.* Col. 421.

4 Communications Satellite Act, s. 305(a) and (b); COMSAT Articles, art. V, s. 5.04(b) and art. VIXX ss. 8.02 and 8.10.

5 For example, as the British Post Office then was.

6 Although peculiar this is not an insuperable difficulty. As I write, the Kuwait Foreign Investment Office has been buying common stock in British Petroleum. However, telecommunications may be more sensitive than oil – note the subjunctive.

7 Cf. Testimony of W.G. Carter, Satellite Communications – 1964, Hearings before a Subcommittee of the Committee on Government Operations, House of Representatives, 88th Cong., 2nd Sess., 1964, Part II at 660, cited above). Since 1983 bilateral agreements have been talked of as a way in which independent US companies could break into the international satellite communications market: however, the question of coordination with INTELSAT remains. See Chapter 4, 'US Separate Systems', p. 187.

8 Cf. the communique issued after the Commonwealth Prime Ministers' Conference, London, July 1964: 'The Prime Ministers also took note of the current international discussions on the establishment of a global system of satellite communications. They endorsed the desirability of establishing such a system and considered how commonwealth countries could best co-operate with each other and with other countries in its development.'

9 CEPT is outlined in Chapter 6, 'EUTELSAT' at p. 266.

10 Satellite Communications Report 1964, p. 95. See also: 'Summary of European Regional Organisations in the Communications Satellite Field' (9 January, 1964, 110 Congressional Record, 175–6, reprinted (1964) III ILM 233), for a statement of the development of the European view and summary of the organizations involved. The meeting of November 1963 is outlined at (1964) III ILM 234.

11 1963–4 690 *HC Deb.* Col 420–1.

12 See *Communications Satellites,* Staff Report Prepared for the Use of the Committee on Aeronautical and Space Science, US Senate, 87th Cong. 2nd Sess. 1962, pp. 56–8, 98–125; and, N.L. Schwartz *et al.,* 'Foreign Participation in Communications Satellite Systems: Implications of the Communications Satellite Act of 1962', RAND Corporation Memorandum RM–2484–RC, February 1963.

13 A USSR view was that the USA had won: I. Cheprov, 'Global or American Space Communications System?' *International Affairs,* December 1964, pp. 69–74.

14 'Agreement Establishing Interim Arrangements for a Global Commercial Communications Satellite System, and Relative Special Agreement, printed, Satellite Communications', 1964 Cmnd. 2436; 51 Department of State Bulletin, pp. 281–90 (24 August 1964): reprinted (1964) III ILM 805; Satellite Communications, 1964 – Hearings, Part II, Appendix 9. See F. Lyall, 'Satellite Telecommunications', *I1 Diritto Aereo,* no. 5, 1966, pp. 315–33.

15 This was the position in the UK at the time the interim INTELSAT arrangements were agreed. The 'Post Office', which had had responsibility for telecommunications since the Telegraph Acts of 1868 and 1869 and the Telephone Transfer Act of 1911, ceased to be a department of government and became a public corporation akin to other nationalized industries in 1969 under the Post Office Act 1969. Telecommunications responsibilities were transferred to the new British Telecom under the British Telecommunications Act 1981, and British Telecom was 'privatized' in 1984. None of these changes has constitutionally affected the UK participation in either the Interim or Definitive INTELSAT Arrangements, although privatization may have affected the UK attitude to the importance of financial considerations in planning and operational decisions within INTELSAT.

16 See note 15 as to the current UK position.

17 Testimony of W.G. Carter, Satellite Communications 1964, Hearings, Part II, p. 663.

18 F. Lyall, 'Law and Space Telecommunications', unpublished LL.M. thesis, (Montreal, Canada: McGill University, 1965).

19 See below, 'Constitutional Structure' subsection 'The Director-General and Executive Organ,' p. 104 and Chapter 2, above, 'The Communications Satellite Corporation'.

20 See Bin Cheng, 'Communication Satellites' 1971 *Current Legal Problems,* 1971, pp. 211–245 at pp. 220–1; Peter B. Trooboff, 'INTELSAT: Approaches to the Renegotiation', *Harvard International Law Journal,* no. 9, 1968, pp. 1–84; Richard R. Colino, *The INTELSAT Definitive Arrangements: Ushering in a New Era in Satellite Telecommunications,* (Geneva: European Broadcasting Union, 1973), pp. 13–21.

21 Cf. Walter A. McDougall, . . . *the Heavens and the Earth: A Political History of the Space Age,* (New York: Basic Books, 1985) particularly ch. 17, 'Benign Hypocrisy: American Space Diplomacy', pp. 34–60 and ch. 20 'Voyages to Tsiolkovskia' 415–35, esp. at pp. 422–9.

22 Agreement relating to the International Telecommunications Satellite Organization (INTELSAT) (1973) UKTS No. 80, Cmnd, 5461; 23 UST 3813, TIAS 7532; (1971) 10 ILM 1909.

23 (1973) UKTS No. 80, Cmnd, 5461; 23 UST 4091, TIAS 7532; (1971) 10 ILM 946.

24 In citing, I adhere to the general practice of using roman numerals for articles in the Agreement and arabic numerals for articles in the Operating Agreement.

25 In the UK this was done by The INTELSAT (Immunities and Privileges) Order 1972, 1972 SI No. 117, pending the Protocol to be mentioned. See now note 27.

26 The Headquarters Agreement would deal with such matters as between INTEL-SAT and the US.

27 Protocol on INTELSAT Privileges, Exemptions and Immunities, Washington 19 May–20 November 1978, (1981) UKTS No. 2, Cmnd. 8103; INTELSAT (nd). Also available as BG–33–50. The UK gives effect to the Protocol by The INTELSAT (Immunities and Privileges) Order 1979, 1979 SI No. 911.

28 Headquarters Agreement between the Government of the United States of America and the International Telecommunications Satellite Organisation: effective 24 November 1976 by exchange of letters dated 31 January and 1 February 1977 between INTELSAT and the US Department of State; 28 UST 2248, TIAS 8542; INTELSAT (nd).

29 Cf. art. 2(1)(d) of the Vienna Convention on the Law of Treaties 1969. A reservation is a 'unilateral statement, however phrased or named, made by a State when signing, ratifying, accepting, approving or acceding to a treaty, whereby it purports to exclude or to modify the legal effect of certain provisions of the treaty in their application to that State'.

30 Documentation accords with the 'Style Manual for INTELSAT Documentation', produced by the Conference Services Department of the Administration Directorate of the INTELSAT Executive Organ in January 1979.

31 Assembly of Parties, Rules of Procedure, 30 September 1976 (AP–2–24 N/9/76) INTELSAT.

32 See below, Chapter 4, section 'Other systems', p. 153.

33 Bin Cheng, 'Communications Satellites', op. cit., pp. 211–45 at p. 228.

34 Meeting of Signatories, Rules of Procedure 19 April 1977 (MS–6–12 S/4/77) INTELSAT.

35 Bin Cheng, 'Communications Satellites', op. cit., pp. 211–45 at p. 228.

36 See below, Chapter 4, section 'Services', p. 149.

37 See below, 'Investment Shares', p. 114.

38 The US 'share' is currently approximately 18 per cent.

39 Board of Governors, Rules of Procedure, 3 February 1977 (BG–26–45 W/2/77) INTELSAT.

40 Richard R. Colino took over from the first Director (and previous Secretary-General of interim INTELSAT), Santiago Astrain at the beginning of 1984. Since April 1987 the position has been held by Dean Burch of the USA.

41 During a vacancy, or when the Director-General is absent or otherwise unable to exercise his functions, the Board appoints a senior INTELSAT officer as Acting Director-General with full powers of the office (art. XI(d)(i)).

42 As was done in the case of Richard R. Colino in December 1986.

43 Cf. 'Office of the Director-General' (BG–28–33), now somewhat out-of-date.

44 The precise procedures for all notices in respect of the withdrawal of a Party are contained in art. XVI(m)(i).

45 The procedures for all notices in respect of the withdrawal of a Signatory are contained in art. XVI(m)(iii).

46 The precise procedures for all notices in respect of the substitution of a Signatory are contained in art. XVI(m)(v).

47 The precise procedures for all notices in respect of the substitution of Party as Signatory are contained in art. XVI(m)(v).

48 The precise procedures for all notices in respect of suspension of a Signatory are contained in art. XVI(m)(iv). Examples are Bolivia, Chad, Honduras, Jordan, Somalia, Uganda, Viet Nam and Zambia.

49 As to financial questions, see Chapter 4, 'Finance', p. 123.

50 The precise procedures for all notices in respect of deemed withdrawal are contained in art. XVI(m)(ii).

51 An interesting example was the settlement consequent upon the withdrawal of the then representative of China (Taiwan) in 1976: see 'Article 21 Financial Settlement' BG–24–51, 26 October 1976; following the decision of the Second Assembly of Parties, September 1976, Agenda item 7, 'Implementation of the United Nations General Assembly Res. No., 2758 (XXVI) Concerning the Representation of China in International Organizations', AP–2–4 at paras 206–24.

52 The Board of Governors has established precise procedures, as it was required to do by art. 6(d)(v).

53 The Board of Governors has estalbished precise procedures, as it was required to do by art. 6(d)(v).

54 See section 'Suspension and Expulsion', p. 111.

55 See also section 'The Board of Governors', p. 97 above.

4 The International Telecommunications Satellite Organization: Its operation

FINANCE

Capital contributions

INTELSAT's finance comes from two main sources: capital contributions and charges for the use of the system. Capital contributions are paid by Signatories as and when required in proportion to their investment shares (art. V(c)). As we have seen, the investment share is fundamentally tied to members' utilization of the system, subject to the minimum investment share of 0.05 per cent (art. V(b)). The Board of Governors advises the Meeting of Signatories which has powers of decision on the capital ceiling appropriate for the enterprise (art. 5; art. VIII(b)(iv)). Capital contributions are used for the direct and indirect costs of the design, development, construction and establishment of the space segment (such as the cost of satellites), and for INTELSAT property costs such as that of building the rather beautiful energy efficient INTELSAT headquarters (art. 4(b)). They may also be used in times of emergency to meet operating costs, or to cover a financial award made following upon a determination of the legal liability of the organization (art. 4(b)). They may also be used in times of emergency to meet operating costs, or to cover a financial award made following upon a determination of the legal liability of the organization (art. 4(b)). The actual sums due from Signatories are contributed as determined by the Board of Governors under art. 4(a) of the Operating Agreement. Initially also, some capital contribution was used for running expenses (see below), but once the system was established and being well used, more and more of the operating costs were able to be referred to revenues from charges for utilization of the system.

Capital contributions may also serve as a fall-back position. Were there insufficient income from utilization revenues to meet the INTELSAT's operating, maintenance and administrative costs, the undertaking would be in acute difficulties. An enterprise which cannot generate sufficient income to pay its normal operating costs is severely threatened. Unless the satellite telecommunications system operates and is maintained, and INTELSAT itself is administered, the whole purpose of INTELSAT fails. To avoid that eventuality the Board is empowered to raise funds by other means in order to keep going. It may meet the deficiency by using any operating funds already created, by overdraft facilities, by raising a loan, by requiring further capital contributions from Signatories in proportion to their investment shares, or by a mixture of such measures (art. 8(f)). Such remedies have not yet been required.

When a Signatory fails to pay a capital contribution within three months of it falling due, the Signatory's rights, as we have seen, are automatically suspended and may eventually lead to its 'deemed' withdrawal.

Utilization charges

The Board of Governors specifies the units by which the utilization of the space segment is calculated for the different services offered and establishes the charge to be paid for that utilization (art. 8(a) and (b)). In so doing it is guided by general rules on the matter which are adopted by the Meeting of Signatories (art. 8(a)). As explained previously the charges set are intended to cover the organization's operating, maintenance and administrative costs, together with necessary operating funds for the future, the amortizing of investments and compensation for the use of capital contributed by Signatories (art. 8(a)). The charges set for specialized telecommunications services are intended fully to cover the cost of their provision, including in appropriate cases that of the separate satellites and other facilities, as well as a proportion of INTELSAT's general costs (art. 8(b)). Under art. V(d), as things stand all system users pay the same rate of utilization charge for each type of service used. This is an important principle, although one which has come under threat during the 1980s. As agreed in 1971, it was intended that the small user of a service should be charged on the same basis as the high traffic generator. Arguably, in terms of system use, it makes no difference whether a link is between the UK and the USA or between two minor telecommunications generators: the satellite costs are the same, the distance served roughly equal, and ground station costs are a matter for the local authority. Nonetheless, the proposals originating in the USA for separate non-INTELSAT international satellite telecommunications systems have attacked the INTELSAT system as being non-competitive because of its policy of worldwide non-discrimination in rates for each service. It is argued that this unnecessarily increases the charge to end-users on major telecommunications routes. One result has been the proposal that art. V(d) be amended to permit INTELSAT to set a different rate for certain services on appropriate routes with the intention of allowing INTELSAT to cope with (and maybe demolish) the threatened competition. This will be discussed later. Suffice it here to say that until the Agreement is modified, non-discrimination in the rate for each service offered by INTELSAT is a matter of international agreement.

In sum, the charges for the utilization of the space segment, and for any specialized telecommunications services, are made on the basis which any prudent commercial operation would employ. This is seen particularly in the provisions for the amortizing of the investment and for compensation for the use of capital. INTELSAT is not in the business of providing subsidized services, although the lesser developed countries do sometimes have special consideration given to their needs in ways which some would interpret as subsidy. Be that as it may, it is noticeable, for example, that the Operating Agreement specifically provides that the Board is to fix a rate for compensation for use of capital as close as possible to the cost of money in the world markets and, in fixing the rate, the Board is also to take into account the risks which are associated with investment in INTELSAT itself (art. 8(c)). Roughly speaking, that translates into the requirement to pay a high dividend on money which may be lost in an extremely risky enterprise. Commercially, high returns are required to produce investment in high risks and, the Board is under a duty to act commercially in this matter.

The Board of Governors has a duty to institute 'any appropriate sanctions' in the case where a Signatory has defaulted for three months or longer in the payment of utilization charges which have fallen due (art. 8(d)). At its most extreme this could involve the commencement of procedures for the suspension and possible expulsion of the Signatory in terms of art. XVI(b) of the Agreement.[1] However, lesser sanctions are occasionally applied, leading, for example, to the exclusion of the defaulting Signatory's ground station facilities from access to the space segment. No pay – no service! It has to be said that there are certain countries which appear on a 'no service' list more often than others. While there are many reasons for such default, in common with other international organizations, INTELSAT finds that frequent problems are encountered by certain states in the obtaining of internationally convertible currency to pay for liabilities.

Revenues from utilization charges

The revenues obtained by INTELSAT through the various utilization charges are dispersed in a fixed order of priority (art. 8(e)). First, operating, maintenance and administrative costs are met. Second, sums are placed in any operating funds which the Board has considered necessary. Third, Signatories are repaid in proportion to their investment shares at the time, sums which go towards the amortizing of the system as that has been determined by the Board. Fourth, a Signatory which has withdrawn from INTELSAT is paid whatever is due to it. Fifth, payments are made towards compensation for the use of capital.

PROCUREMENT

During the discussion of the definitive arrangements for INTELSAT all parties were anxious that the satellite system should be of the highest quality, and should be established as soon as possible. Parties other than the USA which were possessed of, or aspired to, a domestic space-related industry, were also concerned to ensure that procurement contracts in the enterprise would be spread as widely as possible. Some wanted the system to be established as inexpensively as possible, and afford the best value for money. Those countries which had no pretensions to space technology were more interested in getting a good service for as little capital expenditure as possible. The USA, on the other hand, was concerned lest the organization be diverted to become a conduit through which technical skills and knowledge might be syphoned off, even under the guise of 'technical assistance'.

It was, of course, always clear that quality, price and time were factors which would probably favour a US contractor and US suppliers, and that that would conflict with the wish of other Parties to obtain for their own industries a share of an emergent, lucrative market. On occasion, the competition offered by non-US suppliers might be competitive, but the basic position was that the Americans already had an expertise which would give them an edge. The eventual compromise on these matters is contained in art. XIII of the Intergovernmental Agreement and arts. 16 and 17 of the Operating Agreement.

The principles of procurement policies are set out in art. XIII of the Intergovernmental Agreement. The basic position is that contracts for the procurement of goods and services for INTELSAT are awarded on the basis of bids made in response to open international invitation to tender (art. XIII(a)). In general 'the best combination of quality, price and the most favourable delivery time' will secure a contract (art. XIII(a)). Quality, price and time, however, are elements whose balance can vary within several bids, each of which potentially meets the criterion of 'best'. In that case, the sought element of the desirability of dispersing contracts is possible. In such a case the contract is awarded 'so as to stimulate, in the interests of INTELSAT, worldwide competition' (art. XIII(b)).

In that formulation, a factor of great importance is indicated; the desirability that INTELSAT shall be serviced by a number of contractors. If nascent space industries were to be stifled by exclusion from what was to be the largest single group of such contracts in the world, there was a distinct possibility that INTELSAT might find itself with only one single contractor willing to bid for INTELSAT contracts. Such a situation is sometimes known as 'contractor capture' or 'capture by contractor'. Historically, it is clear that such a situation does not permit the purchaser properly to control the activities of such a contractor, to recover for excess charging and to monitor performance on the three crucial factors – quality, price and time. It is different when a government places contracts within its own jurisdiction, since, by legislation or otherwise, a government can claw back what it considers to be excess profits and otherwise enforce its views of fair business.[2] Within the context of an international organization such as INTELSAT, such measures can not be applied and the danger of capture by a single contractor is one which has to be kept in view.

Art. 16 of the Operating Agreement defines procurement procedures. The approval of the Board of Governors is required for the issue of requests for proposals or invitations to tender for contracts which are likely to exceed US$ 500,000 in value (art. 16(b)(i)). Similar approval is also necessary for the award of any contract to a value exceeding US$ 50,000 (art. 16(b)(ii)). This puts the Board in command of all significant procurement contracts, and has allowed it to develop practices which guard against the danger of contractor capture. In addition, the Board has also developed the practice of ordinarily ensuring that large contracts (for example, for the design and procurement of the spacecraft series) contain provisions within them under which certain elements of the contract are subcontracted to other suppliers and usually to foreign suppliers. Thus it has come to pass that the two main companies which act as prime contractors on major INTELSAT spacecraft contracts are Ford Aerospace, and Hughes Aircraft. The tenders made by each of them usually involves the subcontracting of anything up to 40 per cent of the work to foreign subcontractors. Indeed, it seems that it is practice within the Board of Governors to look for such a breakdown of the work involved, and to take that into account when deciding upon the placing of a contract. While it is probably incorrect to say that each of the two main contractors have formed a consortium witht other contractors, the fact remains that each of the main contractors has formed close working arrangements with particular companies working in the space industries, many of them in Western Europe.

That said, such a procedure does assume that competitive bids are received. Of course, this may not always be the case. Art. 16(c)(iv) of the Operating Agreement therefore provides that, where there is only one source of supply which can meet a particular specification by INTELSAT, or where the sources of supply are so severely restricted that it is neither feasible nor in the best interests of INTELSAT to go through the motions of open international tender, the normal processes can be foregone. Direct approach can be made to the single supplier, but where there is more than one potential supplier, all must be given an equal opportunity to put in a bid. Examples of the 'sole source' provision at work are several contracts with COMSAT, largely explained by reason of its location and specialization during the Interim INTELSAT and its tenure of the Management Services Contract to the end of 1978. Thus, to take examples from well in the past, in 1978 having regard to the impending assumption of responsibilities by the Executive Organ, INTELSAT negotiated a Technical Service Contract (BG–34–66), a Management Services Contract (BG–35–55) and a Maintenance and Supply Agreement (BG–35–79) all with COMSAT on a sole-source basis. Early launch agreements with NASA fall into the same category.

In certain other circumstances, the open international tender system may also be departed from, even though bidding procedure would be possible (art. XII(c)). These are:

1 where the estimated value of the contract does not exceed US$ 50,000 (a figure which the Meeting of Signatories can raise) art. 16(c)(i));
2 where procurement of the goods or services is required urgently to maintain the operational viability of the space segment itself (art. 16(c)(ii)); and
3 where what is required is of a predominantly administrative character best suited to local procurement (art. 16(c)(iii)).

The circumstances on which the Board may decide not to go through the open international tender route are therefore circumscribed, but are based on reasonable grounds. If there is only one, or a limited number of potential tenderers, it makes no sense to go through the open tender route, provided that all that might be interested are given opportunity to tender. Again, in INTELSAT terms, a contract for under US$ 50,000 is small. In the case of the 'urgent requirement' the time-consuming tender process is clearly inappropriate – the integrity of the operational system is at stake, and the situation must be rectified as speedily as possible. INTELSAT's economic viability depends on the confidence of its users, and a system which is 'down' for long periods quickly forfeits that confidence. Cable remains an attractive alternative on a number of major telecommunication routes, and might be turned to by distrustful users.

INVENTIONS, PATENTS AND TECHNICAL INFORMATION

Inventions, patents and technical information

Procurement has another side to it. It is not just a matter of countries wanting to be paid for building INTELSAT satellites, providing hardware and doing other work for the organization. It is also a matter of countries wanting to gain expertise, skills and knowledge. The free flow of ideas is not something which happens automatically in international business. There is a complicated law which attempts to protect inventions through the device of patent and other forms of the protection of intellectual property. Short of patent protection, with its associated licensing procedures, there is the simple matter of technical information which can be immensely helpful to a competitor. Such matters could not be ignored in the debate before the adoption of the definitive INTELSAT Agreements. Indeed, the INTELSAT project required new knowledge to be discovered, new processes to be invented, and new devices to make these processes efficient, commercial and usable in telecommunications matters. Such matters had to be coped with before they caused trouble for the new organization.

As has been indicated, in many matters of contract the Board of Governors proceeds by open international tender. Buried within the language then used, however, was reference to a 'request for proposals' (art. 16(b)(i)). This device, the 'RFP', is used by many other international operating agencies, both private and public – among them, INMARSAT and EUTELSAT.

In the RFP, the organization indicates in general terms what it wishes to be done. Thus, it may identify a particular problem which has to be solved, or state a particular result which it wishes to obtain within certain parameters. One example specified that the solution to a particular problem must be capable of being packed into a device of certain proportions and weight occupying a certain position within a spacecraft. The problem was the supply of electric current by battery, which is required to keep the satellite's systems working when the satellite enters the Earth's shadow and the solar panels do not work. The weight, dimensions and positioning of the battery were essentially limited by the configuration of the spacecraft and its other components. If the battery were too heavy it would lessen the weight of propellant which the satellite could carry for stationkeeping and other purposes, and hence would reduce the satellite's effective life. If the battery were badly positioned it might affect the satellite's spin and/or stationkeeping. The supply of electricity had to be adequate in terms of voltage, and had to last for a specified minimum time (measured in years). Nonetheless, a certain constant supply for a specified minimum period from a battery of a specified maximum weight, configuration and dimensions was what the Board was looking for. It got it, although a battery had to be invented, designed and built for the purpose.

Under RFP circumstances, therefore, and sometimes even under the normal 'invitation to tender' process, the successful bidder has to invent something to do the job within the given parameters. On other occasions, a successful contractor may have to obtain, by licence, rights in other inventions in order to meet the requirements of the contract it has entered into. Sometimes both invention and the obtaining of rights to use existing inventions have to occur. The question then arises: what happens to the intellectual property and patent and other rights in such inventions? This matter is dealt with by art. 17 of the Operating Agreement.

The whole area of intellectual property, 'know-how' and rights to property in inventions is, of course, one fraught with the utmost difficulty because of the very nature of the matter and the value which such inventions can hold. In the case of space activities and in telecommunications in particular, the potential value of inventions and of technical information is extraordinarily high. But different legal systems have different approaches to such matters. In the Anglo-American system, work performed under contract usually infers that rights in inventions made go to the placer of the contract, not the inventor. Other contries take a different attitude.

Broadly, under art. 17(a) of the Operating Agreement, INTELSAT is required to acquire rights in inventions made, and in technical information discovered in connection with any work performed by it or on its behalf. But INTELSAT is to acquire only such rights as are necessary in the common interests of INTELSAT and its Signatories in their capacities as Signatories, and no more. In the case of work performed for INTELSAT under contract, such rights are to be acquired on a non-exclusive basis. In short, INTELSAT acquires sufficient rights in inventions and technical information to allow it to carry out its function of establishing and running the telecommunications system efficiently and at minimum cost. It should not itself enter the invention business, acquiring exclusive rights in order to make a profit through further licensing arrangements. Its interest lies in its system and the provision of telecommunications facilities, not in the general area of business and commerce.

Subject to certain exceptions, three sets of rights are contemplated as being those which INTELSAT ordinarily seeks in connection with any work which it performs or which is done for it involving a 'significant element of study, research or development' (art. 17(b)). In ensuring that it possesses itself of these rights, the organization is to take into account 'its own principles and objectives, the rights and obligations of the Parties and Signatories' and 'generally accepted industrial practices' (art. 17(b)).

The first right which INTELSAT seeks in all such cases is a 'right without payment to have disclosed to it all inventions and technical information generated by work performed by it or on its behalf' (art. 17(b)(i)). Second, INTELSAT seeks the right to disclose and to have disclosed to Signatories and others within the jurisdiction of any Party such inventions and technical information (art. 17(b)(ii)). Third, INTELSAT seeks for itself the right to use such inventions and technical information (art. 17(b)(ii)). Fourth, it seeks the right to authorize Signatories and others within the jurisdiction of any Party to use the information and inventions, and to arrange for the authorization of Signatories and these others to do so (art. 17(b)(ii)). In short, INTELSAT seeks knowledge, the right to use it, and the right, directly or indirectly, to pass on both that knowledge and that right.

Whether or not payment is made (in the normal case) for such rights depends upon the material involved. Where the inventions and technical information have been discovered in connection with INTELSAT's space segment or any earth station operating within the system, art. 17(b)(ii)(A) contemplates that the rights will be secured without payment. The reasoning would appear to be the Anglo-American notion that the sums due under contract are sufficient recompense for such invention or information. In other circumstances, however, art. 17(b)(ii)(B) stipulates that 'fair and reasonable terms and conditions' are to be settled with those who have property rights in the invention or information. That said, as noted above the ordinary position of contractors with INTELSAT is that the rights secured by the organization are secured on a non-exclusive basis (art. 17(a)). Contractors normally retain the ownership in any invention or technical information which they generate in carrying out an INTELSAT contract (art. 17(c), although see below as to deviation). That means that, outside the scope of the rights which INTELSAT has, the contractor retains its rights and can deal with them, defend them, license them and so on, as it will. And of course, we must note that, in a given matter, the organization 'taking into account its principles and objectives, the rights and obligations of [its] Parties and Signatories', could decide that it would settle for less than art. 17 might indicate (see below).

Of course, it is also possible that performance of an INTELSAT contract may require the use of inventions and technical information which have been generated elsewhere than on INTELSAT contracts, and hence is not sheltered by the foregoing provisions nor is available under contractual arrangements made in compliance with them. In such a case, INTELSAT is under a duty to acquire on 'fair and reasonable terms' the rights necessary to disclose the content of such inventions and information to Signatories and others within the jurisdiction of Parties and to use and authorize the use of Signatories and others of inventions and technical information directly utilized in the INTELSAT work (art. 17(d)). Naturally, INTELSAT is required to carry out this obligation only to the extent that the contractor has the right to grant such permissions, and the obligation extends only to the extent necessary for performance of the other INTELSAT obligations of disclosure (art. 1l7(d)). The actual couching of these obligations in legal form is somewhat cumbersome, but it is essentially simple, being designed to prevent the acquisition by INTELSAT of rights arising directly out of work done for INTELSAT being rendered nugatory by some other related matter remaining secret.

As can be readily appreciated, the obligations on INTELSAT as to acquiring rights to use and disclose cannot always be complied with. In certain cases, therefore, the Operating Agreement provides for deviation from the ordinary case. In determining whether to deviate from its normal practices, and if so how far to go, the Board of Governors is required to take into account the interests of INTELSAT itself, that of all Signatories, and also the estimated financial benefits to INTELSAT which the proposed deviation would produce (art. 17(g)).

Art. 17(e) deals with the circumstances under which the Board of Governors may approve a contract which deviates from normal practice as to use and disclosure to Signatories and others. First, it clearly states that the Board can only do so in individual cases; a blanket decision as to categories would seem to be ruled out. Second, such deviation can be approved only in exceptional circumstances. In determining whether such circumstances obtain, the Board would seem first to have to tried to stick by its normal contractual practices, for the paragraph speaks of the question arising 'in the course of negotiations'. Two circumstances are envisaged as justifying deviation. The first is that it is demonstrated to INTELSAT that to insist on its normal practice would be detrimental to the interests of INTELSAT itself. The second is that, in the case of onward disclosure of inventions and information to Signatories and others, it must be shown that for the potential contractor to agree would be incompatible with prior contractual arrangements between it and third parties, provided that these arrangements were entered into in good faith and (one assumes) not with the intention of squeezing out the INTELSAT interest (art. 17(e)).

The other circumstance in which a deviation from normal contractual practice is contemplated relates to the retention by a contractor of rights of ownership in inventions and technical information provided for under art. 17(c) (see above). Again in exceptional circumstances, and in terms of art. 17(f), the INTELSAT Board may approve a contract which provides otherwise. A contract may provide for INTELSAT to acquire property rights in such inventions or technical information where each of three conditions are satisfied. First, it must be demonstrated to the Board that failure to acquire such rights would be detrimental to the Organization's interests (art. 17(f)(i)). Second, the Board itself determines that INTELSAT should be able to protect these rights through acquiring patents in any country: that means that the organization would have to be able to prove its standing in the matter (art. 17(f)(ii)). Third, the contractor itself is either unable or unwilling to obtain such patent protection sufficiently early. It seems possible that this condition may operate partially in the sense that INTELSAT may seek to take over rights to the extent that the contractor is unable to secure patent protection (art. 17(f)(iii)).

The most obvious circumstance under which a contractor will be unable to enter into a contract giving INTELSAT the full rights which the Agreement indicates are desirable is one where the national law of the contractor interposes a restriction upon it. Notably, contracts with US contractors – such as COMSAT in its various divisions, or Hughes or Ford Aerospace or other satellite builders, or other US providers of research and development for INTELSAT – are subject to the several restrictions imposed by US law. Thus the International Traffic in Arms Regulations which are administered by the Office of Munitions of the US State Department have caused a problem. The Office is requried to license any export or proposed disclosure to 'foreign nationals' of any item listed on the 'Munitions List' of the Regulation after consulting appropriate other US agencies as necessary (for example, the Department of Defence). The list includes spacecraft, electronic equipment used in communications satellites and relative technical data relating to both. For the purposes of the US Regulations INTELSAT is a 'foreign national'. Extensive discussion with the contractors and the State Department produced in 1980 a 'basic understanding' under which US contractors, the INTELSAT Executive Organ, the Board of Governors and its committees have full access for management purposes, to technical data deliverable under contracts requiring approval from the Office of Munitions. A further review by the Office is necessary for transmission or disclosure of technical data further than the Executive Organ or the Board.[3] The Director-General was not entirely happy with this arrangement, but reckoned that it was the best possible. Occasional difficulties will be encountered, but at least the understanding does, and will, permit INTELSAT contracts to be fulfilled in a reasonable manner, although its effect may be significantly to reduce the availability of technical information to Signatories, Parties and persons under their jurisdiction compared with the position one might deduce from the language of the Agreements. Of course the US internal regulations are not the only source of such difficulty. The Paris-based Coordinating Committee on Exports, which reviews trade in a considerable list of matters between the West and the Communist countries, may also have an impact.

When INTELSAT does acquire rights to have disclosed to it the inventions and technical information generated by work done by it or on its behalf, it remains under an obligation to keep each Signatory which so requests informed of the availability and general nature of such inventions and technical information (art. 17(i)). Where it acquires rights to make the inventions and information available to Signatories and persons under the jurisdiction of Parties, it makes such rights available on their request (art. 17(i)). In the cases of both disclosure and use the terms and conditions which INTELSAT transmits its rights are on a non-discriminatory basis (art. 17(j)).

Disclosure and the transmission of rights of use are dealt with by art. 17(h), which covers materials under both the interim and the definitive arrangements. Disclosure to a Signatory is made subject to the reimbursement of any payment INTELSAT has to make in respect of such disclosure (art. 17(h)(i)). The onward disclosure to others, and the transmission of rights to use, are without payment in the case of inventions and information to do with the operation of earth stations (art. 17(h)(ii)(A)). In other circumstances, 'fair and resonable terms and conditions are to be settled among the different interests concerned, and are subject to the reimbursement of any INTELSAT costs (art. 1l7(h)(ii)(B))).

Data and technical information: procedures

The implementation of the principles set out in the Operating Agreement, of course, require further detail and elaboration, and at its forty-fifth meeting the Board of Governors adopted *INTELSAT Inventions and Data Distribution Procedures* (cited hereafter as *Procedures*),[4] a revision of the arrangements which it had adopted at its twenty-ninth meeting. This has two parts, the first dealing with data and technical information, and the second with inventions and patents. The *Procedures* are, of course, subject to the limitations which its municipal law may apply to a contractor. Thus the restrictions applicable to US contractors under the International Traffic in Arms Regulations and the limited disclosure to INTELSAT for management purposes only have been mentioned above.

As far as data and technical information is concerned, in accordance with the *Procedures* the Director-General has prepared, and maintains, what is called the 'Data Handbook', which is a cumulative listing of 'all significant items of technical data and information which have arisen directly out of or have been utilized in work performed at INTELSAT expense, both under contract and by the staff of the INTELSAT Executive Organ' (*Procedures* Part I, A.1). The Handbook, which is in looseleaf form to permit updating, is cross-indexed and gives a general description of the rights of use held by INTELSAT, any Signatory and any contractor covered by the data (*Procedures*, Part I, A.1). Updating is carried out quarterly for the INTELSAT Advisory Committee on Technical Matters and that feeds through to the other users of the book. Copies of the Handbook are distributed to Signatories as required, and to designees of Signatories on written request by a Signatory (*Procedures*, Part I, A.3). The costs of preparing, maintaining and distributing the Handbook fall to INTELSAT (*Procedures*, Part I, F.1). Signatories are put on notice of the possible need for limiting the dissemination of the information contained in the Handbook (*Procedures* Part I, A.4), and are required each to designate a recipient for data matters (*Procedures* Part I, A.5).

Data contained in the Handbook is usually entered either under 'Foreground Data' or 'Background Data', but if uncategorized data is requested, the Director-General will determine its nature in consultation with any appropriate contractor, and thereafter the procedures appropriate to the category are followed (*Procedures*, Part I, E).

Foreground Data is technical data and information arising directly from contract work (*Procedures*, Part I, B.n.1). This may be acquired by Signatory or its designee in one of two ways depending upon whether INTELSAT has to be involved in the distribution process. Which procedure is required is apparent from the entry about the data in the Handbook (*Procedures* Part I, B.1). Where INTELSAT owns the rights, or has power to grant sub-licenses or to designate recipients of rights of access and use of Foreground Data, a written request for data required can be added to the Director-General by any Signatory. Where the request is for use of the data for a non-INTELSAT purpose, a brief description of that use must also be given 'for the information only' of INTELSAT (*Procedures*, Part I, B.2a). In addition, persons or entities within the jurisdiction of a Signatory may also submit a request for data directly to the Director-General, provided that the appropriate Signatory has previously designated them to receive the information, and, as before, has provided a brief description of the intended use if it is for a non-INTELSAT purpose (*Pro-

cedures, Part I, B.2a). The Director-General complies with the request as speedily as possible, either by making the data available, obtaining it and making it available, or by granting such rights as are required. However, sometimes other procedures, such as patent review or patent applications, may first have to be dealt with. In addition, the *Procedures* allow the placing of explicit restrictions on the use and further dissemination of Foreground Data supplied in accordance with a request in order fully to protect INTELSAT's legal rights (*Procedures*, Part I, B.2b). Where INTELSAT is so involved, INTELSAT's costs and all contractor charges, such as data reproduction and administrative costs, are paid by the Signatory. On the specific request of the Signatory involved, the Director-General will pay contractors' and licensors' costs subject to prompt reimbursement (*Procedures*, Part I, F.2 and 3).

Where INTELSAT is not required to be involved, Signatories can acquire Foreground Data directly from the appropriate contractor or licensor. Terms and conditions are negotiated directly between the parties concerned (*Procedures*, Part I, B.3).

Background Data comprises all technical data and information, other than Foreground Data, which is directly incorporated, directly used or applied in work performed under INTELSAT contracts (*Procedures*, Part I, C.n.1). Whether data falls within that category can usually be established from the entry in the Handbook (*Procedures*, Part I, C.1). As in the case of Foreground Data, Background Data can be acquired through either of two procedures.

A Signatory or its designee can acquire Background Data directly form the contractors (*Procedures*, Part I, C.1) or through the Director-General (*Procedures*, Part I, C.3). If the Signatory acts directly, the Director-General is to assist as he may, and will confirm that the Signatory has been designated to deal with it should the contract require this (*Procedures*, Part I, C.1). If the contract obliges the contractor to disclose such data when requested, the Director-General may also be called in to require compliance with that obligations (*Procedures*, Part I, C.2).

On occasion, however, the Signatory may ask the Director-General to obtain Background Data on its behalf. In that case, the request to the Director-General must include a description of the intended use of the data requested, together with an identification of the Foreground Data with which it is intended to be used (*Procedures*, Part I, C.3). Thereafter the Director-General is to act promplty, seeking an estimate of the royalty the contractor would propose to charge for the specified use of the data and communicating it to the Signatory. A firm royalty is then negotiated on the Signatory's authorization or, if agreement cannot be obtained and the Signatory assents, the matter can be dealt with on the basis that the Signatory gets immediate use of the data and the royalty is adjusted later in terms of the relevant contract (*Procedures*, Part I, C.3). The Signatory pays INTELSAT's costs for obtaining Background Data for a Signatory or its designee, and all contractor charges, such as data reproduction and administrative costs. On the specific request of the Signatory involved, the Director-General will pay contractors' and licensors' costs subject to prompt reimbursement (*Procedures*, Part I, F.2 and 3).

Finally, with regard to the dissemination of data and technical information, the *Procedures* indicate two good practices, which are not, however, binding in the same way as the other practices stated in the book. First, 'it would be useful to the other Signatories and INTELSAT if those Signatories and their designees obtaining Foreground or Background data directly from contractors would notify the Director General' of the start of negotiations and give him, and other Signatories on request, information as to the use to be made of the data and the 'fair and reasonable terms' on which it has been acquired (*Procedures*, Part I, D). Second, it is in

> ... the interests of INTELSAT and other Signatories [that] each Signatory should promptly inform the Director General of any unauthorized use of INTELSAT technical data and information contrary to the terms and conditions of the Operating Agreement or to the provisions of the contract out of which such data arose, of which it has knowledge together with the particulars of such usage as may be available. (*Procedures*, Part I, G)

Inventions and patents: procedures

Part II of *Procedures* deals with inventions and patents and is intended to establish a system by which rights in inventions arising directly from work carried out under INTELSAT contracts can be distributed to Signatories. It therefore involves the protection of inventions through patents and similar forms of statutory mechanisms including 'utility models and registered designs' (*Procedures*, Part II, Preamble, n.3). The arrangements are also designed to gather from Signatories their advice as to the jurisdictions within which such protection would be desirable in the interests of both the organization and the Signatories (*Procedures*, Part II, Preamble).

It is intended that INTELSAT determine where the rights it has in inventions should be protected, and that INTELSAT should, where necessary, distribute such rights to Signatories so that they may act for INTELSAT in the matter. (*Procedures*, Part II, A). Where INTELSAT has acquired only licensed rights in an invention, it is competent for INTELSAT to acquire from the contractor also the right to obtain patent protection for the invention in those jurisdictions where the contractor has chosen not to seek patent protection (*Procedures*, Part II, A.n.1).

It is for the Director-General to review all inventions generated under INTELSAT funding and reported to the organization to see whether, and if so where, the inventions should be protected (*Procedures*, Part II, B.I). The Director-General is also responsible for implementing a decision to file for initial patent protection in accordance with international patent practice (*Procedures*, Part II, B.2). Where INTELSAT itself owns the patent, or has licence rights and the right to secure patent protection, and where the licensor has not filed for any patent protection, the initial filing is to be carried out by the Director-General in accordance with the US patent system (*Procedures*, Part II, B.2). In the case where INTELSAT has licence rights and the licensor has filed appropriately for protection, no INTELSAT action is required (*Procedures*, Part II, B.2). All patent applications by the Director-General, or any Signatory or group of Signatories are made either in the name of INTELSAT, or are held in the name of the applying authority on behalf of INTELSAT. Patents issued are similarly designated. In the case of licence registrations, the registration is in the name of the applying Signatory, INTELSAT and, where appropriate, all other Signatories (*Procedures*, Part II, M).

An 'Inventions Report' is prepared by the Director-General summarizing all decisions and actions taken in accordance with his responsibilities, and this Report is sent to all Signatories at addresses designated by them (*Procedures*, Part II, C.1, 2 and 3). The Report summarizes the technical information on the invention, the Director-General's evaluation – including any recommendation he may make as to where patent or other protection should be sought – the precise scope of Signatories' rights to use the invention in terms of any applicable contract or licence, and the applicant, assignee, serial number and date of filing of any initial patent application (*Procedures*, Part II, C.i). Circulation of the Report is to be 'timely' so that Signatories can advise the Director-General of jurisdictions in which patent or other protection should be sought for the inventions listed (*Procedures*, Part II, C.4). Naturally, such a Report is for limited dissemination and Signatories are notified of that requirement (*Procedures*, Part II, C.3).

Within 90 days of the dispatch of such an Inventions Report, or a shorter period (not, however, less than 30 days) should the Director-General view it as urgent, a Signatory is required to advise whether protection should be sought within its jurisdiction and whether it will undertake to file for protection where the Director-General has so recommended (*Procedures*, Part II, D.1). Where the Director-General has not recommended seeking protection within its jurisdiction, a Signatory may request the Director-General to do so, or may itself undertake the task (*Procedures*, Part II, D.2). In addition, a Signatory or a group of Signatories may recommend the seeking of protection within the jurisdiction of states which are not members of INTELSAT, or may proceed to act in the matter (*Procedures*, Part II, D.3). Where the Director-General is requested to act, he must first re-evaluate his recommendation and decide whether or not such a seeking of protection should be an INTELSAT expense; he then informs all Signatories of his decision (*Procedures*, Part II, D.4). If, after reconsideration, the Director-General is still against filing for patent protection, the requesting Signatory can elect itself to file for protection at its own expense. The fact that a Signatory has elected to act at its own expense does not alter the rights of INTELSAT or any other Signatory (*Procedures*, Part II, D.5).

Further procedures come into play in the case of a protection filing other than the initial filing. Where the Director-General determines that further protection is required in either Signatory or non-Signatory jurisdictions and a Signatory is acting within its jurisdiction on his recommendation, or another Signatory or group of Signatories has decided to act at its own expense, the Director-General provides a copy of the initial filing application and all documentation that he can (*Procedures*, Part II, E.1). Other Signatories are to provide him with information as necessary (*Procedures*, Part II, E.1). If a Signatory recommended to act in the Inventions Report fails to confirm its compliance within the time-limit set out in that Report, the Director-General has the right to file an application within the jurisdiction of that Signatory (*Procedures*, Part II, E.2). If he does so, the Signatory may request the transfer of the application at a later date, and this is done on the authorization of the Board of Governors (*Procedures*, Part II, E.2). The Director-General files applications in the jurisdiction of non-members, unless he determines not to act in which case a Signatory or group of Signatories can act with his and other Signatories' assistance as above (*Procedures*, Part II, E.3).

Where the initial filing is to be within the USA, within 90 days of the mailing of the appropriate Inventions Report, the US Signatory can elect to deal with the matter (*Procedures*, Part II, F.1). In that case, the Director-General passes on all the information which he has, and the Signatory provides any needed documentation (*Procedures*, Part II, F.2). To enable the Director-General to help other Signatories to obtain protection under laws other than those of the USA, the US Signatory transmits to the Director-General copies of its official correspondence with the US authorities (*Procedures*, Part II, F.3). Where, however, the US Signatory does not elect to pursue the matter of protection within the USA within the 90-day period, the Director-General continues the process on behalf of INTELSAT. The process may then be transferred to the US Signatory later at its request, and with the authorization of the Board of Governors (*Procedures*, Part II, F.4).

In all cases where a Signatory is acting, the Director General affords such assistance as he is able, including in the preparation of any application, the investigation of problems as to patentability, and assistance in documentaion (*Procedures*, Part I, I).

Mutatis mutandis, procedures similar to those concerned with patents apply in the case of protection through the registration of patent licences (*Procedures*, Part II, G).

Section H of Part II of the *Procedures* deals with the notification of filing to be given by Signatories to INTELSAT, and the Director-General's powers to intervene and deal with the matter if the Signatory is not acting with necessary speed and within any relevant time-limits. (*Procedures*, Part II, H.1 and 2). Any proposed abandonment of a protection application has to be notified to the Director-General, who may take the matter over (*Procedures*, Part II, F.3). Signatories are to inform him promptly of patents and other protections obtained (*Procedures*, Part II, H.4), and the Director-General makes a report at least annually to all Signatories on INTELSAT's 'Patent Portfolio', giving detail of patents filed – whether by Signatories, licensors or the Director-General himself – patents issued, any registration procedures begun or concluded for both the jurisdictions of INTELSAT Signatories and non-member states, and any applications abandoned or patents deleted from the Portfolio (*Procedures*, Part II, H.5 (a)–e)). This implies a regular review of the Portfolio, which is carried out at least annually in terms of section K of Part II of the *Procedures*. If the Director-General decides on the abandonment or deletion of a patent, the relevant Signatory can elect to take it on at its own expense. Such an election does not, however, alter the rights of INTELSAT or the other Signatories (*Procedures*, Part II, K.2).

Signatories are required to inform the Director-General of any infringements of patents and other protections which INTELSAT has obtained either directly, through licensors or through Signatories, whether or not the notifying Signatory filed, or was associated with, the patent application involved (*Procedures*, Part II, L). The decision on further action then rests with the Board of Governors.

Last, costs of preparing, filing and prosecuting patent applications, the issue of licences and the like are INTELSAT expenses where they are incurred on behalf of the organization, as are the Director-General's expenses in administering the Procedures (*Procedures*, Part II, J). It will, however, be remembered that where a Signatory elects to obtain patent protection without the recommendation of the Director-General, or takes on a matter he has determined to abandon, that is done at the expense of the Signatory alone, although it does not affect the rights of other Signatories or INTELSAT itself.

Comment

Such then are the arrangements which elaborate the provisions of the INTELSAT Agreements with respect to the dissemination of data and technical information, and for inventions, patents and other protections. Of their nature, their effectiveness is difficult to assess, although I am informed, however, that the procedures work fairly well. Problems are not so much with the procedures themselves as with their implementation in certain jurisdictions where INTELSAT business is a minor part of an official's responsibiity. There seems no way round human nature and the problems of too heavy workloads.

SETTLEMENT OF DISPUTES

Introduction

Within any international organization the arrangements made for the settlement of disputes can be important. One has only to look at the United Nations to see how critical such matters can be. Staff disputes are well handled by the UN machineries, but the devices for ensuring that members live up to their obligations under the Charter are inadequate. The same can be said of certain of the specialized agencies of the UN.

Disputes could arise within INTELSAT on a variety of grounds and on a variety of matters. Clearly, there might be dispute about policy and its application, in which case one would hope that the normal decision-making processes within the organization would deal with the matter. Were this not to happen then there would be room for a third party to attempt to use its 'good offices' and otherwise to mediate between the disputing parties.[5] However, should INTELSAT decision-making procedures, negotiations and efforts by third parties produce no solution, three avenues then open. These are:

1   the use of sanctions lesser than 'deemed withdrawal' in the case of non-legal disputes;
2   arbitral proceedings in the case of dispute as to legal rights and duties;
3   if lesser sanctions or arbitration fail, non-compliance with an obligation under the Agreements can ultimately result in the 'deemed withdrawal' of a Party or Signatory, which we have already discussed.[6]

Non-legal disputes

Little was discussed or written within INTELSAT on non-legal disputes and alternatives to arbitration until 1984, when, in the light of
the possible problems emerging from the changed US attitude to the
establishment of international satellite systems separate from
INTELSAT, the Director-General submitted for the information of the
Board of Governors a 'Review of Certain Obligations of INTELSAT
Members Under the INTELSAT Agreements, With Particular Reference to Article XIV(d)'.[7] This memorandum indicates a view that
INTELSAT could taken action in non-legal disputes where the activities of a member or Signatory were considered to be contrary to
the interest of the organization. The hypothetical backgrounds were
the cases where a recommendation of the Assembly of Parties under
art. XIV(d) was sought too late, or where a Party or Signatory failed
either to heed that a finding expressed as a recommendation was
negative on either technical or economic grounds, or failed to comply
with an indication in the recommendation of steps to be taken by
the applicant to meet requirements which the Assembly of Parties
considered necessary in the case. Of course, the Assembly's recommendation is, by definition, not binding in law on a Party or
Signatory, and hence its breach does not *ipso facto* attract legal penalty.
However, the memorandum suggests that the matter may not end
there. The argument depends upon the terms of the INTELSAT
Agreements themselves, general principles of international law, and
upon such international agreements as the Vienna Convention on the
Law of Treaties. It suggests that the subsequent acts of the Party or
Signatory could be made subject to sanction by the organization as a
matter of general law. Thus, while an Assembly recommendation
against the establishment of a separate international telecommunications satellite system might be breached simply by the establishment
of such a system, but not be subject to legal penalty merely for that
breach, the memorandum suggests that if the separate system, when
operational, were to be proven to damage the INTELSAT
organization (by, say, a diversion of traffic and hence revenue), the
organization could lawfully take steps against the Party or Signatory
concerned. It argues that the terms of art. XIV(d) are not exhaustive of
the rights and duties in the matter. Only specific language could
exclude the organization's right to take further action or impose consequences on the breaching Signatory or Party, and that language is
not present in art. XIV.

But what would be the basis of INTELSAT's adoption of further
remedies? By general international law a party to an international
treaty is forbidden from taking steps which prejudice the attainment
of the purposes of the treaty. This principle is encapsulated in

art. 60(3)(b) of the Vienna Convention on the Law of Treaties[8] which includes the violation of a provision essential to the accomplishment of the object and purpose of a treaty among actions constituting a material breach of that treaty. The establishment of a separate system in competition with the INTELSAT system which does withdraw traffic from the INTELSAT system would affect INTELSAT's compliance with the catalogue of purposes in the Preamble to the Agreement. Such an act can therefore only be a material breach of the Agreements, unless INTELSAT itself agreed to the development – which is the point of the art. XIV procedure concerning economic harm.[9]

Assuming that a separate system is set up contrary to a recommendation of the Assembly of Parties, what remedies, are then available to the organization?

With certain limits,[10] failure to comply with any obligation under the Agreements can be made subject to the sanction of deemed withdrawal of a Party of Signatory by the appropriate process, or alternatively in the case of a Signatory by the suspension of its rights.[11] But, by general international law, the statement of these remedies in the Agreements does not close the list of remedies since the statement is not defined as being exhaustive of the remedies available.

The first remedy would, of course, be for a return to the *status quo*, a restitution of the state of affairs as it was before the problem arose. Thus, in an appropriate case, a Party could be required to withdraw certain authorizations or permissions which it has given. A Signatory could therefore be required to cease from acts which damage INTELSAT.

The second remedy (and often a remedy accompanying that of restitution) is reparation. Where the act of a Party or Signatory has caused loss to the organization, reparation for damage caused could be exigible. If that is not forthcoming, then the organization could seek to gain such recompense as it might by limiting the exercise of a Signatory's rights in the organization – for example in financial matters. This would not be a punitive or coercive act, but, as international law would require, would be limited to repairing the loss or damage caused to the organization.

Third, there may be steps which can be taken within the terms of the Agreements, either to secure recompense, or to penalize, or both. The memorandum, 'Certain Obligations', cited above, suggests that options available to INTELSAT in the case of a Party would include suspension of the right to receive documentation, and suspension of the right to participate and/or vote in meetings of the Assembly of Parties. A much wider panoply of rights could be suspended in the case of a Signatory. These include the right to receive documentation, whether or not in connection with meetings; the right to participate

and/or vote in meetings of the Board of Governors (and its committees) and the Meeting of Signatories; the right to repayment of capital;
the right to receive compensation for use of capital; the right to request adjustment of investment share; the right under art. V(d) to
enjoy non-discrimination in rates payable for use of the system; the
right to obtain approval of earth stations; the right to obtain new
allotments of INTELSAT space segment capacity; the right to continue using allotted capacity under existing terms; the right to have
disclosed and use inventions and technical information; and the right
to obtain approval for domestic use of INTELSAT services under
art. III(b)(ii). Of these, those options with financial effects may be of
most interest in a given case.

But, clearly, such steps would produce a dispute about the interpretation of the Agreements. For example, the Agreement prohibits
differential rates being charged for the same service (art. V(d)). There
would also be dispute about the powers and jurisdictions of the
several bodies involved. As such a dispute would be legal in character, it could lead to arbitral proceedings.

Arbitration

Some disputes will fundamentally concern the rights and obligations
enshrined both in the Intergovernmental Agreement and the Operating Agreement. Such disputes might arise between INTELSAT and a
Party or Parties, between INTELSAT and a Signatory or Signatories,
between Parties or between Signatories, or between Parties and Signatories. For all of these, an arbitral mechanism has established in
terms of Annex C to the Agreement, which draws heavily on the
corresponding mechanism existing under the Interim Arrangements.
To date (1988), Annex C has not yet had to be used, and one would
hope that this will continue to be the case. But even if the US-inspired
'separate systems' controversy does activate the mechanism, one can
still hope that its use will be rare and that the more normal
INTELSAT processes will still solve most problems and disputes.

The settlement of disputes arising out of the Agreement or Operating Agreement are dealt with by art. XVIII of the Agreement and
art. 20 of the Operating Agreement. In both the reference is to 'legal'
disputes which eliminates disputes as to policy, and restrict their
application to the interpretation of agreements and law, as indicated
above. Both articles envisage that other steps are to be taken to attempt the solution of the disputes within a reasonable time. Such
steps are not indicated, and presumably cover the gamut of international settlement mechanism short of arbitral or judicial procedures –
that is, negotiation, good offices and mediation.[12] Where such
attempts are unsuccessful within that undefined 'reasonable time',

certain classes of disputes must be referred for arbitral settlement under the Agreements, whilst other disputes may be so.

Acrimonious and long-running disputes could seriously wound an organization like INTELSAT, and the arbitral procedure is provided, compulsory in some instances and optional in others, so that disputes will be promptly settled and not fester, thereby crippling the organization. Obviously no state would accept the compulsory submission of a dispute between it and a PTT (a Signatory) which is neither another state nor an international organization; nor would it have been competent compulsorily to impose the INTELSAT mechanism where other mechanisms exist. These points apart, however, legal disputes are not to be permitted to persist and so weaken the organization.

Assuming that settlement is not achieved in a reasonable time (whatever that is), arbitration is compulsory in disputes as to rights and obligations under the Agreements between Parties, between INTELSAT and one or more Parties (Intergovernmental Agreement, art. XVIII(a)), between INTELSAT and one or more Signatories, or between Signatories (Operating Agreement, art. 20(a)). The arbitral procedure is also compulsory in disputes between INTELSAT and Signatories which arise out of agreements other than the INTELSAT constitutive Agreements which do not contain other mechanisms for their settlement and which are not settled within a reasonable time art. 20(c)).

Disputes betwen a Party and a Signatory are not compelled to be settled by the arbitral procedure, although they may be submitted for arbitration under Annex C provided that the Party and Signatory so agree (art. XVIII(a)). In addition, legal disputes arising out of other agreements between INTELSAT and a Party are subject to whatever procedure they state for their settlement. In the absence of such provision, and subject to the agreement of the disputants, such disputes, if not otherwise settled, may be channelled through the arbitral procedures of the Agreements, Annex C (art. XVIII(c)).

It may be, of course, that a Party of Signatory may have withdrawn from the organization before a dispute surfaces. In this case arbitration is compulsory after a reasonable time, and the arbitral procedure of Annex C is competent, although it is subject to the agreement of the Party or Signatory disputant (art. XVIII(b); art. 20(b)). Where, however, the Party or Signatory withdraws from INTELSAT after the commencement of arbitral procedure in terms of art. XVIII(a) or art. 20(a), the arbitration continues to its conclusion notwithstanding the withdrawal, and the arbitral award is binding on the Party or Signatory (art. XVIII(b); art. 20(b)).

The provisions of Annex C to the Agreement, the 'Provisions on Procedure relating to the Settlement of Disputes referred to in Article

XVIII of this Agreement and Article 20 of the Operating Agreement', are in a form which is fairly standard within international organizations.

Only Parties, former Parties, Signatories and former Signatories, and INTELSAT itself may be disputants in the arbitral proceedings of Annex C (Annex C; art. 1). The tribunal has three members (Annex C; art. 2), all of whom are legal experts selected from a panel of 11 with a further 11 alternates, which is a standing panel in existence between ordinary Meetings and is chosen by the Assembly of Parties. The standing panel elects its own chairman (Annex C; art. 3(b)). A petitioner seeking to invoke the arbitral procedure designates from the panel one member to act in the tribunal for that dispute (Annex C; art. 4(v)), and within 60 days of receipt of the case documents, the respondents similarly nominate another (Annex C; art. 5(a)). If they do not, the chairman of the panel nominates one member from the panel (Annex C; art. 5(b)). Within 30 days, these two then nominate a third from the panel to serve as president of the tribunal (Annex C; art. 5(c)). If they cannot agree that nomination, either may so inform the chairman who is then required to nominate the president again from the panel (excluding himself) within 10 days (Annex C; art. 5(c)). The tribunal is constituted as soon as its president is selected (Annex C; art. 5(d)). Provision is made for vacancies in art. 6. At the request of a disputant, or at its own motion, the tribunal may appoint experts to help it in its work (Annex C; art. 10). The costs of the proceedings, including the remuneration of the members of the tribunal, are ordinarily to be borne equally by each side. Such is the normal case; however, in the particular circumstances of a given case the tribunal may decide otherwise (Annex C; art. 14). Where one side contains several disputants, costs are apportioned within each side's share of the total cost by the tribunal (Annex C; art. 14). If INTELSAT is a disputant, its expenses are an administrative cost of the organization (Annex C; art. 14).

Most of the proceedings are carried through in writing. The process is initiated by a competent petitioner giving a written document to each of the other parties with whom it is in dispute (the respondents) and to the Executive Organ (Annex C; art. 4(a)). This document must contain a statement fully describing the dispute, the reasons for involving the respondents and the relief which is sought (Annex C; art. (a)(i)). The document must show (not merely state) that the matter lies within the competence of the tribunal, and why the relief sought can be granted (Annex C; art. 4(a)(ii)). It must also explain the failure to reach a settlement by means short of arbitration (Annex C; art. 4(a)(iii)), and, in the case of an arbitration which is not compulsory under the Agreements, the evidence of agreement to go to arbitration (Annex C; art. 4(a)()iv)). Copies of the document are promptly sent by the Executive Organ to each Party and Signatory, and to the chairman of the panel of legal experts (Annex C; art. 4(b)).

If one side fails to present its case, the tribunal is not prevented from coming to a decision but it cannot simply issue what, in Scots Law, would be termed a decree in absence – that is, awarding the other party whatever it has asked for. By art. 8 of Annex C the tribunal must satisfy itself that it has competence in the matter and that the case is well-founded in fact and in law before giving its decision in favour of the side calling for the decision in its favour. It must therefore, to some extent, consider the facts and law on the matter.

Usually, however, it is to be expected that the parties will properly present their arguments on fact and law. The proceedings are conducted in writing, with provision for oral argument and testimony if the tribunal considers that appropriate (Annex C; art. 7(d)). The pattern is case, counter-case and reply, with additional pleadings at the discretion of the tribunal (Annex C; art. 7(e)). The competence of the tribunal is the first item on the agenda if that is disputed (Annex C; art. 7(c)). The tribunal may at any time terminate the arbitration if matters proceed beyond its competence (Annex C; art. 7(h)). Counter-claims are competent if falling within the categories of matters within the competence of the tribunal procedures (Annex C; art. 7(f)).

The materials of a dispute may, of course, be extremely complicated, and the information available to a disputant may be insufficient for the proper consideration of the matter. Accordingly, by art. 11, Parties, Signatories and INTELSAT itself may be required by the tribunal to provide any information required for the handling and the solution of a dispute. Such a requirement may be imposed at the instance of a disputant or of the tribunal's own motion.

It is possible for a state Party to the Agreement, whose Signatory is involved in a dispute under Annex C procedure, to intervene and be added as a disputant (Annex C; art. 9(a)). In addition, any other Party or Signatory, or INTELSAT itself, may ask to be added to the disputants if it has a substantial interest in the matter at issue, and will be so added if the tribunal agrees as to the extent of the interest (Annex C; art. 9(b)).

The tribunal decides the date and place of its meetings (Annex C; art. 7(a)). The proceedings are in private and all materials presented are confidential. Only INTELSAT and the Party or Signatory disputant have the right to be present and have access to the materials, unless INTELSAT is itself a party to the arbitration, in which case all Parties and Signatories have right to be present and to have access to the materials (Annex C; art. 7(b)). During the proceedings, the tribunal may indicate such interim measures as will protect the position of the disputants pending the final decision (Annex C; art. 12).

The deliberations of the tribunal are secret (Annex C; art. 7(i)). Its decision is to be given in writing, and be supported by a written opinion (art. 7(j)). Rulings and decisions must be supported by at least two members of the tribunal, and in the case of a dissent in a decision, the minority member may issue a separate written opinion (Annex C; art. 7(j)). During the proceedings the parties may agree to settle the dispute, and the tribunal then records that agreement with their consent as its decision (Annex C; art. 7(g)).

The decision of the tribunal is to be based upon the Agreement and the Operating Agreement, and upon 'generally accepted principles of law' (Annex C; art. 13(a)). This last is an interesting concept, found elsewhere in international law, but of occasionally uncertain content – how 'general' is 'generally', and whose 'acceptance' counts? On the basis of the Agreements and these generally accepted principles of law, as well as the materials and argument presented to it, the tribunal makes its decision. Its decision can be wide-ranging, and, in appropriate cases, the tribunal is able to hold null and void the decision of an organ of INTELSAT; under these circumstances, the tribunal decision binds all Parties and Signatories presumably whether or not such Party or Signatory was party to the arbitral proceeding (Annex C; art. 13(b)). If a dispute arises as to the meaning or scope of the tribunal's decision, it has the power to construe it (Annex C; art. 13(c)).[13] This may seem a somewhat gloomy, if not a demeaning provision, but it would seem to be copied from the Statute of the International Court of Justice, a body which has occasionally been asked to clarify its decisions.[14]

Interestingly, the decision of the arbitral tribunal is binding on all disputants, and is to be carried out by them in good faith (Annex C; art. 13(b)). I am informed that the telecommunications area is the only one where the USA has entered into an agreement under which the decision of an international arbitral tribunal will be binding on the USA. (A similar provision obtains in the INMARSAT arbitral procedure.) It is not normal US practice to enter such agreements, and one may wonder whether, were the Agreements being negotiated now, such a 'prior acceptance' clause would be acceptable to either US negotiators or the US Senate. Given the US withdrawal from the compulsory jurisdiction of the International Court, that uniqueness seems likely to persist.[15]

Finally, however, it must be said that although, as stated, the decision of the arbitral tribunal is binding and is to be carried out in good faith, it is conceivable that a disputant could nonetheless fail to carry out its obligation. Were INTELSAT itself to default, this would surely damage its effectiveness as an organization, and would constitute at least a considerable blow to its standing, making it vulnerable to other calls for change within and outside the organization. Were a

Party or Signatory to default, that would clearly be a case in which the appropriate organ would deem the Party or Signatory to have withdrawn from membership of the organization on the grounds of it having failed to comply with its obligations.[16]

SERVICES, ARTICLE III

The function of INTELSAT is the provision of global telecomunnications services by the use of satellite links. It provides international public telecommunication services, and certain domestic public telecommunications services. It may also provide specialized telecommunications services, and may provide satellites or facilities for any of the above services separate from its own system.

The prime objective of the organization is the provision of international public telecommunications services of high quality and reliability (art. III(a)). These are to be made available to all areas of the world without discrimination (art. III(a)).

Precisely what is an 'international public telecommunications service' is stated in art. I(k) of the Agreement, and, as will be seen below, that category was called in question in 1983 and was the subject of an internal legal memorandum.[17] The memorandum makes clear that it is for INTELSAT to determine whether a proposed service comes within the category of an international public telecommunications service. An international telecommunications service which is 'available for use by the public' or to a section of the public is potentially an international public telecommunications service (art. I(k)), the only exclusion being maritime or air mobile services not provided prior to the signing of the definitive INTELSAT Agreements (an exclusion designed to accommodate the INMARSAT developments) (art. I(k)). International public telecommunications services, within the INTELSAT meaning of the terms, are established by decision of the Board of Governors. This can be inferred from art. III and the enumeration of the functions of the Meeting of Signatories (art. VIII) which is required to approve in advance only one proposed domestic service (art. III(b)(ii)), and the list of the functions of the Assembly of Parties (art. VII) and the fact that it is involved only through it being required to approve certain proposed specialized services at the planning stage (art. III(f)).[18]

Two forms of domestic service are also required by the Agreement to be 'considered on the same basis as international public telecommunications services' (art. III(b)). The first of these is a domestic public telecommunications services between areas of a state which separated by areas not under the jurisdiction of that state or which are separated by the high seas (art. III(b)(i)). In maritime terms, communications here are a matter of cabotage. An example is the Colombian connection between Bogota and San Andres approved by the Board at its thirty-second Meeting in 1978 (BG–32–3).

The second form of domestic service to be taken on the same basis as an international services is the domestic service within the territory of a state between areas which are not linked by any terrestrial wide-band facilities but where there are natural barriers of such an exceptional kind that the creation of viable terrestrial links are impeded. In this case, INTELSAT may provide a domestic service on the same basis as an international service provided that the Meeting of Signatories, advised by the Board of Governors, approves in advance (art. III(b)(ii)). Examples are service within mainland Colombia,[19] Peru,[20] Mexico,[21] and Brazil.[22] For some reason, services between Denmark and Greenland have also been authorized under art. III(b)(ii), although art. III(b)(i) might seem more appropriate.[23]

The initial purpose of the special provision of art. III(b) for domestic services was to ensure that, for the purposes of priority of establishment, the organization would treat the domestic services dealt with on the same basis as international public telecommunications services. However, the final language does not make it clear whether that is the sole interpretation to be put on the words, or whether the effect of the phrase 'considered on the same basis as' is to make a planned separate domestic service system subject to the same rules of coordination as a rival international public telecommunications service. We will come back to this.[24]

International public services, and the domestic services which are equiparated with them by the above provisions are the staple of the INTELSAT responsibility. However, INTELSAT can add to its responsibilities. It is required to make its main system available for domestic services other than the two cases of the split-state and the state with immense problems of topography to the extent that its fulfilment of its prime objective is not impaired (art. III(c)). This benefits INTELSAT through using 'spare' capacity, and benefits states through providing an efficient telecommunications facility.

Beyond that duty, INTELSAT has power to provide facilities for specialized international or domestic (but not military) services, provided that the prime objective is not jeopardized and that arrangements are technically and economically acceptable (art. III(d)). Such services are defined in art. I(l) as including, without being restricted to, radio navigation, broadcasting satellite services for the general public, space research, meteorological and earth resources services. They exclude the normal public telecommunications services indicated in art. I(k). Where a specialized service is required it must be asked for, and appropriate terms and conditions entered into (art. III(d)). In this case, procurement for the facilities cannot begin until the approval of the Assembly of Parties is forthcoming (art. III(f)).

Finally, on request and under suitable conditions, INTELSAT may provide satellites and facilities separate from the normal INTELSAT system for domestic or international public telecommunications within or between Parties, or for specialized but non-military services provided, in all cases, that the efficient and economic operation of the normal INTELSAT system is in no way unfavourably affected (art. III(e)).[25]

INTELSAT is therefore capable of providing a wide range of services, provided always that its prime objective is not prejudiced by other activities. In fact, the international public telecommunications service element of its prime objective is now well served, as is the element of prime objective comprised in the provision of domestic services in awkward terrain or the 'split-state' circumstance. In addition, a good number of countries make use of the availability of INTELSAT facilities under art. III(c) and (d) to meet their internal telecommunications needs, both for public telecommunications and for specialized services either on a pre-emptible lease or on a permanent basis.

Within that potential wide range of services, INTELSAT has pressed on in devleoping services which it is then up to Signatories and other users of the INTELSAT system to market within their marketplace. There has been criticism that INTELSAT has not been sufficiently diligent in developing new services, that being, for example, a main argument of those who would like to see developed satellite systems separate from the INTELSAT system for international telecommunications.[26] Certainly telephony, telex, telegraphy and facsimile were major services provided by INTELSAT from the outset, with certain 'occasional use' services, such as radio back-up, cable restoration and peak telephone traffic diversion from cable, soon being added. The 'occasional use' of the system for television is familiar to all who watch real-time news or sports programmes. But these were available normally on major path transmissions. In 1974 the SPADE system providing services from assigned circuits on demand opened many services to areas where otherwise traffic was small. TDMA (time-division multiple access) added to the availability of services as did the use of the INTELSAT space segment for INMARSAT purposes. INTELSAT Business Service (IBS) was introduced in 1984 providing tailored services of interest to that category. VISTA is a new service for low-density telephony in the Pacific basin. In various of these services, leased circuits operate in a variety of ways ranging from the full lease to the pre-emptible lease which has no priority over other requirements of the system. Whole transponders may be leased, and some countries run their domestic satellite system on such transponders.[27] In autumn 1987 the principle of a transponder right of user (TRU) analogous to the indefeasible right of user (IRU) of the cable system was discussed in the Board of Governors and will be introduced.[28] In short, there are many services now available, and apparently the INTELSAT Tariff Manual lists scores.[29]

But implicit in what has been said is an INTELSAT problem. Not all the services now available are fully, or even adequately, used. Certainly, the threat of competition seems to have spurred INTELSAT in the provision of services, but merely making them available is not enough. INTELSAT provides the space segment for its telecommunications system; it is for the Signatories and others using the system to go out and market the services which are offered. It is also for them to decide how much the end-user is charged for the service he requires. Criticism of satellite charging[30] is not necessarily a criticism of INTELSAT, although, often unfairly, that distinction is not made. Nonetheless, subject to that caveat, INTELSAT has been active in providing new services, particularly within the last decade once its basic system was up and running and the advent of the larger satellite made possible a greater variety of services.

One special case is that of the provision of services to the United Nations. As early as 1971, during the resumed Plenipotentiary Conference on the definitive arrangements for INTELSAT, the UN made a request for a wide-ranging, but free, use of INTELSAT circuits for its requirements.[31] The Conference recommended that this be looked into when the definitive arrangements had come into effect, and this was done. In 1974 a paper summarizing discussions between the UN and the Secretary-General of INTELSAT was put to the Board.[32] This noted the UN requirements as consisting of emergency communications with peacekeeping operations and disaster relief, operational communications with peacekeeping operations, operational communications with major UN Offices (New York to Geneva, Nairobi, Beirut, Addis Ababa, Bangkok, Santiago, Paris, Rome, Vienna, Rome and London – or with two extra New York–Geneva links instead of links between New York and the last four), and, last, occasional informational television links. Such usage was considered by the Secretary-General to have minimal impact on INTELSAT operations, planning or satellite saturation. However, the commercial value of the requested services was high, although the satellite element was low. As also had been the case in the discussions in the Plenipotentiary Conference, various problems were foreseen, ranging from INTELSAT's own duty to equalize charges for the same services and the ITU duty to impose the same charges on all users of a telecommunications service to the point that, since the membership of the two bodies was not identical, it would result in one organization's membership unfairly subsidizing the other. The possibility of the UN itself increasing its own budget and paying the going rate for the services requested was considered as was the question of why other telecommunications modes had not been asked for similar facilities. It was also pointed out that other organizations might also make claims for free or reduced rates.

The Eighth Meeting of the Board of Governors[33] decided that, in principle, two circuits shoule be made available to the UN on a limited 'occasional use' charged basis either to UN Headquarters at New York or to the disaster relief coordinator in Geneva for emergency purposes. INTELSAT was not to consider this offer as a precedent and was reluctant to be regarded as a convenient and free (or relatively so) adjunct to its own operation by the UN and its agencies. Matters dragged on on this basis until 1982 when the UN made a further formal approach asking for a service on a quarter transponder on a full-time leased basis for the purpose of keeping the UN peacekeeping forces in the field in touch with Headquarters.[34] The Board deferred decision, asking for the Director-General to prepare further studies. Following these and further discussion with the UN, the Fifty-Fourth Meeting of the Board of Governors decided to recognize the UN as an entity authorized to ask for service in terms of arts. 14 and 15 of the Operating Agreement but asked for yet more studies on alternative solutions for UN needs for peacekeeping and disaster-relief purposes on the Atlantic region satellite on a long-term basis and on the Indian ocean and Pacific ocean region satellites on an 'as required if available' basis.[35]

OTHER SYSTEMS

Since INTELSAT is not, of course, the only organization offering telecommunications services by satellite, its relationship with other systems and organizations is therefore important. Two categories of system are involved, first the solitary system which is established by Parties which are not members of the INTELSAT Agreements, and second, systems where the membership is partly or entirely made up of INTELSAT members.

The Communist Bloc: INTERSPUTNIK

As to the first, the Russian global satellite system INTERSPUTNIK is the only system which was established with a membership entirely separate from INTELSAT.[36] The legal relationship between the two systems was regulated by negotiation and the ordinary ITU rules as to the coordination of systems. Thus the coordination of the INTER-SPUTNIK and INTELSAT systems was discussed at a series of meetings between the two organizations, and a satisfactory compromise was achieved.[37] Apart from the official coordination requirements, the Soviet Union and other members of INTERSPUTNIK have over the years come to use the INTELSAT system as purchasers of service without becoming members of the organization. This is entirely within the terms of the INTELSAT Agreements.[38] In 1985 such use of the INTELSAT system was placed on a more official basis by a Memorandum of Understanding between INTELSAT and the Soviet Ministry of Posts and Telecommunications, under which earth stations under the Soviet Ministry of Posts and Telecommunications are accorded the same legal treatment as other INTELSAT members.[39]

Systems involving INTELSAT Members

In the 25 years since INTELSAT was set up under the Interim Arrangements, other satellite systems and organizations have been established by some INTELSAT members. Various factors led to such developments. First, certain states and PTTs were anxious to establish a more 'local' system which would provide a more 'tailored' service and which might be more responsive to their views and aspirations than the larger global system. Second, within a smaller 'local' organization there may be an increased opportunity for technological and economic spin-off to national undertakings, and for the development of national space industries through procurement decisions. Third, the forces of simple nationalism and regionalism should not be underestimated. In some instances, a military and security element might also argue both directly for a more immediate national control of communications and also reinforce the other arguments for the establishment of the separate system. In the particular case of INMARSAT, a further impetus was the coincidence of a wish to provide a very specialized service and also to associate the USSR and Communist countries with the development.

The Preamble to the Interim INTELSAT Agreement spoke of the signatory Governments

> . . . [d]esiring to establish a single global commercial communications satellite system as part of an improved global communications network which will provide expanded telecommunications services to all areas of the world and which will contribute to world peace and understanding.

That statement indicates that in 1964 a single global system was envisaged which would have served the needs of all states of the world for satellite telecommunications. However, there were no provisions within the Interim Arrangements for enforcing the 'single system' concept, nor for dealing with the case where an interim INTELSAT member or Signatory contemplated participating in, or setting up, a rival system. The Supplementary Agreement on Arbitration would have been the only recourse in such a case.

Technically, INTELSAT itself might in due course have provided sufficient satellite capacity and services to meet all possible developments in satellite communications but, as time has passed, that option has been excluded for various reasons. One is that INTELSAT could not always make the desired provision within an acceptable time-scale. In the case of the maritime mobile satellite service, another reason was the desire to associate the USSR with the developments which became INMARSAT. The USSR and other members of the Communist bloc are not members of INTELSAT for a variety of reasons. The USSR sent observers to the INTELSAT discussions but was dissatisfied with the eventual proposals.[40] It was uncomfortable with an organization which is only partly based on state membership, which associates telecommunications entities (and some of them highly business-oriented entities) with the very hub of the enterprise, and in which members do not have equality of voting where the important decisions are taken. Whether that view would have been different had the international telecommunications traffic from the USSR and the Communist countries been such as to give them a weight within INTELSAT which they would consider commensurate with their importance in other international affairs is a matter of speculation. Whatever the explanation, the fact is that the USSR and the Communist bloc are not members of INTELSAT. Nevertheless, the USSR's participation was clearly desirable in maritime telecommunications and, accordingly, INTELSAT was not the medium through which a maritime satellite system was to be organized.

The INMARSAT question was typical, and such problems were foreseen. By the time that the definitive arrangements for INTELSAT were being negotiated in 1969–71, it was clear that account would have to be taken of the likely development of other satellite systems and the potential involvement of INTELSAT members and Signatories in them. The USA was itself contemplating domestic systems outside the INTELSAT framework and Europe was of the view that, while it might be true that, separately, no European country would justify having a domestic satellite facility, the European 'domestic' area was comparable to that of the USA, and a European regional system would be as justifiable as a US domestic system.[41] Again, the questions of regionalism and nationalism indicated above were not unimportant in the European arena.[42]

The result was the mechanism set up by art. XIV of the Intergovernmental Agreement, under which satellite systems separate from that of INTELSAT in which members and Signatories are involved are required to be 'coordinated' with the INTELSAT system.[43] This mechanism has operated in all the cases where it has had to be involved, but so far no negative recommendation has been made by the appropriate body or has been advised by the Board of Governors when acting in an advisory capacity. Individual instances are considered below, but it should be stated here that the real test of the art. XIV mechanisms will come when INTELSAT takes a view on a proposal which is not shared by the affected member or members and to which it refuses to defer. This could occur with the requests for coordination of public international systems proposed by US companies under the FCC ruling on separate systems.[44] It is only when a coordination is refused that we will see how strong the coordination procedures of the INTELSAT Agreements are. So far, it might be thought, the Board of Governors has bent over backwards to permit the coordination of other systems with the INTELSAT system.

In all cases under art. XIV, the final findings, which are expressed as recommendations by the Board in the case of a domestic public telecommunications service, or by the Assembly of Parties in the cases of separate systems providing specialized services or international public telecommunications services, are to be made within six months from the date of the commencement of the art. XIV procedures (art. XIV(f)). In order to comply with this time-limit, which exists so that procedures are not necessarily delayed until the Assembly next meets in its normal two-year schedule (art. VII(d)) it is competent for the Parties to meet in an extraordinary Assembly (art. XIV(f)). However, in order to avoid unnecessary extraordinary meetings of the Assembly, a practice has developed, with the consent of interested Parties and Signatories, of proceeding with consultations on an informal basis, converting these to formal consultations six months before the next ordinary Assembly. It should also be noted that, although in one sense procedures begin with a written request to the organization, in all art. XIV cases (those where the Board and those where the Assembly makes the recommendation), that is not the formal commencement of proceedings. In the case of international services, requests matching the first request in terms of traffic and destination must be received from other Parties or Signatories, although not from non-members of INTELSAT. This prevents the use of the art. XIV mechanism as a mechanism to test potential demand, as a means by which to 'peg' a claim, or as a method by which to put down a marker for the future as can happen elsewhere in analogous proceedings. For example, in planning applications, under UK law a development may be proposed and approved to test its viability long before any steps are taken to actually bring it about. INTELSAT has to be given full information in order to deal with a coordination, and all relevant information must be provided to INTELSAT in terms of the appropriate guidance documents which are indicated below. Only when all such requests and information have been received do the proceedings formally commence and the running of the six-month time-limit begins.

Consultation requirements depend upon the type of service which is sought to be provided. They must be gone through in the case of the establishment, acquisition or use by a Party, Signatory or any person within the jurisdiction of a Party, of space segment facilities separate from the INTELSAT system unless these facilities are established and operated for purposes of national security only (art. XIV(g)). Three forms of service are contemplated by art. XIV, and a distinction is made between what is required in the case of international public telecommunications facilities and what is required either for domestic services or for specialized services.

Domestic services: art. XIV(c)

Art. XIV(c) applies in the case of space segment facilities intended to meet the needs of domestic public telecommunications services. The Party or Signatory concerned, or the Party within whose jurisdiction the service is to be established, is required to consult the Board of Governors prior to the establishment, acquisition or use of the facilities. The Board considers the technical compatibility of the facilities and their operation with the radio frequencies and orbital space required by existing INTELSAT satellites and with any planned INTELSAT systems.[45]

There are dissimilarities and similarities between the INTELSAT technical consultation procedures and those which are applied within the mechanisms of the ITU.[46] The two procedures are dissimilar in that third parties are not included in the debate between the Party and INTELSAT – only the interests of the Party and INTELSAT are canvassed. The interests of third parties are dealt with through the ITU coordination process. Again, the consultation parameters of the INTELSAT process are far less detailed or comprehensive than that of the ITU, although by reasons of this characteristic they are flexible and able to change as technology develops in a way that the ITU parameters cannot. The procedures of the two organizations are, however, similar first in their overarching aim, which is the efficient use of the radio frequency spectrum and the best use of geostationary orbital positions. Second, they are similar in that both organizations engage in a two-stage process. In the ITU mechanism there is the requirement of advance notification and then a formal procedure. In

INTELSAT, advance notice is not required but early informal discussions (and sometimes negotiations) are usually conducted, and virtually everything is effectively agreed before the Parties move to the formal consultation required by the terms of art. XIV(c) and the action required of the Board of Governors. The Board's findings must be made within six months of the consultative procedure being commenced (art. XIV(f)) and are expressed in the form of recommendations. They are therefore not binding on a Party or Signatory in the case of the establishment of a domestic public telecommunications satellite service rivalling an INTELSAT service. However, physical constraints, such as the desirability of avoiding mutual radio interference and orbital problems, would seem to indicate that compromise is likely. The unspoken sanction is that an intransigent Party or Signatory may lose out in other INTELSAT decisions; there might also be repercussions of a legal nature.[47] Examples of successful coordinations under art. XIV(c) are: the Colombian SATCOL (Colombian Satellite) network, which was scheduled to be operated from 1981;[48] the Japanese domestic system (which required a reconsultation as the satellite orbital locations and other factors were changed);[49] that for Canada;[50] for Australia;[51] and that for the Indian domestic system.[52] Naturally, there have been consultations for the various US domestic systems, and these have not caused too much difficulty.[52a] There was, however, some difficulty with various UK proposals.[53]

Mention was made above of discussion whether domestic services qualified under art. III(b) had to be dealt with under art. XIV(d)). It seems to be INTELSAT's view that if there is any doubt or question as to whether art. XIV applies, coordination procedures should be carried out. In conformity with that view, art. XIV(c) coordinations have been carried out by members of INTELSAT for the use of the ARABSAT and the PALAPA systems for the domestic public telecommunications requirements.[54]

Specialized services: art. XIV(e)

Art. XIV(e) deals with specialized telecommunications services, which are defined by art. I(1) of the Agreement as:

> . . . telecommunications services which can be provided by satellite, other than [public telecommunications services, which are] those defined in paragraph (k) of the Article, including, but not limited to, radio navigation services, broadcasting satellite services for reception by the general public, space research services, meteorological services, and earth resources services.

Under art. XIV(e) a Party or Signatory intending to establish, acquire or use a satellite system for specialized services, or a Party within whose jurisdiction any person is to establish such a system, is required to consult with INTELSAT whether the intended service is domestic or international. In this case, the matter is dealt with by the Assembly of Parties which is furnished with all relevant information through the Board of Governors. The consultation is concerned only with the technical compatibility of the proposed system with the existing and planned INTELSAT space segment, both as to radio frequencies and orbital space.[55] As in the domestic satellite service consultation under art. XIV(c) dealt with above, a two-stage process has been adopted, starting with a voluntary and early informal consultation which becomes formal in order to comply with the terms of art. XIV(e). The Assembly of Parties is advised by the Board of Governors, which is itself informed by studies done within the Executive Organ, but, within the six-month deadline indicated above, makes its own determination on the matter, expressing this in the form of recommendations which do not bind the Party or Signatory involved. As in the case of a domestic satellite service, however, there are impetuses towards compromise to avoid physical, technical and future problems. Often there is what amounts to negotiation, and sometimes channels may be switched off when INTELSAT enters an area. A good example of the procedure at work is the coordination of the specialized services for television and meteorology available on the planned Indian domestic satellite system, INSAT.[56] Other examples are the Japanese meteorological satellite – established as part of World Weather Watch, the project of the World Meteorological Organization (WMO)[57] – and the Direct Broadcast Satellite systems for West Germany,[58] France,[59] Luxembourg,[60] Australia,[61] and the UK.[62]

International public telecommunications services: art. XIV(d)

The third category of services in which consultation with the INTELSAT system is required is the international public telecommunications service. This is the subject of art. XIV(d) of the Intergovernmental Agreement, and INTELSAT has issued a 'Procedural Manual for Consultation Under Article XIV(d) of the INTELSAT Agreement' which clarifies the stages of the consultation procedure.[63]

The requirements of the coordination of international public telecommunications services are more exacting that those for specialized services. Since more then questions of technical compatibility are considered, it is therefore important whether or not a proposed service falls into the category or public telecommunications. This is a matter for INTELSAT to determine. Article I(k) of the Agreement defines 'public telecommunications services' as:

. . . fixed or mobile telecommunications services which can be provided by satellite and which are available for use by the public, such as telephony, telegraphy, telex, facsimile, data transmission, transmission of radio and television programs between approved earth stations having access to the INTELSAT space segment for further transmission to the public, and leased circuits for any of these purposes; but excluding those mobile services of a type not provided under the Interim Agreement and the Special Agreement prior to the opening for signature of this Agreement, which are provided through mobile stations operating directly to a satellite which is designed in whole or in part, to provide services relating to the safety or flight control of aircraft or to aviation or maritime radio navigation.

That seems clear enough, but the extent of the category 'public telecommunications service' in the INTELSAT Agreements was questioned in 1983. Certain US applicants to the FCC argued that digital data and teleconferencing services which they proposed did not fall within the category since the services proposed were private to the users and not freely available to the public. They were, therefore, not 'public telecommunications services' but rather 'private facilities' requiring consultation under the category 'specialized services'. In making this argument, the applicants relied on concepts developed and used in municipal US law subsequent to the negotiation of the INTELSAT Agreements.[64] If such reasoning were applicable within INTELSAT the classification of these services would have meant that the proposed system would have fallen to be coordinated only as to technical compatibility under art. XIV(e) and not as regards the broader considerations of art. XIV(d). The precise extent of the INTELSAT definition had therefore to be considered. The 'Legal Opinion on the Scope of INTELSAT's "Public Telecommunications Services"' of 13 January 1984,[65] concluded that the proposed services did fall within the 'public telecommunications services' for INTELSAT purposes. That 'Legal Opinion' is right.

The purpose of the definitions of public and of specialized services in the Agreement was to establish categories of service for the INTELSAT's internal constitutional and management purposes. The definitions relate to the responsibilities of different INTELSAT organs. Services available to the public are established by decision of the Board of Governors. Specialized services require the authorization of the Assembly of Parties when they are being planned (art. III). The Parties to the Agreement intended that public services should be established by Board decision alone. Public services are the bread and butter of the INTELSAT system, forming the major part of the facilities which the organization was intended to provide, and which are not intrinsically contentious. Specialized services are of more interest to Parties in their capacities as states and are not, by their nature,

generally available to the public (as a glance at the contrasting definitions in art. I(k) and (l) would substantiate), or if available to the
public are so only as incidental to their governmental function. Particular reference is made in art. I(k) to the maritime and aeronautical
mobile services, excluding forms of service not provided under by
interim INTELSAT. That and the specialized service apart, availability to the public was the test of whether a service was a public
service, and the Board alone was to have competence in the decision
to establish such services.

Other services not available to the public at large might be
expected to be established for state purposes, and be subject to the
direct control of governments as a matter of special national policy or
in the discharge of state obligations (for example, for safety of life at
sea). Given that element of state or governmental interest, the
Assembly of Parties clearly had to be involved if INTELSAT was to
contemplate providing such specialized services. The availability of a
type of service to the public which is intrinsic to the concepts of art.
I(k) and the contrast with the concepts of art. I(l) were thus not
intended to permit a distinction between a 'public service' and a
'private facility' for the requirements of consultation. The distinction
was made for reasons of decision-making within INTELSAT. As said,
the basic services of INTELSAT, its staple, are public telecommunications services which are to be established by the decision of the Board
of Governors alone. Other organs are involved in decision-making
(as opposed to commenting) in areas where there is a special reason
why the Parties to the Agreement wish to retain a standing, having
regard to their other obligations in their capacities as sovereign states.

There are further, separate grounds for rejecting the argument that
'private facilities' are not part of a 'public telecommunications service'. The distinction between private and public telecommunications
services in the form that it is was argued by Orion was one which had
emerged municipally in the USA only after the conclusion of the
Agreements. On general grounds of treaty interpretation it is therefore not competent to read in such meanings to the Agreements,
entirely apart from any question as to the plain meaning of arts. I and
XIV. INTELSAT is concerned with the nature of the service and
whether it is one which members of the public can avail themselves
of as a service. Whether or not they could insist on access to a particular connection within a service provided to other members of the
public – the so-called 'private facility' – is irrelevant. Thus, for example, within the INTELSAT definitions, as illustrated in art. I(k), a
data link between offices of a company is not removed from being a
public service merely because the company can insist that it alone has
access to it. Indeed, the reference to 'leased circuits' in art. I(k) makes
the 'private facility' argument tenuous to say the least. It is therefore

good to see that Davis R. Robinson, Legal Advisor to the US State Department came to the same conclusion on the matter of the first two US applications to the FCC to establish separate systems.[66] Mr Robinson does not consider the matter wholly free from doubt, but is of the view that the proposals in question did not fall into the category of 'specialized services' under the INTELSAT Agreement. He points out, for example, that, were the requirement of consultation with INTELSAT regarding economic harm avoidable by a separate service which was dedicated to users by lease or sale, this would undermine the basic purpose of INTELSAT. The history of the INTELSAT language, the interpretation of 'public telecommunications services' and 'specialized services' all indicate a requirement to consult as to economic harm under art. XIV(d) in the two cases submitted to the Advisor. In his view, only in the case of a wholly-owned private non-commercial space segment being used solely for the internal communication of the entity establishing it would the requirement of coordination with INTELSAT be limited to a technical coordination.

As indicated, the requirement of coordination in the case of public international telecommunications services to be provided through space segment facilities separate from the INTELSAT system is more exacting than that for domestic or specialized services. Like the case of the other separate services, art. XIV(d) requires the Party or Signatory involved, or the Party having jurisdiction over persons who are to establish the proposed system, to provide all relevant information to INTELSAT prior to the establishment, acquisition or use of the separate satellite system. In due course, the Assembly of Parties, advised by the Board of Governors, makes its own findings on the matter, expressing these in the form of non-binding recommendations, if necessary (for example to meet the six-month time-limit for completion of the exercise) meeting in extraordinary Assembly for the purpose (art. XIV(f)). But, unlike the other separate systems cases, consultation under art. XIV(d) has five elements, one technical, and the second and third hinging on INTELSAT's own service and whether the traffic would not be carried on the INTELSAT system in any case. The fourth element concerns prejudice to INTELSAT's establishment of direct links with the participants in the new system. The fifth and most contentious is the question of whether the proposed system will cause 'significant economic harm' to INTELSAT.

First, as in the cases of the domestic and the specialized service system, consultation is required to ensure the technical compatibility of the proposed system with the existing or planned INTELSAT system in relation to both radio frequency use and orbital position. In this element of the procedure the art. XIV(d) consultation is no different from the coordination of the other forms of separate services, and the *Procedures* are used.[67] As in the other cases, the Assembly's recommendations on technical compatibility are not binding, but of course have the weight indicated by the desirability of the avoidance of technical problems.

The second question is whether or not the services which the separate system plans to provide are ones which INTELSAT does, or intends to, provide. This involves comparison of the two systems. If the result is that the new system intends a service which INTELSAT does not, and does not intend to, provide, the further elements of art. XIV(d) are not proceeded with and a favourable recommendation will issue.

Third, there may be circumstances which indicate that the proposed system will carry traffic which would not have gone through the INTELSAT system even if the proposed system did not exist. Thus ground links between proximate countries might be such as to lessen the likelihood of INTELSAT involvement. Again INTELSAT's own 'Procedural Manual'[68] suggests that, because the coverage area of a domestic satellite engaged in video distribution is bound to spill over into adjacent countries, a case could be made out for its extended use being incidental to its purpose and for the traffic not being likely otherwise to have been carried on the INTELSAT system.[69] In all cases, however, the amount of traffic and destinations covered would be important, and the 'Procedural Manual' speaks of other special circumstances being considered on a case-by-case basis.

The fourth element of the art. XIV(d) consideration is that the Assembly of Parties is to

> . . . express, in the form of recommendations, its findings . . . regarding the assurance that the provision or utilisation of such facilities shall not prejudice the establishment of direct telecommunications links through the INTELSAT space segment among all the participants. (art. XIV(d), *ad fin.*)

This is a curious expression, and it has been considered for the guidance of the Board of Governors by David Leive, the INTELSAT Legal Advisor, in a Memorandum, 'Scope of the Article XIV(d) Assurance concerning "Direct Telecommunications Links" ',[70] and also by Richard R. Colino, the then Director-General.[71] From these it seems that the 'prejudice to direct links' language of art. XIV(d) was introduced by a Japanese proposal and was probably originally directed at two questions: that of securing access to the INTELSAT system by participants in the new system; and that of possible economic harm being caused to INTELSAT by the new separate system. The latter aim was largely superseded by the specific provision concerning economic harm to which we are coming, but the question of economic impact may not be entirely irrelevant in considering the 'prejudice to direct links' assurance.

It is, of course, for INTELSAT itself to interpret its Agreements,[72] and three questions emerge from the phraseology of this part of art. XIV(d): 'the assurance that the provision or utilization of such facilities shall not prejudice the establishment of direct telecommunication links through the INTELSAT space segment among all the participants.' Who are the participants concerned? What are the 'direct links' dealt with? And what amounts to 'prejudice' of those links?

As to the first question, the organization acts on the interpretation found in both the Leive and Colino documents cited above, which is that the word 'participants' in the phrase under consideration are the participants in INTELSAT as a whole, and not just those INTELSAT members who happen also to be involved in the proposed separate system.[73]

The 'direct links' language is also similarly easily dealt with at a basic level. Both Leive and Colino agree that, for a favourable recommendation to issue from the Assembly of Parties on this matter, the fundamental element is that the separate system shall not impede interconnectivity. This means that the ability of INTELSAT participants to communicate directly with each other through INTELSAT shall not be prejudiced and that the separate system shall not in the future impede the ability of INTELSAT to establish direct links through its system between a user of the 'new' system and some user outside that system. Thus any new system which, for financial or technical reasons, proposed to interpose itself between INTELSAT members and the INTELSAT system would be proscribed by the language. The INTELSAT system must remain as an alternative link, as it does, for example, in the area served by the EUTELSAT system. Arguably also, all direct links by INTELSAT – whether or not between participants using the proposed system – are also covered by the language.[74] In this case, any effect on the INTELSAT system by the new system would be considered as potentially prejudicing direct links; this makes crucial the exact meaning to be attributed to the qualifier 'prejudice'.

However, difficulty is encountered when we consider the extent of the meaning of 'prejudice' in its precise context. Certainly an interference with connectivity is 'prejudice to the establishment of direct telecommunication links through the INTELSAT space segment'. But can the meaning range more widely – say, to economic or operational questions? Both the Leive Memorandum and the Colino 'Policies' paper mention the question of a separate system used by a small number of countries possibly diverting large traffic streams from the INTELSAT system.[75] A US–Europe link could well fit that description. Consider such a development for the sake of argument. If a new satellite system were established, direct links between INTELSAT participants could well be affected for operational reasons. In the interests of efficiency and arising out of negotiations for the coordination of the new system under the ITU procedures,[76] INTELSAT might have to relocate one or more of its satellites, with a resultant reduction of service in particular areas. Or, by reason of the financial effects of the diversion of traffic from its system to the new facility, INTELSAT might have to raise its rates thereby affecting the participation which some countries could afford. That might affect their participation in INTELSAT, their investment share and hence also their voting weight. The global nature of the INTELSAT system would therefore be affected by the US–European development and the value of the unitary global INTELSAT system diminished.

According to the narration of the negotiating history of the 'direct links' phraseology contained in the Leive Memorandum, the original wording used in the Japanese proposal was that the separate system should not 'prevent' the establishment of direct links through the INTELSAT system.[77] That would have meant that an unfavourable recommendation would result only if the new system would inevitably be an insurmountable obstacle to INTELSAT achieving its goal of a global system. The replacement of the term 'prevent' by 'prejudice' means that the standard is more flexible at INTELSAT's option, an unfavourable recommendation being competent where the INTELSAT system is affected, but not made impossible. However, the terminology adopted also renders the standard more obscure.

Again, it seems that a major element of the thrust of the Japanese proposal was directed towards economic questions, and some would argue that that has been entirely replaced by the 'economic harm' provision also found in art. XIV(d).[78] I would argue to the contrary. What has been set up by international treaty and agreement is an organization whose function, as stated in the Preamble to the Agreement, is the creation and maintenance of a single global satellite telecommunications system, affording efficient and economic access to all peoples of the world. That there is a special phrase dealing with 'significant economic harm' in art. XIV(d) does not necessarily remove economics from the compass of possible 'prejudice' to direct links. On the contrary, it means that economic prejudice to direct links can, at a lesser level than 'significant', be a factor in the Assembly's determination of whether a proposed system assuredly does not prejudice the present or future direct links of the INTELSAT system. As the Leive Memorandum states:

> The word 'prejudice' implies that a system which caused a far lesser degree of harm [than 'prevent'] would not receive Assembly support – that a separate system which merely hampered, injured or damaged

INTELSAT's ability to provide direct links or any participant's ability to access or use those links would not be endorsed. A separate system which resulted in making it more expensive or more difficult, though not impossible, for a participant to acquire and use a direct link via INTELSAT could be said to 'prejudice' that link.[79]

At the Washington Conference therefore, in return for the possibility of separate systems, the proponents of separate regional systems were willing to see language adopted which greatly strengthened the organization's power to fulfil its remit to establish a global and non-discriminatory system as it perceives that to be best achieved. Those, led by the USA, favouring a single global system welcomed that concession.

So, to summarize: this element of art. XIV(d) requires that the Assembly before expressing a favourable recommendation as to a proposed separate system must be assured that the 'provision or utilization of such facilities shall not prejudice the establishment of direct telecommunications links through the INTELSAT space segment among all the participants'. It is a weapon available to the Assembly of Parties and, as such, must be used with care and discrimination. However, its difficulties should not be allowed to render it inoperative in an appropriate case, and argument about it should not be obfuscated by the introduction of concepts and tactics domestic to any member. The interpretation of the INTELSAT Agreements is a matter for INTELSAT, and, if necessary, the provisions made by the Agreements for the settlement of disputes.[80] It is not for a Party or Signatory effectively to impose its view of the meaning of a provision of the Agreements through a decision of a municipal court or administrative agency or by governmental order or fiat.

The fifth and remaining element of the art XIV(d) consultation is directed towards the avoidance of 'significant economic harm' to the INTELSAT global system. The recommendations of the Assembly of Parties must express its findings regarding this consideration as well as findings relating to technical compatibility of the new system with that of INTELSAT and the assurance that the new system will not prejudice direct links among the participants. As can be seen from the attempts of Orion Satellite Corporation and International Satellite, Inc. to avoid the requirement of consultation under this head,[81] 'significant economic harm' is the element of art. XIV(d) most likely to give rise to real difficulty and, in the last resort, could cause a confrontation between INTELSAT and a Party.

The factors that are relevant in arriving at a view on 'economic harm', the respective weighting to be given to each factor, and the import of the adjective 'significant', have been matters of controversy both within INTELSAT and beyond. For many years, the basic document on this form of consultation was 'Inter-system Coordination Procedures: Proposed Procedures for Implementation of Article XIV(d) Requirements Concerning Significant Economic Harm'.[82]

What constitutes 'significant economic harm to the global system of INTELSAT'? We have seen above that it was the Japanese who, at the negotiations of the definitive arrangements, suggested that the development of satellite systems separate from the INTELSAT systems should be permitted, but that safeguards should be established within the INTELSAT arrangements to secure as far as possible that the INTELSAT global system was not prejudiced by such separate systems.[83]

The Japanese proposal was couched in wide terms which conflated questions of direct links and other forms of prejudice to INTELSAT. During the negotiations these elements were separated, and the matter of financial prejudice to INTELSAT operations and efficiency became transmuted to the question of economic harm. In the six-year period, February 1973 to February 1979, from the entry into force of the definitive arrangements there were only two calls for art. XIV(d) coordination, both of these relating to the US MARISAT system.[84] However, after February 1979 the flow of art. XIV(d) coordinations and consultation increased considerably, both because of the establishment of other international organizations such as INMARSAT, EUTELSAT and ARABSAT, and because of other developments such as the US–Canada transborder services and other uses of 'spill-over' facilities. In all these instances INTELSAT proceeded on a case-by-case basis, although naturally a jurisprudence of principle began to emerge. However, with the 'separate systems' question as raised in the form of the US applications to the FCC in 1983,[85] a more coherent approach towards identifiable and weightable criteria for the assessment of 'significant economic harm' became desirable if not necessary. Discussion was helped by three studies commissioned by the Director-General in 1983 from the US consultancy firm, Walter Hinchman Associates. These studies appear as three reports: *The Economics of International Satellite Communications; Significant Economic Harm;* and *International Satellite Competition Impact Analysis* (a country-by-country data analysis).[86]

As far as the definition of 'economic harm' is concerned, the Hinchman reports reckon that the most useful summary indicator of economic harm is increased cost. Given the cost-sharing arrangements which are fundamental to the INTELSAT system, any increase in cost to members and users constitutes harm. If the establishment of a separate system will affect the annual cost per unit of utilized capacity within INTELSAT, then economic harm has resulted. That increase in cost can emerge through all or any of several elements: the reduction in use of the system by the diversion of existing and future traffic, or the loss of potential joint economies or economies of scale. These result in the INTELSAT planned or actual capacity being underutilized and the consequent increase in the cost per unit of utilized capacity.[87] Taking 'cost' as an indicator may not thoroughly exhaust the matter, since it does not take adequate account of the possibility that a future, but as yet unplanned and unforeseen, INTELSAT may be less efficient economically than it could have been in the absence of the separate system. But cost is certainly something which can and must be quantified and which can therefore be used as an indicator of economic harm. Hypothetical developments are much less satisfactory as indicators. Against the relative accuracy of calculable cost, reliance cannot be placed on what might be, if only because it might *not* be. As a working rule, therefore, cost per unit of utilized INTELSAT capacity provides a reasonable yardstick whereby harm is indicated and measured – measured, for it is not all or any 'economic harm' to the INTELSAT system which is proscribed, only 'significant economic harm' drawing an unfavourable recommendation from the Assembly of Parties. A hypothetical development will only marginally affect the appreciation of 'significant economic harm'.

What, then, is the effect of the word 'significant'? The adjective was chosen as a compromise between competing views at the negotiations of the definitive arrangements. Its presence does not render unlawful any harm whatsoever, but allows INTELSAT to indicate the degree of harm which it can tolerate. When adopted, 'significant economic harm' was not a technical term, and it has allowed the Board of Governors and the Assembly to develop and mutate its content. The broad guidelines are contained in BG–28–63, 'Intersystem Coordination Procedures: Proposed Procedures for Implementation of Article XIV(d) Requirements Concerning Significant Economic Harm', but recent developments have, however, forced the systematic consideration of its content.

The result was a document which suggested certain basic questions which INTELSAT will consider in any application for economic coordination, *Policies, Criteria and Procedures for the Evaluation of Separate Systems Under Article XIV(d)*.[88] The US reaction was not, however, favourable to this document.[89] Nonetheless the Board did transmit its recommendations to the Tenth Assembly of Parties, which adopted the proposed criteria, and added as a criterion the cumulative effect of a number of separate developments, a matter which had exercised INTELSAT in relation to the US separate system developments.[90]

Although assessment is still being refined and developed, there are now therefore certain basic questions which are examined in order, concerning the evaluation of a separate system.[91] These are: are the proposed services international public telecommunications services?

can INTELSAT provide those services? is the proposed system likely to prejudice the establishment of direct links through the INTELSAT system? what economic harm will be caused to INTELSAT? are there other factors such as a short-length peripheral service (for example, a limited transborder service) which require to be taken into account? In relation to non-technical consultation, the Tenth Assembly of Parties specifically noted Board practice. The Board document thus approved[92] expands these headings and indicates the questions to be addressed as:

1 Are the services in the proposed separate system public international telecommunications services as defined in art. I(k) of the Agreement?
2 Can the proposed service be provided using the INTELSAT global system comprising, the existing space segment (including normal replacement), any new space segment which is under procurement, and any planned space segment?
3 In the absence of the proposed system, would the traffic have been carried by INTELSAT?
4 How much traffic will be diverted from the INTELSAT global system?
5 What is the estimated effect on INTELSAT utilization charges both in the short and long term?
6 What is the estimated effect on INTELSAT planning and operations?
7 What is the estimated effect on the cost of providing the INTELSAT space segment?
8 What is the estimated effect on the other Signatories' investment of the proposed diversion of traffic in terms of a) changes in space segment investment requirements, and b) variations in the proportion of total investment shares resulting from any decrease in the proposed Signatories' investment share?
9 What other factors are to be considered including a) variables affecting INTELSAT's ability to earn sufficient revenue to cover the cost of providing services, b) its current financial condition, c) its overall growth opportunities and options for responding to competitive systems, and d) the effect of service restrictions that are placed on separate satellite systems [e.g. any obligation to use the system]?

The Tenth Assembly of Parties therefore maintained the former flexible position, its determination as to relevant questions being somewhat less detailed than the requirements of *Policies, Criteria and Procedures* but, as indicated above, allowed the inclusion of cumulative effect as a factor in the INTELSAT decision. Specifically it stated that 'in addition to the procedures above [sc. the above list] the Assembly of Parties will also take into consideration in Article XIV(d) consultations the cumulative economic effect on INTELSAT of one or more systems submitted by a Party or Parties for such consultations over an appropriate period of time. It also reaffirmed 'its commitment to the principle that all Article XIV(d) consultations be conducted on a non-discriminatory basis'.[93]

Of course, it may be that, in its interpretation of the question to be addressed in a consultation, the Board will return to some of the detail indicated in *Policies, Criteria and Procedures*. In particular in considering cumulative economic harm there will be an impetus to setting a limit to cumulation. That this should be a total limit is obvious. It may also, in the interests of equity, have some reference to the extent to which any one country or group of countries should take up the total permissible cumulative economic harm. Should the 'cumulation' not be shared? Surely the whole extent of cumulative harm ought not to be pre-empted by those whom some would call 'vigorous' or 'enterprising' and others 'greedy'? If INTELSAT should decide on something like the 'cap' for the economic harm caused by any one Party's non-INTELSAT activities which is spoken of in *Policies, Criteria and Procedures*, that would be both natural and right. Again, in the interpretation of time in the phrase 'over an appropriate period of time' the Board seems likely to use the INTELSAT 10-year planning cycle and the 10-year expected life of satellites as a convenient benchmark (*Policies, Criteria and Procedures*, paras 74–83). That also is rational.

But the question remains, can a figure be put on the degree of economic harm which INTELSAT will consider as being 'significant'? The final Hinchman report, that on 'Significant Economic Harm' (*supra*), suggests (p. 29) that it might be useful to adopt as an interim measure *de minimis* thresholds of 'significance' at a level of a 1 per cent limit for the individual effect of a proposed separate system and a 5 to 10 per cent limit for cumulative effects. These percentages would be taken on a service by service, region by region and facility by facility basis and could be considered effective within INTELSAT's normal 5 to 10 year planning cycle. The Report, however, also warns at pp. 27–9 that a workable discrete threshold of 'significance' would be difficult to establish and to operate. We shall see as examples and practice accumulates. The 1 per cent limit for the effect of an individual case is awkward for a commentator to grasp. Nonetheless, a 5 to 10 per cent cumulative effect on an INTELSAT operation, whether reckoned by service, region or facility does appear significant, even to an outsider and it may be that these parameters foreshadow a way ahead.

Art. XIV(d) coordinations: practice

It is all very well to have rules as to the coordination of international public telecommunications services. The question is how they are applied and whether their application lessens, or even evacuates, their apparent content. This is particularly the case in art. XIV(d) coordinations. Reasons of politics and economics, the interaction of dogma of competition and doctrines of public service, together with the inexcusable excuse of ignorance of the legal position and a determination to reshape arrangements and those rules to a form more congenial to one's own interests have put the arrangements in the Agreement under strain and the Board of Governors has, as indicated above, been reconsidering the method by which 'significant economic harm to' the INTELSAT systems is assessed and taken account of. In that process of reconsideration, one document usefully lays out and analyses the determinations made in the major instances of art. XIV(d) down to 1984. This is the paper *Policies, Criteria and Procedures for the Evaluation of Separate Systems Under Article XIV(d)*,[94] together with its Addenda, No. 1, 'Tabulation of the Non-Technical Evaluation Criteria (BG–60–69)' and No. 2, 'Application of New Criteria (BG–60–69) to Certain Previous Article XIV(d) Coordinations'. Using the material from that paper (though not in its order, nor all of it) it is possible to outline major categories of art. XIV(d) coordinations.

Use of INTERSPUTNIK Just as certain Communist countries which are members of the USSR-led satellite system, INTERSPUTNIK, use INTELSAT facilities, so it is possible for an INTELSAT member to use the INTERSPUTNIK system. Such use is subject to coordination under art. XIV(d) which specifically speaks of utilization as well as the establishment or acquisition of a separate space segment. In fact, Algeria wished to become a member of INTERSPUTNIK and the coordination processes involved were completed in 1980.[95] The Board of Governors tendered its advice to the Sixth Assembly of Parties (AP–6–20) and that Assembly decided that Algeria's proposed use of INTERSPUTNIK was technically compatible with the INTELSAT system and would not cause significant economic harm to it. The Assembly agreed.[96] However, the Assembly was careful to require that any change from the proposed use would be the subject of yet another coordination, and that its finding was limited to Algeria. Utilization by other INTELSAT Signatories would require new coordination. The Assembly specifically noted that this was the first instance of an INTELSAT member becoming a member of another separate global satellite telecommunications system. This 'development is considered of particular significance in the light of the Preamble to the INTELSAT Agreement' which speaks of its members desiring to achieve a 'single global commercial telecommunications satellite system'. Finally, the Assembly stated its intention to consider in its economic evaluation of the proposal 'any and all uses' of the INTERSPUTNIK system by any INTELSAT Party or Signatories in any future similar requests for coordination. Subsequently, other coordinations involving membership of INTERSPUTNIK and/or the use of that system have been approved.[97]

PALAPA   The PALAPA system began as a domestic telecommunications network designed to meet the requirements of Indonesia, a country where satellite telecommuncations have distinct advantages.[98] Apart from the importance in drawing the country together on a basis of strengthened national and cultural identity,[99] there were other objectives to be served. Indonesia is an archipelago, with dense jungle on many of the islands, and its high rainfall density affects terrestrial radio links. A satellite system was soon seen to solve many problems. Indeed, it was as early as 31 January 1975 that INTELSAT received a request for coordination of a planned system under art. XIV(c), and the advance information on the network was published in the IFRB Weekly Circular (No. 1160) on 29 April 1975.[100] However, there were problems. The initial proposals revealed that an unacceptable level of interference would be caused to the INTELSAT system, particularly through ground stations for the Indian Ocean satellite where equivalent noise temperatures would be increased by more than 2 per cent.[101] Matters continued under discussion, in part because of financial and technical delays. In the meantime the other countries of the ASEAN organization (Malaysia, Philippines, Singapore and Thailand) decided that they too would usefully be served by putting domestic networks and local international traffic through the PALAPA satellite system.[102] Accordingly, the PALAPA system fell to be dealt with under art. XIV(d).

Technical compatibility, which is in a sense an engineering problem, was worked out.[103] More difficulty could have been encountered on the question of 'economic harm', but this was negotiated. In the discussions of the PALAPA system, an important element seems to have been the fact that much of the traffic would originate or end in remote locations where the use of wideband facilities to access INTELSAT facilities was uneconomic. In addition (as had been decided by then for Europe), it was important that INTELSAT did not consider that much traffic would go onto the PALAPA system in preference to the INTELSAT system. The PALAPA use would in a sense be additional traffic, not diverted or 'lost' traffic – although this perhaps undervalues INTELSAT's potential loss of revenue. Accordingly, at the Fourth (Extraordinary) Assembly of Parties in April 1979, a favourable finding on both technical and economic grounds for the

PALAPA B system as described in AP–4–8 was decided upon with due caveats as to any changes, and with a requirement of reconsultation for any alterations and if the networks continued beyond 1990.[104] The PALAPA A network as described in AP–6–13 was approved by the Sixth Assembly of Parties in 1980 with similar caveats but with the omission of the time restriction.[105]

US transborder services   Naturally the US domestic satellite system provides a spillover coverage of many areas contiguous to the USA. Questions were therefore bound to arise regarding possible services to both Canada and to Mexico, Bermuda and the Caribbean islands. Both these matters came up at the Seventh Assembly of Parties, where the Assembly, on the advice of the Board, found that neither the Bermuda development[106] nor the US–Canadian extended service[107] were technically incompatible with nor were likely to cause significant economic harm to the INTELSAT system. Future extensions or alterations of the services from that described in the documentation would, of course, require further coordination processes. So matters were settled for a while.[108] However, one matter ought to be noted: the Board had occasion to 'note the view' during the US–Canadian process that there was serious concern as to whether the requirements of art. XIV coordination were being properly observed by the Parties concerned[109] – a worrying comment.

Notwithstanding, once approval was given the provision of one-way receive only and TV links were soon set up.[110] Limited two-way links followed, and the list of countries to which such service can be provided through US domestic satellites has been extended – most notably in January 1985 including most of the Caribbean and part of Latin America. Other Latin American countries have been added subsequently. The most recent extension came in April 1987 when the Assembly modified the approvals it had given by adding Venezuela to the list, and action was taken by US companies before the FCC for appropriate US permissions.[111] In the most recent of these available to me at the time of writing it is noted that well over 100 transborder authorizations for video, audio, business services and data services had been granted by the FCC to July 1987.[112]

Of course, such developments raise in an acute form the problem that while separately the traffic involved may be minimal and therefore dismissed *de minimis* by those whose interest lies in the result, cumulatively the effect on INTELSAT is to diminish its actual or potential revenues, and hence to affect its stated aim of the provision of a global satellite service at minimal cost and at equal cost to all users for each service (INTELSAT Agreement, Preamble, arts. III and V(d)). The other comment which an outsider must make is that the language of the FCC decisions can be strange, holding, for example,

that INTELSAT will not be harmed by this and that development. That is for INTELSAT to determine. Certainly the FCC and
the State Department genuflect to INTELSAT's role, but the
impression is given that that is increasingly a formality, and there are
signals that compliance with the form of consultation does not require compliance with the INTELSAT finding on the matter.[113]
We will return to this point when considering INTELSAT problems,
below.[114]

Maritime satellite systems   Art. XIV(d) has been involved in two sets
of coordinations regarding maritime satellite systems: the US
MARISAT system established by COMSAT in the hopes that it might
be the system chosen by INMARSAT; and the MARECS system developed by ESA, which was to be a function of EUTELSAT, but
which, as described in the discussion of EUTELSAT, was hived off at
an early stage in the development of that organization. In neither case
were there real problems of technical incompatibility. There might
have been interest centered on the question of economic harm, but
the integration of the systems in the global INMARSAT system elided
much of the problems.

The MARISAT system was of limited life, and hence its effect on
INTELSAT's future development was contained, and its further use
beyond its initial planning was integrated with the MARECS development and the provision of capacity by INTELSAT to
INMARSAT,[115] which again limited the extent to which 'economic
harm' to INTELSAT could occur. INTELSAT was reconciled to the
position that the maritime services were areas which would be served
by a separate organization, and hence were not to be taken account of
in considering whether they were services which INTELSAT itself
would offer. Accordingly, it was relatively easy for the Assembly to
make favourable findings at appropriate points.[116] INMARSAT itself
is discussed separately in Chapter 5.

EUTELSAT   The background of the creation of EUTELSAT is described in the discussion of European developments both in that of
the European Space Agency and of EUTELSAT itself.[117]

When interim EUTELSAT was conceived it was the first regional
satellite system between independent states which was in an area of
major telecommunications traffic. It was therefore the first major test
of art. XIV(d). PALAPA's effect on INTELSAT was minor by comparison. The questions of economic harm being done by EUTELSAT
to INTELSAT and of its effect in prejudice to INTELSAT establishing
direct links between its members was therefore superficially high.
Nonetheless, the Board of Governors and the Assembly of Parties
found it possible to arrive at a favourable finding on all grounds at

the initial stages of consultation and coordination, despite potential for future difficulty.

On the question of economic harm there was much discussion and negotiation, both helped and hindered by the fact that all the members of EUTELSAT are likewise members of INTELSAT, and that certain of the entities involved appoint governors in their own right to the larger body. Much was made of the point that Europe was well served by terrestrial links, and it was argued that, absent the EUTELSAT link, traffic would be placed on the terrestrial links by the telecommunications entities which both owned them and were responsible for the internal end of the telecommunications facilities in the countries involved. INTELSAT links were not likely to be chosen. Accordingly, it was felt that INTELSAT would not suffer 'significant' economic harm by the entry of the new agency into the European telecommunications arena, although some harm was bound to occur.[118]

By the time that the extension of the use of the EUTELSAT system was up for discussion, two matters were beginning to concern IN-TELSAT. The first of these was the question of the diversion of traffic to the EUTELSAT system. Publicity for, and statements about, the new business services indicated that the EUTELSAT service would go beyond that available on terrestrial links and hence would not come within the notion that 'no harm would accrue' to INTELSAT because 'the traffic would not have used INTELSAT circuits in any case',[119] and on the advice of the Board the Seventh Assembly of Parties decided that any material extension of the system after 1988 would require a further new coordination under art. XIV(d).[120] The second was that the new EUTELSAT system was to use the 12 GHz band as a downlink – a band which INTELSAT might use in the future. Accordingly, INTELSAT's view was that a favourable recommendation on technical grounds should be restricted, permitting use of that band until INTELSAT itself wished to use them.[121] Naturally EUTELSAT was disturbed by such a move. If the logic of the INTELSAT finding were followed, EUTELSAT's satellite system might be subjected to major change of waveband after the satellites were 'up' and after only a few years in use.[122] Fortunately, I understand that this latter problem has been elided by a later INTELSAT decision.[123]

ARABSAT   The creation of ARABSAT is described in the discussions of the organization itself.[124] The system is in a sense a regional system like PALAPA B or EUTELSAT, but differs from those in being linearly spread from Mauritania to the Persian Gulf. In the early 1970s the ITU had identified the 'region' as one where telecommunications facilities were deficient. However ARABSAT's organizing principles seem to be political, cultural and religious. It functions as a communications network among the Arab nations in general and the Arab League in particular. Indeed, its cultural and Islamic religious bases seem to be as important as any of the more traditional justifications for the creation of a telecommunications network. As originally coordinated, 17 members of INTELSAT were also members of ARABSAT and there were therefore certain parallels between the case of ARABSAT and that of PALAPA B and the European developments.

When the PALAPA and ECS (later, EUTELSAT) coordinations were in progress, ARABSAT was at a planning stage and attempted to get its affairs dealt with at the same time. This was unsuccessful as will be seen, but there was an Arab intervention at BG–37 in the discussions of these other two systems, which indicated the way in which thought was progressing. Objection was taken to a possible interpretation of 'economic harm' which might have cast severe doubt on any diversion of traffic from INTELSAT, and instead it was emphasized that 'significant economic harm' was the language of art. XIV(d) and that the import of that adjective should be insisted upon.[125] This was an important point. The 'significance' of the 'economic harm' was an important hinge during the discussion of the ARABSAT system.

Informal coordination procedures for ARABSAT began in February 1979 and that for technical compatibility was reasonably and fairly easily dealt with.[126] More difficulty attended the question of economic harm. The information provided by Saudi Arabia as agent for the Arab states was insufficient to allow the Thirty-Seventh Meeting of the Board to make a recommendation on the matter.[127] When matters came properly before the Board the Director-General stated that, while in the cases of PALAPA B and the ECS analysis showed that no traffic was to be carried on either of those systems which would otherwise have gone through INTELSAT, ARABSAT was somewhat different. The first data provided indicated that almost the whole traffic on the planned ARABSAT system was existing INTELSAT traffic which would be diverted or was projected future traffic which would have otherwise gone by the INTELSAT space segment. However, further information was provided. This, based on plans for an increased use of existing terrestrial links and the expansion of these links, showed that the amount of traffic 'removed' from the INTELSAT system was less that had been first indicated. In other words the 'significance' of the economic harm to INTELSAT was prospectively lessened by virtue of an alternative terrestrial telecommunications system which largely existed only on paper. Further the Director-General indicated that, even if the projected traffic 'lost' by INTELSAT was arguably below a 'threshold' of 'significance', that was the case only if the harm was considered in isolation. A cumulation of several such diversions could have a significant effect.[128]

The eventual paper which went from the Board to the Fifth Assembly of Parties recommended a favourable finding,[129] and this recommendation was followed.[130] However, as recommended by the Board, the finding explicitly states that its view of economic harm is based on the existence in 1983 of the terrestrial network details of which had been provided by ARABSAT.[131] I am unaware whether or not the projected terrestrial links have been fully implemented.

PanAmSat-Peru The USA has taken steps towards permitting US companies other than COMSAT to provide public international telecommunications services beyond the 'spill-over' notion of the US Transborder Services dealt with under 'US transborder services' (p. 173 above). The detail of these steps is considered elsewhere in the discussion of future problems for INTELSAT.[132]

The first formal application to INTELSAT for coordination under art. XIV(d) of services planned to be offered by one of these US companies was made on behalf of the Pam American Satellite (PanAmSat). The application followed upon the FCC Separate Systems decision,[133] and the particular PanAmSat proposal was authorized subject to various conditions by the FCC in 1985.[134]

Since the PanAmSat case was clearly going to be the pathfinder for later US requests for this type of coordination, INTELSAT (which had strenuously opposed the whole concept of such developments)[135] had to tread warily. It was helped to a degree by the restrictions placed by the FCC in its decisions following on US government advice on the services to be provided by the satellite system. The proposed PanAmSat satellite (one of the former RCA Astro 3000 series) has 18 transponders using the 4/6 GHz waveband and six capable of operating in the 11/14 GHz band. PanAmSat was authorized by the FCC to use five 4/6 GHz transponders for regional services between the USA and Latin American countries and the remaining 13 4/6 GHz transponders for domestic services by Latin American countries. PanAmSat also had asked for permission to use the six 11/14 GHz transponders for a USA–Europe service but this was refused. By the time the matter went to INTELSAT, what was sought was consultation under art. XIV for five 4/6 GHz transponders to be used on the US–Peru service.

Even so, there were important matters of principle involved. First, there was the whole question of the threat to the unique global purpose of INTELSAT, a point which we will take up elsewhere.[136] Suffice it to say here that clearly any purely commercial satellite system created with the intention (now or later) to offer competition to INTELSAT on its major traffic routes will affect the integrity, costs and finances of the global system. As it happens the PanAmSat proposal that went to INTELSAT did not relate to the North Atlantic, although clearly by the initial application to the FCC the six 11/14 GHz transponders were intended for use on that route.

The second matter of principle was this. Was it competent to consult only as regards a part of a satellite network? When the final PanAmSat application was presented, the question was raised whether what was required under art. XIV(d) was consultation as to the whole satellite, or simply for that portion of the satellite payload which would carry international traffic.[137]

Third, as discussed more fully elsewhere,[138] there was the question as to the effect of the finding of the Assembly. Was compliance with the terms of art. XIV satisfied by the process of consultation, allowing Parties thereafter to proceed as they chose if they did not agree with the terms of the finding?

The decision of the Eleventh (Extraordinary) Assembly of Parties in 1987 defused some of the tension inherent in the situation, but has left awkward possibilities for future disagreement. Following the recommendation of the Board, the Assembly decided that the whole satellite system had to be dealt with. What is coordinated is the satellite network, not the individual transponders. The network was given a favourable finding as to technical compatibility and as to lack of significant economic harm to INTELSAT, and the other elements requried, on the basis of the system as described in the Report of the Board in the matter.[139] However, any material change to the network (and in particular any change as to the countries served) was to require further consultation under art. XIV(d), as was the use of the network beyond 1992.[140] On the basis of this decision by INTELSAT, a Final Permit was given by the FCC for the proposed PanAmSat system.[141]

PROBLEMS

It would be wrong to give the impression that the success of INTELSAT as an organization is entirely unclouded, and that there are no problems on its horizon. In fact, INTELSAT faces problems particularly in the economic field and, at worst, its very integrity may come under threat. The counter-argument is that the organization has been sheltered from proper competition and that the introduction of such competition will refine its activities, producing efficiencies, a greater market sensitivity and better value for money in the services which it provides. Like most similar arguments, both sides have a share of the truth. The materials cited below, particularly in the discussion of the 'separate systems' question will illustrate the argument. But we will begin with a commercial problem which INTELSAT shares with other international organizations providing satellite services.

Relationship with the ITU

INTELSAT has an official relationship with the ITU as it does with other international organizations, and it sends observers to ITU meetings and contributes to its work. However, there are certain areas in which its position might usefully be altered in relation to the ITU. These are discussed separately at the end of Chapter 8 on the ITU itself, but can be summarized here.

First, membership of the ITU itself, and hence official participation and voting at ITU conferences, is open only to states. Organizations such as INTELSAT are dependent upon their member states raising up arguments on their behalf, and upon the necessarily second-hand expression of their views through countries which are in effect designated as their spokesmen. These countries may, however, fail to give the most cogent expression of an organization's view since the spokesman may neither properly understand nor share the same attitude. Worse, the organization's view on a matter may be different from, and perhaps even contrary to, the view of the state which speaks for it.

Second, as an international organization, INTELSAT is dependent on the USA to interact for it with the IFRB. That might be better done direct. Third, INTELSAT does not itself directly participate in ITU coordination and consultation processes for new developments, through, for example, initiating, or having initiated with it, consultations prior to the assignment of frequencies to space and ground stations or in the allocation of an orbital position to a satellite. Fourth, there is some duplication between the ITU coordination procedures and that required for art. XIV consultations within INTELSAT.

It might be thought that there was some room for rationalizing duplicative processes, and perhaps giving the INTELSAT consultations a wider framework. This last possibility has, however, the disadvantge of removing a relatively impartial ITU scrutiny and entrusting highly important and possibly contentious matters to an organization which is itself active in providing telecommunications services. This would obviously be bad practice. Nevertheless, some movement could be brought about on the other three matters by a relatively simple expedient. INTELSAT could be given status within the ITU councils, could interact better with the IFRB and could be brought fully into the consultative and coordinative stages of frequency and orbital assignment by making it – and all other international organizations which provide and operate telecommunications services – eligible for ITU membership. Such treaty-based organizations have limited international personality and that could be founded on for this purpose. Some restriction could be made as to their powers – for example, to nominate and vote on appointments to ITU positions – but, that apart there is no reason why INTELSAT and other like organizations should not be ITU members. Indeed, INTELSAT's revenues are such that it could well afford to subscribe a higher unit class of contribution than many ITU members. However, there is a doctrinal objection in that ITU membership is currently only open to states, and some countries consider that only states ought to be members of international organizations.[142] The ITU Convention wil be revised in 1989: the appropriate steps may be taken then.

That said, the Report of the First Session of WARC-ORB makes special mention of international organizations. It recognizes and recommends to the Second Session of the Conference that the requirements of international organizations which operate satellite telecommunications and other services by given special consideration. That is a welcome statement, justifiable because so many countries pass so much of their international communications through such systems. Nonetheless, it would be useful to give organizations such as INTELSAT an oportunity to participate in the ITU conferences as members, albeit with restricted membership, as indicated above. (But see p. 416).

Fibre optic cables

As indicated in Chapter 1, when satellites were being developed for telecommunications purposes their major justification was that they could provide a better service than the two competitors – radio links and cables. Satellites were not subject to the problems of terrestrial radio and would afford more flexible service than cables, carry a greater traffic density and use a signal transmission band-width which could offer services outside the scope of (for example, long-distance television). The advantages with respect to terrestrial radio remain. Cable technology has not, however, stood still, and the development of the fibre optic cable has allowed cable to plan to offer services which are not dissimilar from those of satellites. In a way the two modalities are now complementary. Telephone transmission by satellite can be affected both by echo, and the effect of delay. Cables are fixed in their route. Satellites can provide a restoration service while cables are out of order for one reason or another, and there is likely to be a higher degree of 'redundancy' in the satellite service by reason of deliberate planning, making it that much more able to cope with fluctuation in demand.

But, whatever one may say by way of mitigation, the INTELSAT system is a commercial enterprise in competition with alternative methods of telecommunication. That some of its members are active also in the cable business serves to complicate, rather than ameliorate, the position. The application by Tel-Optik (now Nynex Corporation) working with Mercury Communications of the UK (a wholly-owned subsidiary of Cable & Wireless Plc) for authority to lay a high-capacity private fibre optic cable across the Atlantic and the related decision by the FCC shows the potential for damage to INTELSAT.[143] In addition, plans were announced on 1 October 1986 by Cable and Wireless with Japanese associates for a round-the-world fibre optic communications link for public telecommunications. The state of planning of both these enterprises – the private and the public cables – show that such cable developments are no mere science-fictional dreams: these or similar cable developments are likely to happen soon. Certainly the FCC has taken the advent of fibre optic technology very seriously and has incorporated it into both its Atlantic and Pacific Ocean regional plans.[144]

Competition from cable-borne services will have an effect on INTELSAT, both in relation to the satellite services which are in direct competition and also in the possible loss of some other business from countries which decide that cable has attractions which satellites do not share. However, as said, cables are static, and once laid cannot be shifted. Like satellites they are ultimately dependent on land links, but they remain less flexible than satellite links. The argument will not, however, depend on such platitudes. Commercial pressures (increased since the privatization of British Telecom and Nippon Telegraph and Telephone Company) mean that the end-user, the telecommunications entities, will closely scrutinize the value which they will get from their investment in, and use of, either system as against the other.[145] The cost of establishing a cable or satellite link together with the capacity each offers, and the projected life-span of each are questions affecting decisions. Cable is likely to have a life-span of 25 years, satellites of 10 years. The cost of the transAtlantic fibre optic cable may be approximately US$ 350 million, as opposed to the approximate US$ 180 million of the INTELSAT VI series. Circuit costs by satellite are about one half of the cable circuit. Cable costs are distance-dependent. Relatively speaking, satellite costs are close to uniform – the distance up and down swallowing the minor element of surface distance. The evidence so far to hand indicates that cable and satellite (and in particular the INTELSAT system) will remain competitors in international telecommunications, and it remains to be seen how much damage each does to the other's profitability. Both are cost-competitive with the other.

But although there is certainly room for both systems of telecommunications in the future, the advent of cable will add to a possible glut of telecommunications capacity due to an overestimate of demand. The entry of the INTELSAT VI series, along with regional systems, will provide more capacity than will probably be necessary for the world's telecommunications needs for some years. Cable will accentuate that problem, and make investment in either modality less profitable than had been anticipated. Until recently the FCC imposed a 'balanced loading' policy on US carriers, requiring them to share their international traffic equally between INTELSAT and the cables on the Atlantic route. The cables, being part-owned by US common carriers such as AT&T would otherwise have tended to favour the cable route at the expense of INTELSAT, and the satellite share of the transatlantic market did require that initital protection.[146] That policy was terminated for most cases in 1985, following upon discussions between the FCC and the members of the Conference of European Postal and Telecommunications Administrations (CEPT),[147] and was completely abandoned by the FCC in 1988.[148] It remains to be seen what will be the precise effect in the telecommunications marketplace of this change in policy, and of the commencement of operation of the new cables, together with the interaction of the new cables and the new satellites.

US Separate Systems

More danger to INTELSAT comes from the potential development of separate international telecommunications satellite services by US companies. As we have seen above, procedures exist within INTELSAT for the expression of 'views' on proposals for separate systems involving INTELSAT members through the requirements of art. XIV. We have also seen that developments such as INMARSAT, EUTELSAT and ARABSAT were relatively easily dealt with, INTEL-SAT having had the first-named in contemplation when it itself was established, and the other two being relatively easy to foresee and to harmonize with the INTELSAT system.

Greater difficulty accompanies the US separate systems. Although, at present, internal US steps have been taken to reduce the impact on INTELSAT there is no doubt that these developments foreshadow systems planned with the direct intention of competing with INTELSAT services on certain of INTELSAT's major service routes. Some results of the US steps have been noted above, particularly in the discussion of art. XIV(d), but the threat implicit in the moves already within the USA should be more fully indicated by rehearsing them.[149] We will follow through the internal US developments before outlining the INTELSAT response. These developments are: first, the authorization subject to restrictions of certain separate systems; and, second, legislative changes which clear the ground for other changes in the US attitude to INTELSAT.

Authorizations of separate systems

We deal first with the authorization of separate systems by the FCC. The first legal steps in the development of the US separate systems were taken by Orion Satellite Corporation (Orion) which on 11 March 1983 filed an application with the FCC[150] for authority to establish a communications satellite network to provide international services. This was followed by similar applications by International Satellite Inc (ISI) on 12 August 1983,[151] by RCA American Communications, Inc. (RCA) on 13 February 1984,[152] by Cygnus Satellite Corporation (Cygnus) on 7 March 1984[153], and by the Pan American Satellite Corporation (PanAmSat) on 31 May 1984.[154] On 12 June 1984 Systematics General Corporation filed two applications, but withdrew them six weeks later. On 20 July 1984 Western Union Telegraph Co. asked for a modification to its existing permission for a domestic satellite in order to provide coverage to Central and South America.[155]

These applications showed that US companies, other than COMSAT, were anxious to set up satellite systems separate from INTELSAT for the provision of public international telecommunications services. To an extent, the way had been paved by the trans-border decisions of INTELSAT and the FCC noted above, but the 1984 applications posed a direct challenge to the INTELSAT system. Accordingly, following on the Orion application, the US State Department and the US Department of Commerce wrote to the Chairman of the FCC asking that no final action be taken on the application until the US government could study and determine the impact of the introduction of separate systems on US national interest and foreign policy. The Department of Commerce repeated this request following the filing of the second application, that of ISI, in August 1983, which the Department felt raised additional questions also requiring governmental review.

A Senior Interagency Group on International Communications and Information Policy (SIG)[156] was set up to formulate the US government's position and duly reported in a 'White Paper' in late autumn 1984.[157] This document and its conclusions appear to be the basis on which on 28 November 1984 President Reagan signed Presidential Determination No. 85-2, 'International Communications Satellite Systems. A Memorandum for the Secretary of State and the Secretary of Commerce'.[158] This formally determined that 'separate international communications systems are required in the national interest', ordered consultation with INTELSAT and instructed the Secretaries of State and Commerce to inform the FCC of 'criteria necessary to ensure' that the USA met 'its international obligations and to further its telecommunications and foreign policy interests'. The reference to the national interest activated that part of s.102(d) of the US Communications Satellite Act 1962 (47 USC 701(d) which states the legislative intent of Congress in the Act not 'to preclude the creation of additional communications satellite systems . . . if . . . required in the national interest'. But this takes us into the realms of Humpty Dumpty's approach to words indicated in *Alice in Wonderland*.[159] One would have thought that the 'national interest', though separate from the 'unique governmental needs' for which also other systems might be established, was something more than the interests of certain US nationals seeking a profitable enterprise.

In accordance with the Presidential Determination, the Secretaries of State and Commerce informed the Chairman of the FCC (the Schultz–Baldrige letter of 28 November 1984) of the necessary criteria directed towards ensuring US telecommunications and foreign policy interests and the meeting of US international obligations. The FCC should ensure that:

1   each system it authorized should be restricted to providing service through the sale or long-term lease of transponders or space segment capacity for communications;
2   that the service provided be not interconnected with public-switched message networks (except for emergency restoration service;
3   that one or more foreign authorities are to authorize use of each system and enter on art. XIV(d) consultation with INTELSAT about it.

The letter also enclosed a Memorandum of Law on art. XIV of the INTELSAT Agreement which had been prepared by Davis R. Robinson, Legal Advisor to the State Department.[160]

Having received this data from the appropriate parts of the Executive Branch on matters under their jurisdiction, the FCC issued a 'Notice of Inquiry and Proposed Rulemaking on Separate Systems'.[161] To permit it to deal with the matters at issue without the pressure of further applications, on 31 May 1985 the FCC issued a 'Freeze Order' on accepting further requests for authorizations to establish separate international satellite systems.[162] That device worked and after much debate, argument, representations and counter-representations, analysis and cogitation the 'Separate Systems Report and Order' as it is usually called, was issued on 3 September 1985.[163]

It is long, complex and detailed, so much so that, unlike most FCC reports, it starts with a Table of Contents. Its main sections contain discussions of general policy including the benefits of competition and the economic impact on INTELSAT as well as dealing with more technical domestic US matters such as policies on the qualifications of those who should be licensed to provide such services. The report goes through the elements of art. XIV(d) including the question of 'prejudice to direct links' (para. 192). In particular, it discusses in detail the various Hinchman reports to INTELSAT on the question of 'significant economic harm' (dealt with above) and specifically states that, given the restriction to be placed in the FCC licences that the services shall not interconnect with public switched message systems, 'we do not believe that entry by separate international satellite systems would inflict significant economic harm on INTELSAT' (para. 188). On the contrary, INTELSAT will be able to compete effectively (para. 189) and move into the markets served by the 'new' systems before the separate systems, a development which would itself benefit INTELSAT (para. 190). Further, the new systems, by 'forcing INTELSAT to compete for some of busines, would put pressure on INTELSAT to operate its satellite system more efficiently, minimize its costs consistent with reasonable quality of service, and set prices to reflect its costs in those markets where entry occurs'. As a result, its economic efficiency would improve (para. 191). 'Many users are dissatisfied with the inefficient service choices offered by INTELSAT in the past and with the inflexible and inadequate arrangements and a lack of responsiveness' (para. 82). Competition would spur INTELSAT on. Again, the FCC considered that other INTELSAT members would not be disadvantaged by the introduction of separate systems into the satellite telecommunications market (paras. 200–2, 201). Nor would new systems prejudice the establishment of direct links (paras. 192–3). Given such reasoning, and an extensive discussion of the economic and public benefits foreseen, it is not surprising that the FCC was able to find that the establishment of separate systms within certain restrictions would benefit the (US?) public without harming INTELSAT (paras. 68–86).

To comment here briefly, it has to be said that INTELSAT did not then, and still does not, agree the detail of the FCC analyses, nor its conclusions. Nor does it seem right that such detail was entered into by the FCC prior to INTELSAT itself considering the questions which each new service would raise. The FCC's detailed discussion of INTELSAT criteria, opinion, documentation and performance sounds like a pre-emptive strike by a Party which was uncertain that, in the absence of such a demonstrated determination to get its own way, an unfavourable finding might issue from the Assembly of Parties. Certainly the FCC had to authorize the service before another country could be sought as the other end of the proposed link and a request for art. XIV consultation could go to INTELSAT. But, first, what of the usual informal discussions which is INTELSAT practice? Second, why was INTELSAT so discussed in the FCC proceeding? Last, there is a non-barking dog – the question of profit for those serving a more restricted high-traffic route.[164]

Be that as it may, the Separate Systems Report and Order is fundamental to the US development of international telecommunications satellite systems separate from that of INTELSAT. It sets out the financial qualifications any applicant for authorization must meet and integrates the orbital policies for such developments with its other orbital decisions (for example, on domestic satellites).[165] The Report and Order further specified the mechanisms through which the FCC policies the restrictions that Presidential Determination imposed (paras. 108–14). These include that the operators of the separate systems shall not operate as common carriers (para. 117), although common carriers and enhanced service providers can resell capacity provided that the other restrictions on the systems are observed and that that use by a common carrier is authorized by the FCC in each case (paras. 126–7). The Report and Order lays down the technical requirements to prevent interconnection of the authorized systems with the public switched message services (paras. 132–8), although some doubt whether the so-called 'leaky PBX', the exchanges that permit such interconnection, can ever be excluded (cf, para. 130 and n. 84 reporting Norway and Sweden's comments). To comply with the terms of the Determination, the Report and Order stipulates the services to be dealt with by the US separate systems. These are restricted to 'customized services', a term which was used in the Executive Branch White Paper. However, the FCC treated the term as being indicative and not as definitive. Thus, while a 'customized' service might be defined as a service which a corporation uses for its own internal message requirement, the FCC chose to consider as 'customized' services which are tailored to the requirements of their users (para. 115). Finally the FCC decided that no minimum unit of capacity be specified for sale or lease purposes on a separate system (paras. 116–18), but that the 'long-term lease' restriction in the Presidential Determination should be interpreted as requiring a minimum one-year lease arrangement – this apparent departure from the normal meaning of the words being justified by comparing that period with the lease terms offered by domestic systems and INTELSAT practice (paras. 117–23).[166]

On the same day as the Separate Systems Report and Order was adopted, the FCC conditionally granted two of the applications it had before it,[167] granted another,[168] and deferred action on the two remaining, requiring in these cases the production of further financial data.[169] These applications were later granted.[170] In addition, the FCC has subsequently granted later applications on terms similar to the initial three as to financial qualification, service restriction and the requirements of INTELSAT consultation.[171]

In the meantime, different parts of the Separate Systems Order were appealed. In the first Separate Systems (Reconsideration) Order,[172] minor modifications were made to a number of elements relating to questions of interconnection and services, including affirming that final authorization would only be given to a separate system under the 1985 Order only where the 'foreign' end of the link was willing to comply with the restrictions on interconnectivity (paras 30–4, 36). Two major matters were decided. First, following upon clarification from the Executive Branch,[173] the FCC determined that a 'long-term lease' within the meaning of the Presidential Determination applied to the term of the contract, not to the use being made of the facility contracted for: therefore occasional use of any service, including television, might be contracted for in a lease whose term was one year although one year's use would not be made of the facility (paras. 30–2). Second, and technically very important, an attempt to use FCC procedures to deal with a possible conflict of orbital positions was avoided, it being clearly stated that such matters are for the ITU and INTELSAT procedures (paras. 44–8). Six months later the FCC reaffirmed its holding as to 'occasional use' and 'long-term' lease in a further reconsideration of its Separate Systems Order.[174]

The Separate Systems Order, as modified and reconsidered in the two subsequent proceedings, therefore provide the framework within which the USA will proceed to authorize systems separate from the INTELSAT system, subject to restrictions it considers necessary to protect INTELSAT. It is, however, clear that providing competition for INTELSAT is seen desirable. Indeed, one has the superficial impression that the principal aim of the USA is the betterment of the INTELSAT system. It appears to be unintended and entirely coincidental that, by providing such 'necessary competition' to the international organization on selected routes, certain US companies and individuals may make a substantial profit. Be that as it may, one separate system is now on the threshold, having satisfied both the FCC and the INTELSAT requirements. This is the service which PanAmSat will provide between the USA and Peru, and within Peru.[175] (See also Chapter 9, p. 412.)

US legislative activity

Interconnected with the FCC and governmental activity on separate systems, in 1985 the US Congress was also at work in two measures, altering the law so that the prevailing US views of INTELSAT might be given legislative force. The technicalities are detailed elsewhere.[176] Suffice it here to say that by the Statement of Managers accompanying the Supplemental Appropriations Act 1985, Public Law 99-88 of 15 August 1985, Congress urged the Executive Branch and the FCC, in dealing with separate systems, to avoid causing significant economic harm to INTELSAT, that the requirements of the Presidential Determination be adhered to, and laid out a procedure for action should a proposed system acceptable to the USA be found against in the INTELSAT procedures. That procedure was overtaken by the amendments to the Communications Satellite Act of 1962 made by s.146 of the Foreign Relations Authorization Act, Fiscal Years 1986 and 1987, Public Law 99–93 of 16 August 1985. That Act repeats the requirements of compliance with the Presidential Determination of 1985 and establishes a procedure for dealing with art.XIV(d) matters. Under this procedure, if the INTELSAT Assembly of Parties makes an unfavourable finding on a separate system under art. XIV(d) and the USA is nonetheless minded to proceed with the system, the President must determine that it is in the US national interest to proceed and the Secretary of State must send to Congress a detailed report on the matter including the foreign policy reasons for the Presidential Determination, and submit 'a plan for minimizing any negative effects of the President's action on INTELSAT and on United States foreign policy interests' (s.146(d)). Thereafter Congress has 60 days in which to take action before the Secretary of State may inform the FCC that US obligations under the INTELSAT agreement as to art. XIV(d) have been met (s.146(e)).

INTELSAT reaction

These legislative and FCC steps were not taken in a vacuum. Throughout the discussions of the FCC, and during and preceding the legislation, INTELSAT, INTELSAT sympathizers and INTELSAT Parties and Signatories actively lobbied the Congress and in made their views on the US developments well known: indeed, unusually, the Director-General designate gave evidence to Congressional committees on INTELSAT matters.[177] The matter was also discussed within legal literature.[178] But while there was clearly some effect on the legislative steps, the fact remains that the FCC has proceeded to license separate systems.

Obviously the fact that the USA has proceeded as it has done has had an impact on INTELSAT. Questions of new services and of pricing flexibility will be discussed shortly but, first, something must be said about the legalities involved.

Earlier, when outlining the rights and obligations of members, I wrote that the cynical might think it possible to go through the motions of consultation under art. XIV(d) and then proceed at will in defiance of what would merely be a negative finding expressed in the form of a recommendation by the Assembly of Parties.[179] Certainly the provisions of the Foreign Relations Authorization Act 1985, with its requirement of a solemn determination by the President, a full report by the Secretary of State to Congress and a 60-day period within which Congress might act is such a case, comes as a welcome chink of light. Yet one wonders whether that light is dawn or sunset. It was the US government which first indicated to the FCC that compliance with the terms of art. XIV(d) did not mean that the recommendation of the Assembly of Parties be followed: it was enough that the process of consultation had been engaged in, and two sovereign state might 'in good faith' determine to proceed to establish a service separate from INTELSAT.[180]

What does 'in good faith' mean under such circumstances? To repeat what has been said earlier,[181] INTELSAT 'findings' are recommendations, not decisions. However, art. XIV(a) and the Preamble to the Intergovernmental Agreement clearly express a determination to achieve a 'single global commercial telecommunications satellite system' and 'a single global commercial telecommunications satellite system as part of an improved global telecommunications network'. Should this 'good faith' be broken, there is the ultimate sanction of art. XVI(v)(i) by which a party may be deemed to have withdrawn from the organization.

There was some talk of whether or not the USA was standing in danger of these waters in its early explorations of the separate systems questions. Certainly an outsider has the impression that the US government of the day was dissatisfied with INTELSAT, that it was interventionist in regard to COMSAT's role within the organization (see the discussion of COMSAT and the US government) and that it was willing to coerce rather than persuade INTELSAT as to improvements through the use of competition. An outsider must also say that different elements of the US government having responsibilities within the telecommunications arena appeared to be pulling in different directions that while some of that conflict was genuine difference of opinion as to telecommunications questions, some of it was jockeying for position within a rather loose governmental structure; that international law has been conveniently ignored; and that some saw a chance to make money amid the welter and confusion. The result has been that the plain meaning of words has been evaded and the spirit of agreements strained. Unfortunately one has also to say that these problems do not seem to be confined to questions of international telecommunications.[182]

US action and 'initiative' in telecommunications have not ceased. One awaits with a major degree of concern the outcome of the FCC review of US regulatory policies relating to international telecommunications which has begun.[183] Comment was sought on the extent to which the US public interest requries that the telecommunications

policies for foreign governments should be considered in the formation of US policies on telecommunications. The Notice of Inquiry
also indicated that the FCC will seek to establish a model to represent
an ideal to be sought in international telecommunications and a
benchmark for international policies and practices (para. 1). This
model will have four bases:

1   open entry to the telecommunications market;
2   non-discrimination;
3   technological innovation; and
4   international comity.

Information will be sought on the penetration of the US market by
foreign-owned firms, and attention will be paid to the extent to which
foreign governments permit access to their national markets for US
firms. A competitive market structure in telecommunications will be
sought, and obviously reciprocity will be important. This proceeding
has been carried further by a Report and Order and Supplemental
Notice of Inquiry of March 1988, under which more extensive information is to be required to be filed by 'foreign carriers' to enable the
purpose of the initial Notice to be fulfilled.[184]

Such proceedings interact with the whole matter of US participation in INTELSAT, INMARSAT and the ITU, as well as with its
bilateral arrangements. The FCC may be a useful arena in which to
develop ideas but, in terms of international law and practice, it may
be doubted whether a domestic tribunal is the place to conduct international negotiations. There is a danger that, once more, the USA is
taking steps which it construes as leadership, but may be construed
by others as bullying.

New services

The controversy surrounding the 'separate systems' has had two
effects – one practical and the other legal. INTELSAT has been stimu-
lated to introduce a variety of new services, and has begun steps to
alter its pricing policies.

As to the first, various new services have been developed by
INTELSAT over the years. The major development of the 1980s has
been the ''Business Services' category, the establishment of which is
not unconnected with the criticisms of INTELSAT referred to in the
FCC Separate Systems Report and Order of 1985, and which is to an
extent a pre-emptive reaction to the possibility of competition from
the separate systems.[185] Of course, although INTELSAT provides a
service, it is for the common carriers providing telecommunications

services within national jurisdictions to decide whether and how to market the new services, and that affects the success of a new service. The potential of pricing flexibility discussed below is significant, and the introduction of INTELSAT Business Services has not been without effect. PanAmSat sought to influence the US development of a position within INTELSAT on business services through an FCC application which was, however, unsuccessful,[186] and has sought to challenge COMSAT's participation in the necessary modification of INTELSAT satellites.[187] PanAmSat did, however, succeed in challenging some of COMSAT's (and through it, INTELSAT's) lobbying expenses.[188] That sort of skirmishing is indicative that INTELSAT's response has affected possible competitors, and it is clear that the new services offered by INTELSAT has attractions for common carriers which may well prefer to use it rather than the new systems.[189]

The other point that is relevant here is that there is likely to be a glut of telecommunications capacity afforded by the existing international satellite organizations, particularly with the entry of INTELSAT VI series and the new fibre optic cable links. This also will affect the economics of the separate satellite systems, and may well result in many plans remaining just that,

Pricing flexibility

By art. V(d) of the INTELSAT Agreement the 'rates of space segment utilization charge for each type of utilization shall be the same for all applicants for space segment capacity for that type of utilization'. Once a particular service or utilization has been established, all users, without discrimination on a geographical or personal basis, are to be changed the same rate for each unit of utilization. Certainly this logical so far as geographic questions are concerned, for access to and through a satellite makes distance an insignificant variable. As a matter of principle it is also important that there is no discrimination on the basis of the nationality of the user or origin of the use. Of course, the actual end-cost to the user will vary, for different telecommunications authorities making use of the INTELSAT segment may charge differently for the total 'call' or 'connection', incorporating different elements into their own pricing. But the INTELSAT element remains the same in all utilization.

The White Paper on International Telecommunications which was the basis of the Presidential Determination on Separate Systems took account of pricing questions,[190] and correspondence between the Secretary of State and the Secretary of Commerce indicated that the White House had directed these Departments to examine the scope of INTELSAT's pricing flexibility as an element in the criteria to be adopted under the Presidential Determination.[191] However, pricing did not appear as a criterion, but was reserved for further discussion. In brief, under US law the pricing of telecommunications is supervised by the FCC and is closely related to cost. 'Whip-sawing' (the subsidization of one service by another or one route by another) and the unjustifiable passing on of costs to particular end-users is not permitted. The US government on the one hand is concerned that INTELSAT price its services in accordance with US practice so that US end-users are benefited. If necessary competition with INTELSAT will aid that purpose.[192] On the other hand, it is concerned that INTELSAT does not cripple the new separate systems through offering unduly competitive rates.[193]

Much therefore turns on the extent to which INTELSAT can adjust its prices. Although what eventually counts for the user is the charge which the carriers make for services, the INTELSAT segment rate is an element of that charge. An INTELSAT view has been that it cannot compete effectively with the proposed separate systems because of the requirement implicit in art. V(d) that it average its rate for particular services on worldwide basis. It cannot, for example, separately set its transatlantic rate on telephone traffic in order to offset telephone costs elsewhere. A separate system, by restricting its operation to the transatlantic route, would be able (perhaps) to offer lower rates.

As to the law on the matter, David Leive, INTELSAT's Legal Advisor, produced a Legal Memorandum on the matter, which was backed up by three later Opinions from leading Washington law firms.[194] In short, the view of the Memorandum and the Opinions is that under art. V(d) INTELSAT must apply the same charges for the same type of utilization on a worldwide basis for all users. The computation of the charge for a type of utilization is not mechanical. As art. 8 of the Operating Agreement requries, and as developed by the First Meeting of Signatories (MS-1-6), INTELSAT charges are cost-related so far as practicable, each type of utilization making, however, an appropriate contribution to INTELSAT's revenue requirements. However, a charge need not fully recover the cost of providing that type of service: 'cost-related' does not mean 'cost-determined'. Such flexibility allows INTELSAT to depart from basing the charge for a particular utilization of the system on a strict calculation of the cost of its provision. It is permitted to take into account what is practicable in the case of a given type of utilization and what is desirable in overall INTELSAT terms. The Board's power to establish types of utilization is therefore crucial. Working from the Agreements and the Minutes of the Plentipotentiary Conference, the Leive Memorandum demonstrates that the Board's practice is well-founded. By this, many different types of utilization have been established, and operational parameters have been taken into account – for example, through restricting certain utilizations to or outside Primary or Major Paths, and through the use of conditions as to whether a service is pre-emptible or non-pre-emptible.[195] Not all services need be offered in every ocean region although, when offered, a service is available without discrimination to all users. Further utilization categories can be created through reference to voice, TV, data, power, band-width, transponder use and so on. However, a differential charge within an offered service, based on the geographic nature of a link or the user, is not possible within the INTELSAT structure. Nor could INTELSAT abuse its power to establish types of utilization to nullify the prohibition on discrimination.

The Leive Memorandum therefore indicated that INTELSAT has a cetain flexibility in setting its charges through the use of different types of utilization subjected to different restrictions and conditions. INTELSAT could not, however, offer a special price on a particular service over a particular route in order to respond to competition from a separate satellite system. To give an exact example: had PanAmSat's initial request to provide transatlantic service been acceded to by the FCC, INTELSAT could not alter its transatlantic prices on the same services, unless these services were available only on the transatlantic route (which they are not). It had to average its rate on a worldwide basis. PanAmSat, offering a more geographically restricted service and not sharing INTELSAT's global revenue requirement, might have been able to offer a cheaper rate, and INTELSAT would have been unable to respond by reducing its transatlantic rate in order to compete.

Thus, the obvious question is whether this lack of flexibility (which does not amount to total inflexibility) is a defect of the INTELSAT Agreements which should be changed. It is therefore interesting that the US State Department issued a paper 'Flexibility to Compete: INTELSAT in an Era of Separate Systems' which, selectively using data from the Leive Memorandum and the Opinions, sought to demonstrate that INTELSAT had sufficient flexibility to cope with competition as things were.

The INTELSAT Executive Organ picked substantial holes in the paper.[196] The State Department considered INTELSAT was clearly using market principles in setting of rates for a vastly increased range of types of utilization on individual routes, and that this amounted to sufficient flexibility for the organization to meet competition from the proposed separate systems. The response of the Executive Organ stated that this was not what was done in INTELSAT rate-setting, and that the principle of non-discrimination, explicitly incorporated into the Agreements, remained and would not improperly be circumvented. I must say I agree with INTELSAT. Whether INTELSAT *should* be able to offer different rates for the same service is a separate question. As matters stand, INTELSAT is bound in law to offer the same rate for the same service to all destinations within the range of a satellite's coverage, although it is not bound to offer the same services from each satellite. Whether INTELSAT should be able to offer different rates for the same service on different routes in order to meet competition is a different matter.

Steps have been taken to alter the legal position. On 5 July 1985 Cameroon and Tanzania submitted proposals for the amendment of art. V(d), that of the former two Parties specifically dealing with the question of discrimination, as did Colombia three days later. Their proposal would have deleted the current art. V(d) and replaced it with the following:

> The Board of Governors shall establish utilization charges taking into account factors such as ocean region areas, satellite(s), time of access, priority of access, characteristics of satellites, earth station characteristics, length of lease. The Board shall establish space segment utilization charges at the same rate for each type of utilization, except when it determines that it is in the best interests of INTELSAT to deviate from this charging principle including, but not limited to, instances where this is necessary to meet competition in various ocean regions or traffic routes.[197]

That statement does reflect, to a degree, previous Board practice in the establishment of new types of utilization, and clearly also copes with differential pricing in response to competition.

The proposal went before the Tenth Assembly of Parties in October 1985.[198] That Assembly also had on its agenda various documents relating to the FCC proceedings on separate systems and the US legislative developments outlined above. 'Following a lengthy and detailed discussion of the issues involved during which nearly every representative present spoke',[199] the matter of the amendment of art. V(d) was deferred to the Eleventh Assembly, the Director-General and Board being asked in the interim to take account of existing flexibilities, the Meeting of Signatories being asked for its views, and others given the opportunity to submit proposals on the matter.[200] No doubt developments continue.

NOTES

1   See Chapter 3, subsection 'Withdrawal' p. 109.
2   Cf. What has happened to various defence contractors in the UK and the USA in the 1980s, following on charges of excess profits.
3   'Report on Application of US Munitions Control Regulations to INTELSAT Programs' BG–42–52, Contribution of the Director-General, 4 June 1980.

4 *INTELSAT Inventions and Data Distribution Procedures*, BG–29–61E (Rev. 1, 18 March 1981.

5 In international law 'good offices' and 'mediation' are terms of art indicating different procedures, the former being largely the conveying of messages in their attempt to solve the matter and the latter involving a more active attempt to bring a solution through making proposals and engaging in persuasion and argument.

6 See Chapter 3, subsection 'Suspension and Expulsion', p. 111.

7 'Review of Certain Obligations of INTELSAT Members Under the INTELSAT agreements, with particular reference to Article XIV(d)', BG–60–62, 13 August 1984.

8 Convention on the Law of Treaties, Vienna 23 May 1969 (1980) UKTS 58, Cmnd. 7964; (1969) 8 ILM 679).

9 See section 'Other Systems' below, p. 153.

10 Non-payment of capital contributions is dealt with specially by art. XVI(c).

11 See Chapter 3, section 'Membership, withdrawal, suspension and expulsion' p. 109, especialy the subsection dealing with 'Withdrawal' and 'Suspension and Expulsion'. A Party is dealt with by the Assembly of Parties, which has a discretion to expel (art. XVI(b)(i)). A Signatory's rights may be suspended by the Board of Governors (art. XVI(b)(ii)), or it may be deemed to have withdrawn by the Meeting of Signatories (art. XVI(b)(ii)).

12 Cf. J.G. Merrills, *International Dispute Settlement* (London: Sweet and Maxwell, 1984).

13 One text in my possession has an interesting misprint: the tribunal has power to 'construct' its judgement.

14 The power of the International Court is substantially contained in art. 60 of its Statute.

15 Cf. in relation to the USA's obligations under the UN Headquarters Agreement: Applicability of Obligation to Arbitrate under Section 21 of the United Nations Headquarters Agreement of June 26, 1947, 1988 ICJ Rep. 12.

16 See sec. Chapter 3, subsection 'Suspension and Expulsion', p. 111.

17 See below section 'Other Systems', particularly subsections, 'Domestic services', 'Specialized Services', and 'International public telecommunications services' and the 'Art. XIV(d) coordinations: practice' pp. 157–170. See also David Leive, Legal Advisor to INTELSAT, 'Legal Opinion on the Scope of INTELSAT's Public Telecommunications Services' (Leive Memorandum), 13 January 1984, Attachment No. 1 to 'Scope of INTELSAT's Public Telecommunications Services', 10 August 1984, BG–60–48.

18 Cf. Leive Memorandum, op cit.

19 Cf. *inter alia* 'Request from the Signatory of Colombia for an Article III(b)(ii) Determination', DG, BG–38–30, 15 May 1979 regarding the Bogota–Leticia service (Leticia is across the Andes and down in the Amazon basin), and the same captioned BG–48–46, 9 November 1981, which deals with connections to other regional centres.

20 'Request from the Signatory of Peru for an Article III(b)(ii) Qualification', DG, BG–35–49, 21 November 1978, and the same captioned BG–47–77 of 8 September 1981. The main services requested are between centres of population in the jungle areas.

21 'Request from the Signatory of Mexico for the Provision of Space Segment Capacity', BG–41–69 and relative discussion BG–41–3, 172, and the same captioned DG, BG–45–57, 26 February 1981, together with 'Request from the Signatory of Mexico for an Article III(b)(ii) Determination', BG–45–34, 17 March 1981.

22 See 'Advice tendered by the Board of Governors to the Second Meeting of

Signatories Pursuant to Article X(a) (xxiv) of the Intergovernmental Agreement in Respect of Domestic Public Telecommunications Services in Brazil', BG–8–44. MS–2–12. 26 March 1974.

23 See 'Request from Denmark for an Article III(b)(ii) Qualification', DG, BG–37–39, 22 February 1979.

24 See 'Applicability of Article XIV(d) to Domestic Services falling under Article III(b) of the Agreement' with its Attachment on 'Legal Aspects', DG, BG–15–52, 24 April 1975. The history of the provision is summarized at para. 3 of the Attachment. Cf. subsections 'International public telecommunications services, art. XIV(d)' and 'Art. XIV(d) coordinations: practice' pp. 159 and 171 respectively.

25 Cf. 'Separate Services under Article III(e) of the INTELSAT Agreement', DG, 8 March 1982, BG–50–63; 'Establishment and Operation of a Separate Space Segment under Article III(e), DG, BG–52–45, 24 August 1982.

26 See the 'Separate Systems Decision': 'In the Matter of Establishment of Satellite Systems Providing International Communications', Report and Order, CC Docket No. 84–1299, adopted 25 July 1985, released 3 September 1985, (1985) 101 FCC 2d 1046, at paras. 52–55, 64–81, 82–86.

27 Cf. 'Provision of Domestic Services', DG, BG–42–39, 4 June 1980, which indicates that pre-emptible leases of transponders had been approved for Algeria, Brazil, Chile, Colombia, France, India, Malaysia, Nigeria, Norway, Oman, Peru, Saudi Arabia, Sudan and Zaire. Leases had been approved in principle for Argentina, Australia, Colombia, Denmark (Greenland) and Egypt. Inquiries had been received from Bolivia, Cameroon, Chad, Chile, Libya, Mauretania, Mexico, Morocco, Niger, Pakistan, Portugal and Thailand.

28 Under the Indefeasible Right of User (IRU) concept, a right is in effect bought to use a particular channel of communication by contributing a sum approximating to a proportionate share of the cost of construction of the channel, plus a sum for maintenance and (sometimes) administation. The IRU holder is not responsible for the construction, nor in law is a part-owner of the system. Its advantage is that an IRU is not 'defeasible' – in satellite terms the right to use a transponder cannot be 'pre-empted' if the system requries additional capacity on a short or long-term basis.

29 See M. Perras, 'Development of New INTELSAT Services' in J.R. Alper and J.N. Pelton (eds), *The INTELSAT Global Satellite System*, vol. 93, 'Progress in Astronautics and Aeronautics', (New York: American Institute of Aeronautics and Astronautics, 1984) pp 255–68.

30 As, for example, in the case of argument about COMSAT before the FCC: see the discussion of COMSAT, Chapter 2 above at n. 115.

31 The UN had attempted to get favourable charging from the ITU at the Buenos Aires ITU Conference: cf. 'Use of the United National Telecommunication Network for the Telegraph Traffic of the Specialised Agencies', Res. No. 35 of the Malaga-Torremolinos ITU Plenipotentiary Conference.

32 'United Nations Request for Free Access to INTELSAT Space Segment', Contribution of the Secretary General, BG–7–29, 17 January 1974.

33 Minutes of the eighth meeting of the INTELSAT Board of Governors', BG–8–3, April 1974, para. 53.

34 'Provision of Telecommunications Services to the United Nations', DG, BG–52–60, 26 August 1982 and relative discussion, BG–52–3).

35 See 'Provision of Telecommunications Services to the United Nations' with Attachment No. 1, 'The United Nations as a Duly Authorised Telecommunications Entity', DG, BG–53–74, 7 December 1982; 'Provision of Telecommunications Services to the United Nations' DG, BG–54–67, 23 February 1983, together with relative Board discussions, BG–53–3,122 and BG–54–

3,122–3. The UN was negotiating service through the Symphonie satellite system.

36 INTERSPUTNIK is more fully discussed in Chapter 7.

37 Cf. Status Report(s) on Intersystem Coordination: USSR Networks, SG/DG, BG–19–51, BG–25–27, BG–29–22, BG–34–36. The Report on the Twelfth Intersystem Coordination Meeting between INTELSAT and the USSR, DG, B6–76–23, 29 April 1988, shows the process continuing.

38 Cf. arts. 14 and 15 of the Operating Agreement.

39 Memorandum of Understanding between INTELSAT and the Ministry of Posts and Telecommunications of the USSR, Attachment 1 to the same titled INTELSAT document, DG, BG–64–59, 27 August 1985. See also Chapter 7 below.

40 Cf. I Cheprov, 'Global or American Space Communications System?' 1964 (December) International Affairs (Moscow) pp. 69–70.

41 Donald Fink, 'Europe Unifying Policy for INTELSAT Talks', 87 *Aviation Week and Space Technology*, no. 87, 27 November 1967, pp. 69–70.

42 See also Chapter 6, 'Europe' and its section, 'The European Telecommunications Satellite Organization', pp. 264.

43 As to the negotiation of art. XIV and a discussion of its main elements, see R.R. Colino, *The INTELSAT Definitive Arrangements: Ushering in a New Era in Satellite Telecommunications*, (Geneva: European Broadcasting Union, 1973) 88-89.

44 'In the Matter of Establishment of Satellite Systems providing International Communications', CC Docket No. 84–1299: September 2 1985 Released; Adopted July 25, 1985, (1985) 101 FCC 2d 1046. See below, subsection, 'US Separate Systems', p. 182.

45 The Board deals with such matters initially through its Technical Committee (BG/T). The procedures are contained in 'Article XIV Intersystem Coordination Procedures (Technical Compatability)' (BG–28–70), of 29 June 1977, as revised particularly by 'Review of Article XIV Technical Consultation Procedures' 3 November 1982 (BG53–111), producing the composite BG–28–70 (Rev. 1; 14 December 1982). BG–53–111 also modifies 'INTELSAT Guidelines for Intersystem Coordination', 18 September 1980 (BG–43–71). 'Transmission Characteristics of the INTELSAT System for Use in Article XIV Consultation Procedures' are contained in BG–45–83 of 23 March 1981.

46 See Chapter 8, section 'Space Frequencies', p. 357.

47 See above, section 'Settlement of Disputes', p. 141.

48 See 'Article XIV(c) Consultation for SATCOL Satellite Network', DG, BG–34–37, 29 August 1978.

49 Cf. 'Article XIV(c) Consultation for the CS–2 Satellite System', DG, BG–44-30, 11 November 1980; 'Article XIV(c) Re-Consultation for the CS–2 Satellite System', DG, BG–53–38, 8 November 1982, for part of the process. On the experimental stage see, "Coordination of a Japanese Domestic Experimental Communications Satellite System with the INTELSAT System', Contribution of Japan, BG–6–38, 7 December 1973 which antedates the setting-up of the Executive Organ.

50 Cf. 'Intersystem Coordination Status Report, ANIK B-1 Space Station of the Telesat B–1 Satellite Network', DG, BG–29–23, 16 August 1977; 'Article XIV(c) Consultation, ANIK C–1 and C–2 Space Stations of the Telesat C–1 and C–2 Satellite Networks' DG, BG–30–14, 24 October 1977; 'Article XIV(c) Consultation for the Telesat D–1 Satellite Network', DG, BG–44–48, 14 November 1980, for part of the process.

51 'Article XIV(c) Consultation for the AUSSAT Satelite System', DG, BG–48–63.

52 Cf. 'Intersystem Coordination Status Report, INSAT-1A Network', DG, BG–42–34, 3 June 1980, which contains as Attachment 1 the 'Agreement between the Government of India and INTELSAT concerning Coordination of INSAT and

INTELSAT Satellite Networks in the Indian Ocean Region', and 'Consultation for the INSAT 1B Network of the INSAT Satellite System under Article XIV(c) and Article XIV(e)', DG, BG–43–28, 22 August 1980. The INSAT system was also coordinated for specialized services; see below.

52a Cf. 'Article XIV(c) Consultation for the WESTAR III US Domestic Satellite Network', DG, BG–41–27.

53   'Article XIV(c) Consultation for the UNISAT-1 Domestic FSS Network', DG, BG–52–52, 25 August 1982; 'Further Consultations under Article XIV(c) Between the UK Signatory and INTELSAT Concerning the UNISAT-1 Domestic FSS Network' DG, BG–53–60, 19 November 1982. Cf. 'Offer of Space Segment Capacity by the UK Signatory', DG, BG–53–69, 7 December 1982; 'UNISAT-1/ INTELSAT Operating Agreement' [a draft], DG, BG–54–25, 28 January 1983.

54   See 'Coordination of ARABSAT System under Article XIV(c)', DG, BG–41–68, 12 March 1980; 'Article XIV(c) Consultations for the Malaysian Domestic Satellite System using PALAPA B Space Segment Facilities', DG, BG–43–50, 10 September 1980; 'Article XIV(c) Consultations for the Philippines and Thailand Domestic Satellite Systems using PALAPA B Space Segment Facilities', DG, BG–43–61, 15 September 1980 on technical compatibility, and BG–43–48 on economic harm. BG–37–38 dealt with the use by these countries of the PALAPA A satellite). On ARABSAT see subsection on 'ARABSAT' p. 175 below and Chapter 7, Part II, p. 303. On PALAPA, see subsection 'PALAPA' p. 172 below.

55   The procedures are contained in the documents cited above, n. 47.

56   See 'Intersystem Coordination Status Report, INSAT-1A Network', DG, BG–42–34, 3 June 1980, which contains as Attachment 1 the 'Agreement between the Government of India and INTELSAT Concerning Coordination of INSAT and INTELSAT Satelite Networks in the Indian Ocean Region'; 'Coordination of the INSAT Satellite Network (Meteorological Services) Article XIV(e), DG, BG–34–6, 31 August 1978; 'Article XIV(e) Consultation for the Television Broadcasting Service of the INSAT-1A Satellite Network, DG, BG–42–35, 3 June 1980; and 'Consultation for the INSAT 1B Network of the INSAT Satellite System under Article XIV(c) and Article XIV(e)', DG, BG–43–28, 22 August 1980. The INSAT system was also coordinated for domestic public telecommunications services, see above.

57   'Consultation for the GMS-2 Satellite Network under Article XIV(e)', DG, BG–43–27, 18 August 1980.

58   'Consultation under Article XIV(e) for the TV-SAT Broadcasting-Satellite System', DG, BG–43–56, 4 September 1980, and 'Intersystem Coordination Status Report, TV-Sat Network', DG, BG–47–86, 9 September 1981.

59   'Article XIV(e) Consultation for the SARIT Broadcasting Satellite Network', DG, BG–52–47, 24 August 1982.

60   'Article XIV(e) Consultation for the LUXSAT Broadcasting Satellite System', DG, BG–52–25, 12 August 1982.

61   'Article XIV(e) Consultation for the AUSSAT Broadcasting Satellite Network', DG, BG–51–85.

62   'Article XIV(e) Consultation for the UNISAT-1 Broadcasting Satellite Network', DG, BG–52–21, 5 August 1982; 'Article XIV(e) Technical Consultations for the BSB Broadcasting Satellite Network', DG, BG/T–63–37 and BG–73–58 (Rev.1) 11 August 1987.

63   The Manual is issued by the External Relations Department of INTELSAT, and makes use of the Attachments to AP–10–35 on 'Article XIV(d) Coordination.' Cf. 'Policies, Criteria and Procedures for the Evaluation of Separate Systems under Article XIV(d), 22 August 1984 (BG–60–69) (which narrates procedures and attempted to systematize criteria found in previous practice) and its Addendum No. 2 of 14 December 1984, 'Application of New Criteria (BG–60–69)

to Certain Previous Article XIV(d) Coordinations' which shows that, with a minor qualification relating to lack of data on parts of the ECS and PALAPA systems, these and the US – Canada transborder coordinations would have received favourable findings under the BG–60–69 formulations of the appropriate criteria. See also subsection 'Art. XIV(d) coordinations: practice' p. 171 below.

64 See applications by Orion Satellite Corporation and International Satellite, Inc. discussed in subsection 'US separate systems; p. 182.

65 Attachment No. 1 to 'Scope of INTELSAT's Public Telecommunications Services', BG–60–48, 10 August 1984.

66 Davis R. Robinson Legal Advisor, US State Department, Memorandum of Law, 'The Orion Satellite Corporation and International Satellite, Inc. Applications for International Satellite Communications Facilities', attached to letter by G.P. Schulz, Secretary of State, and M. Baldrige, Secretary of Commerce, to Mark S. Fowler, Chairman, FCC, 28 November 1984, concerning Presidential Determination No. 85–2: available as Appendix B to Attachment A of 'US Government Executive Branch Paper on the Subject of Separate International Communications Satelite Systems', BG–62–40; MS–15–14, 12 March 1985; and, Attachment No. 2 to 'US Federal Communications Commission Notice of Inquiry and Proposed Rulemaking: Establishment of Satellite Systems Providing International Communications', CC Docket No. 84–1299, AP–9–24, 21 January 1985.

67 Procedures, BG–28–70 (Rev. 1) 18 March 1981. See above n. 46–7 for a discussion of technical coordination under art. XIV.

68 The 'Procedural Manual for Consultation Under Article XIV(d) of the INTELSAT Agreement' is issued by the External Relations Department of INTELSAT, and makes use of the Attachments to AP–1–35 on 'Article XIV(d) Coordination'.

69 'Procedural Manual', op. cit., Attachment no. 2 to AP–10–35, p. 2.

70 Attachment No. 1 to 'Proposed Procedures for Implementation of Article XIV(d) Requirements Concerning Prejudice to Direct Telecommunications Links,' BG–60–61, 15 August 1984.

71 'Policies, Criteria and Procedures for the Evaluation of Separate Systems under Article (XIV(d)),' BG–60–69, 22 August 1984, paras. 51–64.

72 Subject to an appeal to the procedures which the Agreements set up for the settlement of disputes. See section 'Settlement of Disputes', above.

73 Leive Memorandum op. cit., p. 4: Colino, 'Policies, Criteria ...' op. cit., p. 20, para. 52.

74 Leive Memorandum, op. cit., p. 4: Colino, 'Policies, Criteria . . . op. cit., p. 20, para. 56.

75 Leive Memorandum, op. cit., p. 4: Colino, 'Policies, Criteria . . . op. cit., p. 20, para. 56.

76 For detail see Chapter 8 sections, 'Space Frequencies' p. 357, 'The Geostationary Orbit' p. 387, 'WARC–ORB 1985–88' p. 393.

77 Leive Memorandum, op. cit., p. 7.

78 Cf. I.B. Schwartz, 'Pirates or Pioneers in Orbit? Private International Communications Satellite Systems and Article XIV(d) of the INTELSAT Agreements', Boston College International and Comp. Law Review, vol. 9, 1986, pp. 199–242, 220 n. 203.

79 Leive Memorandum, op. cit., p. 7–8.

80 See section 'Settlement of Disputes' above, p. 141.

81 See above at n. 64; and below subsections 'Art. XIV(d) coordinations: practice' p. 171 and subsection 'US Separate Systems', p. 182 and articles there cited at n. 178.

82   Board of Governors BG–28–63. That document was couched in very general terms and has now been superseded by AP–10–33 (BG–64–78)(Rev. 1).
83   See 'Negotiating History', s.III (pp. 2–3) of David Leive's 'Legal Memorandum: Scope of the Article XIV(d) Assurance Concerning 'Direct Telecommunications Links', Attachment No. 1 to BG–60–61, 'Proposed Procedures for Implementation of Article XIV(d) Requirements Concerning Prejudice to Direct Telecommunications Links.' Cf. I.B. Schwartz, 'Pirates or Pioneers in Orbit? Private International Communications Satellite Systems and Article XIV(d) of the INTELSAT Agreements', (1986) 9 *Boston College International and Comp. Law Review*, no. 9, 1986, pp. 199–242 at 219–20.
84   See AP–1–5 and Summary Minutes, AP–1–4 para. 19; AP–2–11 and Summary Minutes, AP–2–4 paras. 93–98.
85   See below, subsection 'US separate systems', p. 182.
86   Walter Hinchman Associates, *The Economics of International Satellite Communications*, BG–59–34, 18 May 1984, *Significant Economic Harm*, BG–60–63, 15 August 1984; *International Satellite Competition Impact Analysis*, BG–62–20, 25 January 1985.
87   See 'The Economics of International Satellite Communcations', BG–59–34, pp. 62–6; 'Significant Economic Harm', BG–60–63, pp. 21–2.
88   Board of Governors, BG–60–69, 22 August 1984.
89   Cf. the US Signatory's reaction, 'Comments on proposals for the Establishment of Policies, Criteria and Procedures for the Evaluation of Separate Systems under Article XIV(d) of the INTELSAT Agreement (BG–6–69), BG–62–54, 22 March 1985.
90   Tenth Assembly of Parties, AP–10–4, paras. 34–5.
91   The procedure is contained in 'Procedural Manual for Consultation under Article XIV(d) of the INTELSAT Agreement', issued to all Parties and Signatories, May 1986.
92   Procedures Adopted by the Board of Governors for Non-Technical Consultations under Article XIV(d) of the INTELSAT Agreement, AP–10–33, with Attachment 1, 21 September 1985.
93   Tenth Assembly of Parties, AP–10–4, para. 35.
94   DG, BG–60–69, 22 August 1984.
95   'Article XIV(d) Consultation concerning Potential Economic Harm to INTELSAT by Algeria's Planned use of the INTERSPUTNIK System', DG, BG–43–43, 28 August 1980; 'Article XIV(d) Technical Coordination for the use of the INTERSPUTNIK System by Algeria', DG, BG–43–51, 4 September 1980.
96   Sixth Assembly of Parties, AP–6–1, pp. 23–4, and relative minutes, AP 6–4, pp. 108–16.
97   Cf. 'Technical Compatibility Between the Use of the INTERSPUTNIK System by the UK and INTELSAT Networks', BG/T–65–22; 'Assessment of Technical Compatibility Between the Use of INTERSPUTNIK System by the Syrian Arab Republic and INTELSAT Networks', DG, BG–73–34, 5 August 1987.
98   The system is fully described in its original form in the Indonesian entry in the *Eighteenth Report by the International Telecommunications Union on Telecommunication and the Peaceful Uses of Outer Space*, (ITU Report) ITU Booklet no. 25 (Geneva: 1979) pp. 128–48.
99   The ITU Report, op. cit., states in a note at p. 131 that 'PALAPA' is taken from the Javanese phrase 'Amukt Palapa', an oath sworn by Prime Minister Gajah Mada (1319–1364). Palapa means the fruit of the effort and 'amukti' means enjoyment. The oath means that the Prime Minister would rest only when Indonesia unity was accomplished. PALAPA therefore refers to the unity of Indonesia.
100 'Status of Intersystem Coordination Under ITU Regulations – Indonesian

Domestic Satellite System', Contribution of the Secretary-General, BG–16–48, 1 July 1975.

101 'Coordination Between the INTELSAT System and the Indonesian Domestic Satellite System', MSC, BG–16–50, 2 July 1975. Cf. 'Indonesian Domestic Satellite System Status', Contribution of the Southeast Asia Group, BG–16–67, 11 July 1975.

102 We have noted the coordinations for Malaysia, the Philippines and Thailand under art. XIV(c) which are part of that development: see above, Article XIV(c).

103 Cf. 'Article XIV(d) Consultation Concerning Technical Compatibility of the Proposed PALAPA Network (Indonesia)', DG, BG–37–38, 23 February 1979.

104 Fourth Assembly of Parties, AP–4–1, p. 9 with Minutes, AP–4–4, pp. 30–5 and 59.

105 Sixth Assembly of Parties, AP–6–1, p. 12 with Minutes, AP–6–4, pp. 64–90.

106 Seventh Assembly of Parties, AP–7–1, p. 30 with documentation, AP–7–22.

107 Seventh Assembly of Parties, AP–7–1, p. 34 with documentation, AP–7–25.

108 See also 'Article XIV(d) Consultation concerning Potential Economic Harm to INTELSAT by the Planned Use of Domestic Satellite Systems to Extend Telecommunications Services between Canada and the United States', DG, BG–52–17, 7 September 1982; Article XIV(d) 'Consultation Concerning Potential Economic Harm to INTELSAT by the Planned Use of the RCA SATCOM Satellite System by Bermuda', DG, BG–52–64, 2 September 1982; 'Article XIV(d) Consultation for the Use of US SATCOM III–R and US SATCOM IV Satellite Networks to provide Service to Bermuda and the Cayman Islands', DG, BG–52–54, 25 August 1982.

109 Board of Governors. BG–51–3, p. 180.

110 See Transborder Satellite Video Services (1981) 88 FCC 2d 258.

111 Cf. 'In the Matter of Hughes Communications Galaxy, Inc.; For Authority to make available to Turner Broadcasting System, Inc. Television Channels of Communication via the Hughes Communications Galaxy, Inc. Satellite Galaxy I for Transmissions to Various Transborder Locations', Order and Authorization, (1987) 2 FCC Rcd 4536. adopted 23 July 6 1987, released 4 August 1987.

112 In the Matter of Overseas Telecommunications, Inc.: For Authority . . . etc', Memorandum Opinion, Order and Authorization, (1987) 2 FCC Rcd 5438, adopted 19 August 1987, released 4 September 1987, at para 12. The note to that paragraph recites the 13 prior major transborder decisions as being Transborder Satellite Video Services, 88 FCC 2d 258 (1981) (Transborder I); Eastern Microwave, Inc., I–P–C–81–049 et al., Mimeo No. 2617, released 1 March, 1983 (Transborder II); American Telephone and Telegraph Company, File Nos. I–P–C–82–048 et al., Mimeo No. 6119, released 26 August 1983 (Transborder III); Bonneville Satellite Corp., et al., File Nos. I–T–C–83–148 et al., Mimeo No. 1554, released 29 December 1983 (Transborder IV); Western Union Telegraph Co., el at., File Nos. I–T–C83–068, et al., Mimeo No. 3286, released 4 April 1984 (Transborder V); Eastern Microwave, Inc., et. al., File Nos. I–T–C–84–095 et al., Mimeo No. 6425, released 11 September 1984 (Transborder VI); Western Union Telegraph Co., et al., File Nos. I–T–C–83–136, et al., Mimeo No. 3643, released 8 April 1985 (Transborder VII); American Satellite Company, et al., File Nos. I–T–C–85–187 et al., Mimeo No. 7299, released 30 September 1985 (Transborder VIII) RCA American Communications, Inc., et al., File Nos. I–T–C–002 et al.,Mimeo No. 2578, released 14 February 1986 (Transborder IX); Hughes Galaxy, Inc., File No. I–T–C–83–101, Mimeo No. 2472, released 11 February 1985; Satellite Signals Unlimited, Inc., et al., File Nos. I–T–C–81–069 et al., Mimeo No. 3519, released 29 March 1985; RCA American Communications, Inc., et al., File Nos. I–P–C–81–038 et al., Mimeo No. 4294, released 6 May 1985; and National Cable and Satellite Corp., et al., File Nos. 417–DSE–ML–86 et al., Mimeo No. 5220, released

20 June 1986 (Transborder X); and American Telephone and Telegraph Company, *et al.*, File Nos. I–T–C–-86–105 *et al.*, DA No. 86–26, released 9 October 1986 (Transborder XI); and RCA American Communications, Inc., *et al.*, File Nos. I–T–C–86–175 *et al.*, DA 87–273, released 23 March 1987 (Transborder XII); GTE Spacenet Corp., *et al.*, File No. I–T–C–86–103, DA 87–444, released 20 April 1987 (Transborder XIII); Washington International Teleport, Inc., File No. I–T–C–87–086, DA No. 87–555, released 10 July 1987, Cf. BG–73–17 to 21 and 24–31 comprising art. XIV(c) and XIV(d) technical consultations for Galaxy II, Spacenet II, Westar, Comstar III, Satcom II–R, Telstar 303, Galaxy B, Galaxy IV, Comsat Gen–B, Fedex A and B, Gstar III and Westar B satellite networks.

113   See, for example, the discussion in the Transborder I proceeding, Transborder Satellite Video Services (1981) 88 FCC 2d 258 at paras. 45–52 and the letter dated 23 July 1981 from James L. Buckley, Under-Secretary of State for Security Assistance, Science and Technology to Mark Fowler, Chairman of the FCC, Appendix A to that proceeding (1981) 88 FCC 2d 258 at 287–89. Cf. for a US Court's View: *Communications Satellite Corporation, Petitions v. FCC and USA, Respondents*, (1988) 836 F. 2d 623.

114   Subsection 'US Separate Systems,' p. 182 below.

115   See 'Provision of Maritime Services by INTELSAT', DG, BG–43–42, 28 August 1980 and its attached draft Lease Contract between INTELSAT and INMARSAT. The final contract was not precisely on those terms.

116   See for MARISAT, First Assembly of Parties, AP–1–3,19 with AP–1–5, Second Assembly of Parties, AP–2–3,17–18, with AP–2–11, Seventh Assembly of Parites, AP–7–3,27–29, with AP–3–14 and AP–7–21: for MARECS, Seventh Assembly of Parties, AP–7–3,24–26 with AP–7–20. The Minutes of the discussions (AP–7–8) make interesting reading.

117   Chapter 6 below.

118   See with appropriate Minutes, Fourth Assembly of Parties, AP–4–3, paras. 6–8 with AP–7; Sixth Assembly of Parties, AP–6–3, paras. 17–19 with AP–6–17. See also 'Article XIV(d) Consultation Concerning Potential Economic Harm to INTELSAT by the Planned European Communications Satellite System (ECS)', DG, BG–34–27, 30 August 1978; 'Intersystem Coordination Status: Regional European Communications Satellite System (ECS) (Technical Compatibility), DG, BG–34–41, 31 August 1978; 'Article XIV(d) Consultation Concerning Potential Economic Harm to INTELSAT by the Planned European Communications Satellite System (ECS)', DG, BG–35–62, 27 November 1978; 'Article XIV(d) Technical Consultation: ECS Primary Network of the ECS System', DG, BG–37–16, 27 February 1979; 'Article XIV(d) Consultation Concerning Potential Economic Harm to INTELSAT by the Planned European Communications Satellite System (ECS)', DG, BG–37–34, 22 February 1979; 'Report of the Board of Governors to the Assembly of Parties Pursuant to Article XIV(d) Concerning Coordination of the European Communications Satellite System', AP–4–7, 16 March 1979.

119   Cf. 'Article XIV(d) Consultation Concerning Potential Economic Harm to INTELSAT by the Planned European Communications Satellite System', DG, BG–52–41, 20 August 1982.

120   Seventh Assembly of Parties, AP–7–3, paras. 31–3 with AP–7–24.

121   Seventh Assembly of Parties, AP–7–3, paras, 31–3 with AP–7–24. See also 'Article XIV(d) Consultation for the EUTELSAT 1–2 (Spare) Network of the ECS System', DG, BG–43–17, 22 August 1980; 'Status of Article XIV(d) Technical Consultation Concerning the Expansion of the European Communications Satellite System', DG, BG–52–46, 24 August 1982.

122   See 'Further Information on the EUTELSAT/INTELSAT Article XIV(d) Con-

sultation', Interim EUTELSAT Council, Contribution of the General Secretariat, ECS/C 23–58, 21 September 1982.

123 'Extension of Article XIV(d) Technical Consultation for the EUTELSAT Network', DG, BG–73–82 (Rev. 1), 21 August 1987.

124 See Chapter 7, Section 'ARABSAT' p. 303 below.

125 'Intersystem Coordination under Article XIV(d)', Contribution of Arab Group II, Arab Group III and United Arab Emirates,' BG–37–51, 12 March 1979.

126 'Article XIV(d) Informal Technical Consultation for the ARABSAT System', DG, BG–38–21, 20 April 1979.

127 'Article XIV Consultation Concerning the Planned ARABSAT System', DG, BG–37–23.

128 'Article XIV Consultation Concerning the Planed ARABSAT System', DG, BG–40–54, 19 November 1979.

129 'Report of the Board of Governors to the Assembly of Parties Pursuant to Article XIV Concerning Coordination of the Arab Communications Satellite System (ARABSAT),' AP–5–8 (BG–41–51), 14 March 1980.

130 Fifth Assembly of Parties, AP–5–3, 5–7 with relative Minutes, AP–5–4.

131 Fifth Assembly of Parties, AP–5–3,5(a)(ii). The 'Terrestrial Transmission Links in the Arab Network' is Attachment No. 4 to AP–5–8.

132 See below, subsection, 'US Separate Systems', p. 182.

133 ''In the Matter of the Establishment of Satellite Systems Providing International Communications', (CC Docket No. 84–1299, adopted 25 July 1985, released 2 September 1985; (1985) 101 FCC 2d 1046.

134 Pan American Satellite Corporation, (1985) 101 FCC 2d 1318; later modified, FCC 86–257, released 21 May 1986.

135 See sub-section INTELSAT reaction p. 188 below.

136 See below, subsection, 'US Separate Systems', p. 182.

137 See 'Request for Partial Consultation of a Satellite Under Article XIV of the INTELSAT Agreement', DG, BG–67–64, 17 June 1986.

138 See subsection 'International public telecommunications services: art. XIV(d)', p. 000 above.

139 'Report of the Board of Governors to the Eleventh Assembly of Parties Pursuant to Article XIV(d) on the PanAmSat Network', AP–11–10, 20 February 1987.

140 Eleventh Assembly of Parties, AP–11–3, and relative Minutes, AP–11–4.

141 'In the Matter of Pan American Satellite; Request for a Final Permit for the Construction, Launch and Operation of a Subregional Western Hemisphere Satellite System', Order, Slip Opinion, FCC 87–310, adopted 28 September 1987, released 30 September 1987. See also p. 412 for subsequent developments.

142 Cf. the attitude of the USSR to INTELSAT itself and at the initial stages of INMARSAT, both described elsewhere.

143 'In the Matter of Tel-Optik Ltd: In the Matter of Submarine Lightwave Cable Co., Memorandum, Opinion and Order, adopted 1 March 1985, released 5 April 1985, Release no. FCC 85–99; 100 FCC 2d 1033 (1985) (The Private Submarine Cable decision).

144 See: For the Atlantic: FCC Docket No. 79–184, the most recent in series being: 'Inquiry into the Policies to be Followed in the Authorization of Common Carrier Facilities to meet North Atlantic Telecommunications Needs during the 1991–2000 Period', (1987) 2 FCC Rcd 2088; For the Pacific: FCC Docket No. 81–343, the most recent in the series being: 'Inquiry into the Policies to be Followed in the Authorization of Common Carrier Facilities to meet Pacific Telecommunications Needs during the period 1981–1995', (1985) 102 FCC 2d 353. For prior considerations see FCC Docket No. 18876, 'Policies to be Followed in Future Licensing of Facilities for Overseas Communications', the last in series being (1979) 73 FCC 2d 326.

145 Cf. Yang-Soon Lee, 'Competition between Fibre-optic cables and satellites on Transatlantic Routes', *Telecommunications Journal* no. 34, 1987, pp. 809–14 for a recent analysis, and FCC proceedings for the North Atlantic and Pacific and the Tel-Optik cables cited above.

146 'Policies for Overseas Common Carriers', (1980) 82 FCC 2d 407.

147 'In the Matter of Inquiry into Policies to be Followed in the Authorization of Common Carrier Facilities to Meet North Atlantic Telecommunications Needs During the 1985–1995 Period', FCC Docket No. 79–184, Second Report and Order, adopted 7 August 1985, released 22 August 1985, FCC Release No. 85–456.

148 'In the Matter of Policy for the Distribution of United States International Carrier Circuits Among Available Facilities During the Post-1988 Period', CC Docket No. 87–67, Notice of Proposed Rulemaking, (1987) 2 FCC Rcd 2109; Report and Order, adopted 24 March 1988, released 14 April 1988, Release No. FCC 88–122 (1988).

149 Cf. Richard R. Colino, "A Chronicle of Policy and Procedure: The Formulation of the Reagan Administration Policy on International Satellite Telecommunications', *Journal of Space Law*, vol. 13, 1988, pp. 103–56.

150 FCC File no. CSS–83–002–P.

151 FCC File no. CSS–83–04–P (LA).

152 FCC File no. I–T–C 84 085.

153 FCC File no. CSS–84–002–P (LA).

154 FCC File no. CSS–84–004–P (LA).

155 FCC File no. 114–DSS–P/LA–84.

156 The SIG was composed of representatives of the Departments of State, Commerce, Justice, Defense; the Offices of Management and Budget, Science and Technology Policy, Policy Development and the US Trade Representative; the National Security Council; the Central Intelligence Agency; the US Information Agency; the Board for International Broadcasting; the Agency for International Development; and the National Aeronautics and Space Administration.

157 'A White Paper on New International Satellite Systems', released February 1985, Attachment No. 1 to 'US Government Executive Branch Paper on the Subject of Separate International Communications Satellite Systems, DG, BG–62–40, MS–15–14, 12 March 1985.

158 Presidential Determination No. 85–2, Public Papers of the Presidents, 20 Weekly Compilation of Presidential Documents 1853, 49 Federal Register 46987, 30 November 1984.

159 Cf. Dissenting Opinion of Lord Atkin in *Liversidge v. Anderson* [1942] AC 206 at 245.

160 The Schultz–Baldrige letter and the Memorandum is Appendix B to 'US Government Executive Branch Paper on the Subject of Separate International Communications Satellite Systems', DG, BG–62–40, MS–15–14, 12 March 1985.

161 'In the Matter of Establishment of Satellite Systems providing International Communications', CC Docket No 84–1299; Release No. FCC 84–632, (1985) 100 FCC 2d 290, adopted 19 December 1984, released 4 January 1985. Also available as Attachment No. 2 to 'US Federal Communications Commission, Notice of Inquiry and Proposed Rulemaking: Establishment of Satellite Systems providing International Communications', CC Docket No. 84–1299, DG, AP–9–24, 21 January 1985.

162 'In the Matter of Processing of Pending Applications for Space Stations to provide International Communications Service', FCC 85–296, adopted 31 May 1985, released 6 June 1985.

163 'In the Matter of Establishment of Satellite Systems Providing International Communications', Report and Order, CC Docket No. 84–1299, (1985) 101 FCC 2d 1046, adopted 25 July 1985, released 3 September 1985.

164 Cf. A. Conan Doyle, *Silver Blaze* – the case where Sherlock Holmes notes as important that the dog did not bark in the night.
165 Cf. 'The Orbital Plan – Domestic Satellites: In the Matter of Assignment of Orbital Locations to Space Stations in the Domestic Fixed-satellite Service', (1980) 84 FCC 2d 584; Release No. FCC 80–711, adopted 4 December 1981, released 30 January 1981: as revised by Licensing of Space Stations in the Domestic Fixed-Satellite Service, (1983) 48 Fed Reg. 40233, which reduced orbital spacing in the light of the ITU Administrative Conference for Region 2 (See Chapter 8, sec. I subsections 'The World Broadcasting-Satellite Administative Radio Conference, Geneva, 1963', p. 360 and 'Developments regarding frequencies to WARC–ORB 1985–88, p. 380.
166 'Long-term' was to be further interpreted: see below at note 173.
167 International Satellite, Inc. (1985) 101 FCC 2d 1201; Pan American Satellite Corporation, (1985) 101 FCC 2d 1318.
168 RCA American Communications, Inc., (1985) 101 FCC 2d 1342. RCA later gave up this permission.
169 Cygnus Satellite Corporation, (1985) 101 FCC 2d 989; Orion Satellite Corporation, (1985) 101 FCC 2d 1302.
170 Orion Satellite Corporation, Mimeo No. 6871, released 6 September 1985; Cygnus Satellite Corporation, Mimeo No. 0362, released 22 October 1985.
171 See Financial Satellite Corporation, FCC Release No. 86–13, unpublished, released 15 January 1986; Columbia Communications Corporation, FCC Release No. FCC 86–536; In the Matter of McCaw Space Technologies Inc., Application for Authority to Construct, Launch and Operate a Two-region Satellite System, adopted 5 Jaunary 1987, released 15 January 1987, Release No. FCC 87–-1, (1987) 2 FCC Rcd 259.
172 'In the Matter of Establishment of Satellite Systems Providing International Communications . . ., Memorandum Opinion and Order, CC Docket 84–1299, adopted 3 April 1986, released 17 April 1986, Release No. FCC 86–144.
173 Letter dated 10 January 1986 from Diana Lady Dougan, Coordinator and Director of the Bureau of International Communications and Information Policy, Department of State and from Rodney L. Joyce, Acting Assistant Secretary for Communications and Information, Department of Commerce (the 'Dougan–Joyce' letter).
174 Establishment of Satellite Systems Providing International Communication . . ., CC Docket no. 84–1299: Memorandum Opinion and Order, adopted 23 October 1986, released 5 November 1986, Release No. 86–471, (1986) 1 FCC Rcd 439.
175 See the discussion of PanAmSat above in relation to art. XIV(d) (subsection 'PanAmSat–Peru', p. 177), and: the PanAmSat Initial Authorization Order, 'In the Matter of the Application of Pan American Satellite Corporation; for Authority to Construct, Launch and Operate a Subregional Western Hemisphere Satellite System Consisting of Two In-Orbit Satellites', Memorandum Opinion and Order, adopted 25 July 1985, released 3 September 1985, FCC Release No. FCC 85–398: the PanAmSat (Modification) Order, 'In the Matter of an American Satellite Corporation; Application for Modification of Conditional Authority to Construct a Subregional Western Hemisphere Satellite System', Memorandum Opinion, Order and Authorization, adopted 16 May 1986, released 21 May 1986, FCC Release No. 86–257: and the PanAmSat (Final Permit) Order, the PanAmSat (Modification) Order, 'In the Matter of an American Satellite Corporation; Application for a Final Permit for the Construction, Launch and Operation of a Subregional Western Hemisphere Satellite System', Order, adopted 28 September 1987, released 30 September 1987, FCC Release No. 87–-310, (1987) 2 FCC Rcd 7011.

176 Richard R. Colino, 'A Chronicle of Policy and Procedure: The Formulation of the Reagan Administration Policy on International Satellite Telecommunications', (1985) 13 *Journal of Space Law*, vol. 13 1985, pp. 103–56 at 116–19, 130–2, 136–43, 146–53. See also 'Events in the United States Relating to Separate International Communications Satellite Systems', DG, BG–64–76, 30 August 1985, with its Attachments Nos. 1 and 2; and the similarly titled document, BG–65–17, AP–10–39, 4 October 1985, and its Attachment No. 1.

177 'Foreign Policy Implications of Competition in International Telecommunications', *Hearings before the Subcommittees on International Operations and on International Economic Policy and Trade of the Committee on Foreign Affairs, US House of Representatives*, 99th Cong. 1st Sess., 19 February and 6 and 28 March 1985; 'International Communication and Information Policy', *Hearings before the Subcommittee on Arms Control, Oceans, International Operations and Environment of the Committee on Foreign Relations, US Senate*, 98th Cong. 1st Sess., 19 and 31 Oct. 1983. Cf. the letters printed as Attachment 3 to Events in the United States Relating to Separate International Communications Satellite Systems, BG–65–17, AP–10–39, 4 October 1985, and Attachment 2 to the similarly titled BG–64–76 of 30 August 1985. Cf. also the list of 69 objectors (mainly INTELSAT members and PTTs) Appendix A to the PanAmSat Initial Authorization Order (1985) 101 FCC 2d 1318. Cf. also the unanimous resolution of the Thirteenth (1983) and Fourteenth (1984) INTELSAT Meeting of Signatories and those of the Eighth (1983) and Ninth (Extraordinary (1985) Assembly of Parties. Cf. also a stream of INTELSAT Comments on US Policy Statements some of which are printed as Attachment 3 to BG–64–76 (above) and as Attachment 3 to AP–10–39 (also above), Attachment 1 to BG–63–37 of 7 June 1985 which attacks seventeen 'myths' about INTELSAT and competition. Cf also 'A Critical Review of Responses' in the US FCC Proceeding on Separate Satellite Systems, Attachment 1 to 'FCC Proceeding on Separate Satellite Systems' DG, BG–63–35, 31 May 1985.

178 See L.A. Caplan, 'The Case For and Against Private International Communications Satellite Systems', *Jurimetrics Journal*, 1986, pp. 180–201, J.C. Glassie, "Analysis of the Legal Authority for the Establishment of Private International Communications Satellite Systems', *George Washington Journal of International Law and Economics*, vol. 18, 1984, pp. 355–91; J. McKenna, 'Bypassing INTELSAT: Fair Competition or Violation of the INTELSAT Agreement?', *Fordham International Law Journal*, vol. 8, 1985, pp. 479–512; B.W. Rein, B.L. McDonald, D.E. Adams, C.R. Frank and R.E. Neilsen, 'Implementation of a US "Free Entry" Initiative for TransAtlantic Satellite Facilities: Problems, Pitfalls, and Possibilities', *George Washington Journal of International Law and Economics*, vol. 18, 1985, pp. 459–536; S.Z. Chiron and L.A. Rehberg, 'Fostering Competition in International Telecommunications', *Federal Communications Law Journal*, vol. 38, 1986, pp. 1–57; I.B. Swartz, 'Pirates or Pioneers in Orbit? Private International Communications Satelite Systems and Article XIV(d) of the INTELSAT Agreements', *Boston College International and Comp. Law Review*, vol. 9 (1986), pp. 199–242.

179 Chapter 3, section 'Rights and Obligations of Members', pp. 106–108.

180 See letter of 23 July 1981 from James L. Buckley, Under-Secretary of State for Security Assistance, Science and Technology to Mark Fowler, Chairman of the FCC, Appendix A to the FCC Transborder Satellite Services Decision, Transborder Satellite Video Services, etc., Memorandum Opinion, Order and Authorization, adopted 22 October 1981, released 30 October 1981, Release No. FCC 81–492, (1981) 88 FCC 2d 258.

181 See Chapter 3 Section, 'Rights and Obligations of Members, particularly pp. 106–109.

182   Cf. the argument as to the ABM Treaty, exemplified in *ABM Treaty Interpretation Dispute,* Hearing before a Subcommittee on Arms Control, International Security and Science of the Committee on Foreign Affairs, US House of Representatives, 99th Cong. 1st Sess., 22 Oct. 1985.

183   In the Matter of Regulatory Policies and International Telecommunications, CC Docket No. 86–494, Notice of Inquiry and Proposed Rulemaking, adopted 23 December 1986, released 30 January 1987, Release No. FCC 86–563, (1987) 2 FCC Rcd 1022. Two attempts to extend the period for the filing of comments were denied: that by the Information and Telecommunications Technology Group of the Electronic Industries Association (1987) 2 FCC Rcd 2425; that by COMSAT, (1987) 2 FCC 2d 2977 and 3232.

184   'In the Matter of Regulatory Policies and International Telecommunications', CC Docket No. 86–494, Report and order and Supplemental notice of Inquiry, adopted 25 February, released 25 March 1988, Release No. FCC 88–71, (1988) 64 Rad. Reg. 2d (P & F) 1976. See also Chapter 9, p. 420 and n. 31.

185   M. Perras, 'Development of New INTELSAT Services' in J.R. Alper and J.N. Pelton, (eds) *The INTELSAT Global Satellite System,* vol. 93, 'Progress in Astronautics and Aeronautics', (New York: American Institute of Aeronautics and Astronautics, 1984) pp. 255–68.

186   'In the Matter of Communications Satellite Corporation Participation in INTELSAT's Planed Domestic Services', Memorandum Opinion and Order, adopted 25 November 1987, released 30 November 1987, FCC File No I–S–P–88–001.

187   'In the Matter of Pan American Satellite Corporation, Complainant v. Communication Satellite Corporation, Defendant', Memorandum Opinion and Order, adopted 5 August 1986, released 20 August 1986, FCC Release No. FCC 86–347, FCC File No. I–S–P 85–003, an unsuccessful application by PanAmSat for review of Mimeo No. 1685 released 31 December 1985.

188   Pan American Satellite Corporation, v. Communications Satellite Corporation, Mimeo No. 4988, released 6 June 1985; Memorandum Opinion and Order, adopted 6 October 1986, released 15 October 1986, FCC Release No. FCC 86–422, (1986) 1 FCC Rcd 111.

189   Cf. 'In the Matter of RCA American Communications, Inc.; Application for Authority to Establish Channels of Communication between the US and INTELSAT Atlantic and Pacific Satellites for Provision of INTELSAT Business Services and International Television Service', Memorandum Opinion, Order and Authorization, adopted 5 August 1987, released 13 August 1987, (1987) 2 FCC Rcd 4751.

190   'A White paper on New International Satellite Systems', February 1985, available as Attachment 1 to BG–62–40, MS–15–14, at pp. 27–9 and 49–50.

191   Letters printed as Appendix A to BG–62–40, MS–15–14, above.

192   Cf. the Separate Systems Report and Order, op. cit., (1985) 101 FCC 2d 1046 paras. 195–9.

193   It is amusing to find that, the US having 'encouraged' INTELSAT to be competitive, PanAmSat has complained that INTELSAT has been so: 'In the Matter of Policy for the Distribution of United States International Carrier Circuits Among Available Facilitites During the Post-1988 Period, CC Docket No. 86–87, adopted 24 March 1988, released 14 April 1988, Release No. 88–122, (1988) 3 FCC Rcd 2156 at para. 14.

194   David M. Leive, 'Legal Memorandum: Determination of INTELSAT Space Segment Utilization Charges', Attachment 1 to 'Legal Opinion Concerning the Determination of INTELSAT Space Segment Utilization Charges', DG, BG–61–67, 19 December 1984; AP–9–18, 4 January 1985. Attachments 2, 3 and 4 are the

opinions respectively of Messrs Arnold and Porter, Messrs Wiley and Rein and Messrs Ginsburg, Feldman and Bress. The opinions are on more restricted questions, such as must be asked in an opinion.

195   A pre-emptible service is one which INTELSAT, at its option, may interrupt if the circuits are needed for another purpose. Obviously it costs less than a non-pre-emptible service.

196   The State Department paper is Addendum 1 to 'INTELSAT's Legal Flexibility to Compete', DG, BG–63–39, 5 June 1985.

197   'Proposed Amendments to Article V(d) of the INTELSAT Agreement', DG, BG–64–51, 23 August 1985. See also the same titled AP–10–37, 2 October 1985 where explicatory letters by the proposers are available. The Colombian proposal had to do with special rates for programmes of high humanitarian and social content, including education in developing countries.

198   See also 'Proposed Amendments to Article V(d) of the INTELSAT Agreement: Substantive Aspects', DG, AP–10–40, 1 October 1985, which also contains as Attachment 1 the Leive Legal Memorandum and Opinions on the determination of utilization charges BG–61–67, together with the State Department's paper on INTELSAT's flexibility to compete and the Executive Organ's response, BG–63–39.

199   Quoted from 'Report to the Board of Governors on the Tenth Assembly of parties', DG, BG–65–47, 21 November 1985, para. 3.

200   Tenth Assembly of Parties, AP–10–4, paras. 32–3.

5 The International Maritime Satellite Organization

INTRODUCTION

The International Maritime Satellite Organization, known as INMARSAT, is much smaller than INTELSAT although it similarly provides a global telecommunications service. As implied by its name, however, at its inception the service INMARSAT was to provide was intended to meet the communications needs of maritime traffic, in effect as an application of satellite technology to the existing terrestrial maritime mobile services. The pending extension of the INMARSAT service to air mobile and perhaps in the future also to land mobile services will not require any major growth of the organization nor will it necessarily make it more complex. Because it deals with a smaller constituency INMARSAT will always remain the junior international organization providing global satellite telecommunications. Of course, this comparison ignores INTERSPUTNIK, but that is justifiable. While INTERSPUTNIK offers a global service, in practical operation its system is apparently restricted.

INMARSAT came into being on 16 July 1979. The next 30 months or so were a hectic period of discussion and negotiation before the organization began to operate commercially on 1 February 1982. It is now well on the way to owning (as opposed to leasing) its own satellite system.

INITIAL DISCUSSIONS

The history of INMARSAT is intriguing.[1] Although there had been discussion within INTELSAT about the provision of a maritime service through INTELSAT facilities, and although such a service would have lain within INTELSAT's competence,[2] the initiative which lies behind the establishment of INMARSAT originated within what was then the Intergovernmental Maritime Consultative Organization (IMCO), now the International Maritime Organization (IMO), the specialized agency of the United Nations concerned with maritime affairs.[3] IMO's function is the fostering and facilitation of cooperation between governments, and in the regulation of international shipping and maritime transport.

In 1966, following upon the work of an expert committee, and in the light of Interim INTELSAT, IMCO came to the conclusion that a satellite communications system devoted to the needs of marine transport would be of significant benefit. IMCO was encouraged to continue in its studies by the 1967 World Administrative Radio Conference which dealt with maritime matters,[4] and in 1971 the World Administrative Radio Conference for Space Telecommunications specifically allocated certain frequencies[5] to a Maritime Mobile-Satellite Service.[6]

In 1972 IMCO established a panel of experts to take matters further. In the six years since the first formal suggestion of a maritime system had been made within IMCO, the global communications system created by Interim INTELSAT had demonstrated the potential of satellite communications. There is also a curious analogy with the development of 'normal' radio. The advantages of 'wireless' communications with ships were underlined by the *Titanic* disaster in 1912, and much progress was made as a result of that tragedy.[7] Safety of life at sea remains a major international preoccupation,[8] and clearly a satellite system would have many advantages in providing a distress and emergency facility. It was therefore not surprising that it was the IMCO Maritime Safety Committee which, in March 1973, decided to recommend that an international conference be held in 1975 to decide upon the principle of establishing a Maritime Communications Satellite System, and, if possible, to agree more specifically on the structure of any organization needed for the implementation of such a service. The IMCO Assembly agreed to this recommendation in November 1973,[9] and the IMCO Panel of Experts, which was already busy at work, was able to proceed in the knowledge that such a conference would be held.

The actual conference was held in three sessions, from April 1975 to September 1976, with major contributions being made by inter-sessional working groups.[10] The result was agreement on the basic instruments of INMARSAT. Also adopted were other Resolutions and Recommendations which have significance in that they express opinions, which although not binding on the new organization, are nonetheless indicators of how the organization's founders considered that matters should develop.

Two distinct sets of questions had, and have, to be considered in relation to the setting up of INMARSAT. First, why is a maritime

satellite system required at all? Second, why is a separate organization needed for the operation of the system?

First, why have a maritime satellite system? Maritime communications serve a variety of functions. Some are technical, relating to the
safe and economic use of the seas. Others are matters of ordinary
public communications.

As indicated in the treatment of the creation of the ITU, the first
impetus towards putting radio on ships was directed towards the
safety of life at sea. Safety continues to be an important purpose of
the satellite system: the distress message, summoning help in emergency, the provision of search and rescue, are all matters where a
good radio link is essential.[11] Short of extremity, the provision of
meteorological data and of radiolocation, and the monitoring and
enforcement of rules for separation of traffic at sea and the traffic
advice (not to say instruction) that is required of shipping lanes in
congested areas such as the English Channel, all these are largely
dependent on radio communication. The satellite service can often
provide a much clearer link than ordinary terrestrial radio.

But, if radio were to be available through satellites, needs other
than navigational could be met by the facility. When radio was introduced, shipowners and the shippers of cargo found the ability to
control the destination and arrival of ships, together with up-to-date
news of any problems and deviation from plan, was useful. Satellites
would augment that facility.

Of course, all these matters could to an extent be served by the state
of the art in radio prior to the coming of satellites. However, the non-
satellite maritime radio system had, and still has, defects. Both the
high-frequency and the medium wavebands are congested. Despite
the best efforts of technology the frequencies set aside for maritime
commerce are close to saturation. Despite the efforts of the ITU and
the maritime nations, there were problems of interference. The reception of messages could be very difficult, and in some areas of the high
seas virtually impossible. These difficulties were compounded in the
short waveband (the only band suitable for longer-range communication) by natural hazards. Frequent radio blackouts, interference
and fading are caused by changes in ionospheric propagation
characteristics. Voice communication, requiring a better signal than
that needed for data traffic (telex and morse code), always suffers
most under these conditions. Living in a fishing port as I do, one has
only to listen in to what passes for communication between trawlers
and their head offices down at the harbour to understand that the
radio service available in the days before satellite was not good.

Furthermore, only a limited service is in practice available for ships
at sea through the non-satellite systems. The ship-to-shore communications network is by no means global in its coverage and where it is

provided, it is often not available on a 24-hour basis. A full service is maintained by shore stations only in the areas of major shipping traffic. Outside these areas a ship can communicate or be contacted only within specified hours. Once a time difference is added, the inconvenience of the sporadic availability of the ordinary radio service of, say, the early 1960s, can be marked.

Many of these problems are rendered irrelevant by a properly designed and operated satellite system, which therefore provides significant benefits. At an early stage in the IMCO discussions it was clear that a satellite system would relieve congestion in both the medium and high frequency bands. Such a system would improve the reliability, quality and speed of communications, improve geographic coverage, and provide a continuously available service. Obviously it would also particularly improve the distress and emergency communication system. Satellites could also be used to make provision for ships to fix their position by radio determination with a very high degree of accuracy. And, in addition, because these prior needs of the maritime radio services would be better met, further developments were possible, both on board ship and between ship and shore. For example, the radio telephone services could be automated and unskilled persons could use the system as simply as one dials a telephone number on-shore.[12] Again, the teleprinter service could be enhanced, and high-speed data transmission between ship and shore and vice versa could be established. With such a system the performance of a ship, and of anything on board, could be monitored from on-shore through the transmission of data from sensors on board the ships. Thus engineering problems could be diagnosed, surveyed and dealt with at an early stage by specialists who would not need to be carried on the ship itself. For example, the condition of refrigerated goods could be kept under constant review from head office, and so on.[13] Transmitting in the opposite direction, constant and detailed data on weather and sea conditions could also be fed to a ship so that it could be advised of minor course corrections to avoid storms which would both delay it and increase its fuel consumption (a matter of increasing importance since the oil crisis of the mid-1970s).[14]

Of course, the weighting afforded to the different factors, problems and benefits of a maritime satellite system varied greatly amongst the many participants in the 1975–76 IMCO conference. But, even if many considered the extent to which the running of a ship could be left to supervision by data link to be a somewhat hypothetical advantage (although it has proved not to be so), it was clear that a satellite system would better provide for the then existing needs of maritime communications. These benefits alone were considered quite sufficient to justify going ahead with the establishment of such a system.

That left the second question, whether a separate organization was the best vehicle for the development of maritime services by satellite. INTELSAT was already in existence, its definitive arrangements having entered into force in 1974 and it having operated in its interim form from 1964. It had the experience. Elegance and Occam's Razor would have dictated that it provide and operate the new system. A logical alternative was for some existing commercial communications entity to provide a system which shipping companies would simply use, like any communications common carrier charging for its services on a time/distance/mode/traffic basis. The decision actually taken resulted in a separate international organization, INMARSAT. This was not a decision taken lightly or without much discussion and argument, but with hindsight there does appear to have been an inevitability about it. It has its own logic, that of securing the divergent interests (or the perceived interests) of a large group.

Despite the logical alternatives, it was apparent throughout all the preparatory studies for the maritime satellite system that some special institutional arrangement or structure would have to be worked out. As had been the case of the negotiations prior to the setting up of Interim INTELSAT, it was clear that states were unwilling to buy or to permit shipping companies under their jurisdictions to buy a communications service from a single commercial communications company, even though they recognized that a company such as the COMSAT might well have the expertise to provide such a service.[15] Indeed, following on both the interim and definitive INTELSAT negotiations, this point was so clear that there was never a stage in the formal discussions which led to the INMARSAT Agreements at which the provision of a maritime telecommunications service by a single company incorporated in any state was considered as a viable or even a formally discussable proposal.

Chronologically, the earliest discussion within the IMCO Panel of Experts centred on whether there was a need to have an international organization charged with the matter, or whether a solution might be the provision of the new facility by a consortium of national and international entities including existing telecommunications carriers. At the time, of course, the interim INTELSAT arrangements were providing a model, and there was provision within INTELSAT's structure of that time for the INTELSAT system to continue as a consortium. Not surprisingly, the USA therefore suggested that the provision of the maritime satellite system should be entrusted to a similar consortium.[16] However, even as that idea was being floated, the model had itself chosen a different development. The definitive arrangements for INTELSAT had progressed to that curious mixture of intergovernmental organization and an international commercial telecommunications organization, a hybrid within which existing telecommunications entities (ranging from departments of state to purely commercial entities) play major roles. It would have been strange indeed if INMARSAT had adopted a solution rejected by its pathfinder. But, that apart, it should be acknowledged that the involvement of governments and telecommunications administrations in the interim structure of INTELSAT had left them both with a feel for such matters, and an unwillingness to let the reins of such developments pass from them.

Within the IMO Panel of Experts, therefore, there was no support for the consortium solution. That solution could not have been made acceptable to the intended participants in the system which were departments of state, or in other cases to the governments which stood behind the telecommunications entities. And in any event, as in the case of INTELSAT, there were advantages in governmental participation in an international organization. First, governments would be less likely to take exception to, or be obstructive in, the implementation of decisions in which they had been involved. Second, with government involvement, and therefore backing, finance for the new system might be somewhat easier to come by. This was an important factor in such a high-cost enterprise, which was also made high-risk, at least in its initial years, by the chances of launcher and satellite failure.

In the light of the attitudes of members of the Panel of Experts, the USA then suggested that a suitable way of proceeding would be to establish the system through some sort of 'users' group' which could be contained within the constitutional structure of IMCO itself. Such a users' group would have a supervisory role with limited powers, while the actual contracting for, and the construction and operation of, the system would be dealt with by entities with a proven record in such matters. This proposal had attractions, although transparently it meant that the technology and spin-off advantages would be confined to those countries with an already healthy space capability. (In short, COMSAT would have had the inside track.) The suggestion, however, ran into the more basic problem of the nature and purposes of IMCO itself. The primary purpose of IMCO was indicated in its then title, the Intergovernmental Maritime Consultative Organization. It was a consulting organization and its main thrust was within the area of marine safety through the fostering of cooperation and harmonization. The new facility would certainly have an impact on safety, but clearly safety was a minor category of wide-ranging elements. The satellite link, although important for safety, was likely to be used far more for other purposes. Indeed, there was even a fear that the nascent organization could prove to be a cuckoo within the IMCO nest, forcing out other worthy IMCO causes in a competition for finance. The majority therefore considered it inappropriate that IMCO should house such an organization.

The third option then proposed by the USA was that the new service should be provided as a specialized service by INTELSAT. The power for INTELSAT to provide such specialized services was one of the fruits which the USA had obtained in the renegotiation of the INTELSAT Agreements.[17] It was also certainly true that by this time INTELSAT had proved itself to be an efficient provider of satellite telecommunications services and clearly had the operational and structural capability to move into the provision of such services had it so wished. Nonetheless there was a widespread feeling that INTELSAT, an organization providing public telecommunications facililties between countries, was perhaps too broad an organization properly to exercise financial and other policy controls over the specific needs of a maritime satellite system. Tucked away within this argument, although not made properly explicit, was the simple point that the USSR and many other communist countries were not (and are not) members of INTELSAT. These countries are major shipping nations, and obviously their participation in the maritime satellite system was desirable. The system would lose much of its point were ships of these major nations not to use the system because, for one reason or another, their states of registration did not wish to be members of organization providing the system. Had the system been provided by INTELSAT that boycott might have been a real possibility.[18] Indeed, the material of this difficulty went even more broadly than the attitudes of the communist countries and could affect even INTELSAT members.

The problem lies with the structure of INTELSAT. For all practical purposes, that structure vests the highest authority in the Board of Governors. Were INTELSAT to move into the maritime satellite business, major decisions would be taken by a Board on which non-INTELSAT members (notably the USSR), would not be represented, and on which some INTELSAT members with a major interest in maritime communications would play a very small role. Norway, for example, has a large fleet but a small INTELSAT share. Again, many developing countries are building up their maritime commerce and their own fleets.[19] None of these countries, however, has sufficient international telephone traffic originating within their jurisdiction to give them a large investment share in INTELSAT. The addition of their maritime traffic would not significantly alter that fact. Consequently, their voice within INTELSAT would be insignificant. One has only to look at the list of investment shares in INTELSAT to see the problem – a problem which is only marginally alleviated by the provision within the INTELSAT Agreement for representative Governors. In their view, the interest of such countries in maritime matters would not be properly reflected within an organization for which maritime messaging is a secondary responsibility. Finally there was the point that many interests other than those of telecommunications had to be given some place. Maritime communications involve shipowners, ship operators, maritime unions and a whole variety of sea users, ranging from pleasure craft to ocean liners, tankers to fishing vessels; and above these interests there is a superstructure of regulation and supervision ranging from the appropriate national government organ to the International Maritime Organization itself.

As discussions continued, it was therefore apparent that the interests of the various states, telecommunications entities and those active in maritime communications were sufficiently different from their interests in ordinary public international telecommunications to justify different solutions to the organizational problems presented by the new services. Accordingly, the creation of a new intergovernmental organization came to be seen as the best solution. The proposal for such an organization was made by the USSR early in the discussions of the panel of experts, and the draft Convention it submitted became the basis of negotiations. The advantage for the USSR could be argued to be its place amid the maritime nations which secure it a role and a voice in the new organization, which it could not have had within INTELSAT. No doubt that element was not insignificant, just as the potential for leadership for the USA was not irrelevant in the proposals it made. But be that as it may, the fact is that the special nature of both the problems and the clientele for a maritime communications satellite system were seen sufficiently cogent to result in what superficial reaction might dismiss as yet another international organization of dubious justifiability.

Naturally the existence of the INTELSAT Agreements, and the negotiations which had gone into both the interim and the definitive INTELSAT arrangements provided both models and warnings for the solution of particular problems in INMARSAT. Indeed, it is interesting to see that many of the difficulties which were encountered in the INTELSAT negotiations emerged again in the INMARSAT negotiations. For example, there was much argument as to the division of responsibilities between the organs of the international organization. Again, there were problems as to the division and allocation of financial burdens and the calculation of investment shares.[20] And, of course, procurement policy remained a matter of sensitivity.

Another question was fundamental. What should membership of the organization consist of? Some participants in the negotiations considered that only states should be members of an international organization. For example, the USSR and other communist countries initially had doctrinal objections to an international organization which was a mixed state–company enterprise. Non-state entities (the telecommunications administrations) could be given a status akin to that of the 'recognized operating agencies' which exist within the structures of the ITU.[21] That view could not, however, be sustained. It is true that in both the USA and Japan – countries whose participation in the new organization was essential – the provision of telecommunications facilities was in the hands of private enterprise. Indeed, in the USA the Communications Satellite Corporation had been created by government at a private company separate from government for the very purpose of being the USA's participant in arrangements for the provision of international telecommunications by satellite. Such countries could take part in the work of a regulatory organization such as the ITU with the state as the member of the Union, while the telecommunications entities provide personnel for the various ITU study groups and soon. But such a division of responsibilities was not a possibility when dealing with an organization which was itself to have operational competence. The appropriate telecommunications entities had to be more closely built into the decision-making structures.

In short, therefore, when the alternatives were thoroughly explored, there was really no viable alternative to the INTELSAT model. The eventual solution had to provide an arena for satisfactory cooperation between the diversity of constitutional arrangements found amongst states' provision for telecommunication. This lay once more in the linking of an intergovernmental agreement dealing with constitutional structures and the operating agreement between telecommunications entities. There is room for much variation of species, but the genus is clear.

THE CONVENTION AND OPERATING AGREEMENT

The result of the Preparatory Conference was the drawing up of the INMARSAT Convention and its relative Operating Agreement which were opened for signature at London on 3 September 1976.[22] It is notable that an extremely high level of ratification was required for the Convention to come into effect. States holding no less than 95 per cent of the initial investment shares had to deposit ratifications with the depositary for the convention (the Secretary General of IMCO (now IMO) (art. 35(1)) before the Convention could enter into force. Mathematically, the failure of any of the USA (initial investment share 17 per cent), the UK (12 per cent), the USSR (11 per cent), Norway (9.5 per cent), Japan (8.45 per cent) or of any two of Italy (4.37 per cent), France (3.5 per cent), West Germany (3.5 per cent), Greece (3.5 per cent), Holland (3.5 per cent), Canada (3.2 per cent) and Spain (2.5 per cent) could have blocked the coming into force of the INMARSAT agreements,[23] and the potential for a fatal refusal to ratify was also present in various configurations of the small shareholders.[24] But this did not happen: the 95 per cent level was reached in May 1979 and the Convention came into force on 16 July 1979 in accordance with art. 33 of the Convention, only some 45 days before the Convention and Agreement would have lapsed in terms of art. 33(2).

No reservation can be made either to the Convention or to the Operating Agreement (art. 32(50)). In the period between the signing of the two Agreements and the organization becoming formally operational, a maritime satellite system giving partial global coverage was established and operated by COMSAT, the US Communications Satellite Corporation. That system proved the utility of the embryonic organization and helped to eliminate any remaining doubts. Whether it had commercial effects as well as a matter of argument.

The amendment process for the Convention is contained in art. 34.[25] It begins through a proposal by any Party to the Convention being communicated to the Directorate, which informs other Parties. Three months' notice is required before a proposal is considered by the INMARSAT Council, which then has up to six months from the date on which the proposal was circulated by the Directorate to Parties in which to submit its views to the Assembly of Parties. The Assembly considers the matter within a further six months, or in a lesser period if it so decides, such a decision being a substantive decision for the purposes of its Rules of Procedure (art. 34(1)). Naturally, since the Convention is an international treaty, it is necessary that the various Parties to the original document ratify any changes. If the Assembly adopts the proposal it enters into force 120 days from the time that the Depositary (the Secretary-General of the IMO (art. 35(1)) receives notices of acceptance from two-thirds of the Parties holding two-thirds of the total investment shares at the time of the adoption of the proposals (art. 34(2)). There is no time-limit beyond which an amendment adopted by the Assembly expires if it has not received the appropriate level of acceptances. On entry into force, an amendment binds all Parties and Signatories including those which have not accepted it (art. 34(2)). This moderately unusual provision is of course necessary for the constituent document of an organization such as INMARSAT.

The process and timetable for the amendment of the Operating Agreement is contained in art. XVIII and reflects that for the Convention with minor variation. Thus a Party or Signatory may propose an amendment to the Operating Agreement – only Parties may deal with the Convention. Proposed amendments are circulated to Signatories, not Parties. The Assembly confirms the proposal; it does not adopt it. There is no time-limit for the ingathering of acceptances of an amendment. Only the appropriate Party may transmit the consent of its designated Signatory to a proposed amendment which has been confirmed by the Assembly, and that transmittal signifies the Party's acceptance of the amendment. There is the same requirement of acceptance by two-thirds of the Parties holding two-thirds of the total investment shares, and the same binding of all Signatories by an amendment which has entered into force.

The two agreements have been modified under these procedures extending the organization's responsibilities to the provision of an aeronautical mobile service. Such an extension was a matter of debate for some years, and the necessary amendments to the Convention and the Operating Agreement were respectively adopted and confirmed by the Fourth Session of the INMARSAT Assembly meeting in London from 14 to 16 October 1985.[26] The processes of ratification and acceptance for these amendments are now being gone through.

THE ORGANIZATION

General

One's first impression on looking at the Convention on the International Maritime Satellite Organization (INMARSAT) and the relative Operating Agreement is that the INMARSAT package was either a simpler matter to draft than that of INTELSAT, or that the drafters did a better job. There is perhaps a grain of truth in the latter suspicion for some of the provisions in the INTELSAT Agreements are tortuous. But the main explanation of the difference must lie in the relative complexities of the organizations required to deal with the purposes served by the two sets of Agreements. INMARSAT was created for a more concise set of purposes, and it will always be the smaller of the two organizations because of the specialized nature of its services. The extension of INMARSAT's remit to the provision of an aeronautical mobile service, as decided by the Fourth Session of the INMARSAT Assembly in October 1985, has not required an increase in structural complexity. The commercial and procurement decisions, and the volume and nature of the telecommunication traffic involved in maritime or aeronautical services, are not such as to cause states to hedge about the procedures of the organization in the way in which they have done in INTELSAT. Even if at some future stage land-mobile services are added (and that is under discussion), no institutional changes would be required unless there is a considerable consequential increase in membership, and a demand from new members for a greater share in the organization. The INTELSAT model suggests that many states would be more likely simply to buy land-mobile services without asking for membership.

INMARSAT is a fully-fledged international organization, having legal personality in terms of its constituent Convention (art. 25), and being entitled to the normal privileges and immunities of such organizations (art. 26).[27] Its headquarters are in London, and the Headquarters Agreement provides for the organization's privileges and makes the taxation arrangements for its employees which are usual in such cases.[28]

Membership of INMARSAT

As in the case of INTELSAT and EUTELSAT the division between the Convention (which is an agreement between states) and the Operating Agreement (which is between telecommunications entities) makes the question of membership of the organization legally complex. For practical purposes, however, membership of INMARSAT is open to all states, either through signature followed by any appropriate national procedure such as ratification prior to the entry into force of the Convention or thereafter through accession (art. 32), but no state may so become a Party to the Convention until either it, or a telecommunications entity designated by it, has signed the Operating Agreement (art. 32(4)). The questions of the allocation of an investment share to a Signatory and of the withdrawal and termination of membership are dealt with below after the structure and functions of the organization's organs have been outlined. The Signatory for the UK is British Telecom, that for the USA is COMSAT,[29] while the USSR created a special entity 'Morsviaz' sputnik' for the purpose of its participation.[30]

STRUCTURE

The International Maritime Satellite Organization created and governed by the Agreements of 1976, as amended, has three organs. Specified in Article 9 of the Convention, these are a) the Assembly; b) the Council; c) the Directorate, which is headed by a Director-General.

The Assembly

The Assembly is composed of all the Parties to the Convention (art. 10(1)). It holds regular meetings every two years, but extraordinary session may be convened on the request of one-third of the Parties, or on the request of the INMARSAT Council (art. 10(2)). The quorum for any meeting of the Assembly is formed by a majority of the Parties (art. 11(4)). Each Party to the INMARSAT Convention has one vote in the Assembly (art. 11(1)). Decisions on matters of substance are taken by a two-thirds majority, and decisions on other matters, including matters of procedure, are taken by a simple majority of Parties present and voting (art. 11(2)). Abstention from voting is considered as not voting (art. 11(2)). The decision as to whether a matter of substantive or procedural is made in the first instance by the Chairman of the Assembly. The overturning of his decision is itself a substantive decision, to be taken by the Assembly itself, and is subject to the two-thirds majority rule (art. 11(3)).

Functions

The functions of the Assembly are laid out in art. 12(1) of the Convention. In performing these functions it is the duty of the Assembly to take into account any relevant recommendations made by the INMARSAT Council (art. 12(2)), but the Council has no power to bind the Assembly by them. In form, therefore, the position of the Parties as sovereign states is preserved within the organization. They form the Assembly. It is noticeable, however, that, as in INTELSAT, the Assembly of Parties is not in practice the main power within the organization. It meets too infrequently to exercise great control over the organization's activities, and its functions are supervisory rather than executive.

The first enumerated functions of the Assembly are to 'consider and review the activities, purposes, general policy and long term objectives of the organization and express views and make recommendations' on these matters to the Council' (art. 12(1)(a)). It is noticeable in this list that the power of the Assembly is limited to expressing views and making recommendations. Its power to make substantive decisions is very limited. Thus, although only the Assembly has the power to authorize the establishment of additional space segment facilities for the special or primary purpose of radio determination, distress or safety services, that power is exercised only on the recommendation of the Council (art. 12(1)(c)). Further, while that power may presumably be used in the negative and the recommendation of Council be not followed, nonetheless the 'power to authorize' operates only in the realm of additional facilities. The ordinary facilities created to provide maritime public correspondence services can themselves be used to provide for distress, safety and radio determination purposes without the authorization of the Assembly being required (art. 12(1)(c)).

The other decision-making functions of the Assembly are similarly relatively unimportant in practical terms. Thus the Assembly decides on questions of formal relationships between the organization and states, or with other organizations (art. 12(1)(f)). It has a role in deciding upon amendment to the Convention of the Operating Agreement (art. 12(1)(g)). It decides whether membership of the organization should be terminated in terms of art. 30 of the Convention (art. 12(1)(h)). It decides on 'other recommendations of the Council' and expresses views on the reports of the Council (art. 12(1)(d)). Comprehensively, the Assembly ensures that the organization's activities are consistent with the basic documents of the organization and with the United Nations Charter and any other treaty to which the organization becomes bound (art. 12(1)(b)). This last duty is very high-sounding, but in modern international law is tautologous. Nonetheless it is perhaps just as well to have it stated in the constituent document so that, if necessary, the duty may be pointed to during discussions within or outside the organization.

Finally, the Assembly elects four representatives to the Council both in order to secure geographic representation within that body, and to ensure that the interests of developing countries do not go by default (arts. 12(1)(e) and 13(1)(b)). In pragmatic terms, because the Assembly meets infrequently compared with the Council, this may well be the Assembly's most important substantive power and duty. The Assembly meets ordinarily every two years; the Council meets every four months. The Council therefore carries greater weight in the organization's affairs than the Assembly. We will shortly turn to its precise remit, but it suffices here to point out that the Council is necessarily more closely in touch with the organization activities and plans. It is therefore crucial that the membership of the Council should suitably reflect geography and interest – factors which might be overlooked by a simple allocation of power on the basis of use or contribution to cost. The election of these four representatives to the Council by the Assembly is the vehicle through which these ends are achieved.

So at least an argument may run; but it may be based on a false appreciation of facts. The position is even more interesting than it sounds. During the negotiation of the Convention and Agreement there was contention as to the relative balance of power between the Assembly and the Council. The argument arose because the communist countries and certain of the developing countries wanted to elevate the role of states and governments Parties to the Agreements to major importance. However, they failed to get that bias written into the organization's constitution. Nonetheless, in all of these countries telecommunications is a governmental matter, and the entity signing the Operating Agreement on behalf of each of these states is in fact a government department. We are therefore left with the curious position that states which sought, but failed, to enhance the role of the Assembly, may, through representation by an arm of government in the Council, be more directly be able to influence INMARSAT decisions and development than states which wished to have the Council pre-eminent, but which do not have similar direct governmental representation on the Council. That said, it must nevertheless be recognized that the states whose interests are represented by the four members of the Council elected by the Assembly have only a shared representation. Countries such as the USA, Japan and (now) the UK, where the entity signing the Operating Agreement is a private telecommunications agency, do have individual membership of the Council. Opinion varies as to the degree to which INMARSAT affairs are affected by governmental instructions or advice.

The Council

Membership

The INMARSAT Council consists of 22 representatives of the Signatories of the Operating Agreement (art. 13(1)). Eighteen of these represents the largest investment shares in the organization (art. 13(1)(a)). [31] For the purpose of that calculation an investment share may be the investment share of one Signatory, in which case that Signatory is entitled to representation itself (art. 13(1)(a)). Alternatively, the investment share for the purpose of the calculation may be the aggregate investment share of a group of signatories having smaller investment shares which have banded together for the purpose of having a single representative on the Council (art. 13(1)(a)).[32] If a group of Signatories and a single Signatory have the same total investment share, the single Signatory is entitled to representation, taking priority over the group of Signatories in competition for a seat on the Council (art. 13(1)(a)). No such mechanism exists for solving the problem of a straight competition between single Signatories. If two or more Signatories have equal investment shares then, exceptionally, each is entitled to one representative, even though that may result in the total membership of the Council exceeding 22 (art. 13(1)(a)).

The remaining four seats on the Council are reserved for representatives of Signatories not otherwise represented on the Council. These are elected by the Assembly without regard to investment shares, but with the intention of ensuring that an equitable geographic representation is present on the Council, and also 'with due regard' to the interests of developing countries in the work of the organization (art.

13(1)(b)). A Signatory elected to serve on the Council in one of these four 'special' seats is elected to represent each Signatory within its geographic area which has agreed to be represented in this way and which is not otherwise represented on the Council (art. 13(1)(b)). The four members of the Council so appointed take office at the Council meeting immediately following the Assembly which elects them, and serve until the next ordinary meeting of the Assembly (art. 13(1)(b)). On convening, each Assembly holds elections for the four reserved seats, and the ordinary tenure of one of the 'special' seats is therefore something under two years.[33]

The result of such provisions is that not all members of INMARSAT are on, or are represented on, the INMARSAT Council. When investment shares were redistributed on 31 January 1987 no new representative groups were notified to the Director-General and the resultant Council represented members holding 97.18112 per cent of the investment shares in the organization. That seems an adequately high ratio. The only oddity might be thought to be the omission of Korea, Poland and the major flags-of-convenience countries like Panama, but the facts rule against these, their investment shares being insufficient for representation and they not being part of a group nor voted on as one of the four 'special' elected seats. Their representation is through the appropriate 'special' geographic representative.[31]

Procedure and voting weights

The Council meets as often as is necessary, but not less than three times per year (art. 14(1)). The quorum for any meeting of the Council is a numerical majority of its members so long as these represent at least two-thirds of the total voting participation of all Signatories and groups of Signatories which are represented on the Council (art. 14(4)). A gap in the Council membership pending the filling of a vacancy does not mean that the Council is improperly constituted (art. 13(2)).

The Council is instructed to 'endeavour to take decisions unanimously' (art. 14(2)). If such agreement cannot be achieved, a vote is taken. Decisions on substantive matters are taken by a majority of the members of the council representing at least two-thirds of the total voting participation of Signatories and groups of Signatories represented on the Council (art. 14(2)). Decisions on procedural matters are taken by a simple majority each Council member having one vote (art. 14(2)). The decision whether a matter is substantive or procedural is taken by the chairman, but he may be overruled by a two-thirds majority of representatives present and voting, each in this instance having one vote (art. 14(2)). It is open to the Council to adopt a different voting procedure for the election of its officers (art 14(2)), and this has been done. By Rule 8 of the Rules of Procedure for the Council of INMARSAT and its Subsidiary Organs,[34] the chairman and vice-chairman of the Council are to be elected unanimously, if possible, failing which they are elected by secret ballot by a two-thirds majority of representatives present and voting, each representative having one vote (Rule 8A). If two rounds of voting does not produce a result, then the requirement drops to a simple majority (Rule 8B).

Lurking amid the language of the previous paragraph are references to voting participation being dependent upon investment shares. The question of voting in the Council being related the investment share was naturally an important one within the INMARSAT negotiations, various countries for doctrinal and other reasons favouring a one-member one-vote system. Practical exigencies triumphed. It was eventually recognized that some relationship between use, contribution to costs and voting weight was necessary, and the investment share method was adopted. However, as is the case within INTELSAT, steps have been taken in the Agreements to ensure that the impact of any large single investment share within the most important decision making-body is subject to limitations. By art. 14(3), an upper limit of 25 per cent of voting participation in the Council is made possible, but unlike the equivalent case in INTELSAT, that limit is not absolute. This may reflect the realization that, within the smaller organization, while such a limit may well be desirable in many instances as a way to prevent a large holder overbearing others, yet there could be times and circumstances in which such a limit might operate unfairly. The large user, and therefore holder of a large investment share, should not be penalized in its voting potential merely because others are unwilling to take on the financial burden of accepting its excess investment allocation.

Art. 14(3)(a) therefore makes the ordinary provision, by which each representative on the Council is entitled to a voting participation equivalent to the total investment share or shares which he represents. However, no representative may cast, on behalf of one Signatory, more than 25 per cent of the total voting participation in the Council unless certain procedures have been gone through (art. 14(3)(a)). Where a Signatory is entitled, as a result of its investment shares, to a voting participation of more than 25 per cent, it may choose to offer to other Signatories part or all of the excess above that figure. Any of that excess not offered to other Signatories is distributed for the purposes of voting participation equally among all the other members of the Council (art. 14(3)(c)).[35] Where a Signatory does offer all or part of its 'excess', other Signatories notify the organization of the extent to which they are prepared to accept that excess as additional to their own responsibilities (art. 14(3)(b)(ii)). Where the total of the shares offered to be taken up by other Signatories does not exceed the amount offered by a Signatory, the Council distributes the excess amongst the willing Signatories in accordance with their notifications. Where the total which Signatories have notified the organization of their willingness to accept exceeds the offered amount, a division has to occur. The 'excess' is distributed by the Council either by the agreement of the notifying Signatories, or, failing agreement, among the notifying Signatories proportionately to the total of the notifications received (art. 14(3)(b)(ii)). Only in the case where there is more 'excess' than other Signatories are willing to absorb is the Signatory holder of the original investment share entitled to a voting participation within the Council of more than the 25 per cent norm (art. 14(3)(b)(iv)). Any distribution made of 'excess' may not increase the investment share of any Signatory above 25 per cent of the voting participation within the Council (art. 14(3)(b)(iii)). A Signatory cannot, therefore, by the judicious buying-up of investment shares increase its own investment share above the maximum for voting purposes, nor thereby acquire a disproportionate financial interest in the organization.

Functions

The functions of the Council are mainly laid out in art. 15 of the INMARSAT Convention. With 'due regard for the views and recommendations of the Assembly', the Council is responsible making provision for the space segment (the satellites) necessary for the working of the organization 'in the most economic, effective and efficient manner' consistent with the INMARSAT Agreements. In particular, it determines the organization's requirements and adopts 'policies, plans, programmes, procedures and measures for the design, development, construction, establishment, acquisition by purchase or lease, operation, maintenance and utilization' of the space segment, including procuring launch facilities (art. 15(a)). It also deals with the criteria and procedures for the approval of earth stations, a term which includes land stations as well as the mobile ship and aircraft stations (art. 15(c)). The criteria should be sufficiently detailed to permit national licensing authorities to use them for the purposes of their own type-approval purposes, if they wish (art. 15(c)). The Council adopts procurement procedures, regulations and contract terms and approves procurement contracts (art. 15(f)). It also deals with financial matters, adopting financial policies, approving financial regulations, annual budget and financial statements; periodically determines charges for use of the space segment; and decides all other financial matters including investment shares and capital ceiling (art. 15(g)). It consults regularly with organizations and groups representative of system users, shipowners, aircraft operators, and maritime and aeronautical personnel. It adopted and now implements the management arrangements under which the Directorate works (art. 15(b)), directs the Director-General and periodically submits to the Assembly reports on the organization's activities, including its financial affairs (art. 15(e)). The Council appoints and may remove the Director-General (art. 16(1) and (2)) and approves appointment of senior officials reporting directly to the Director-General (art. 16(5)).

The Council's powers are therefore considerable and, as indicated, central to the organization. Meeting every four months or so, it is in a much better position to control the ordinary work of the organization than is the Assembly. Indeed, as already noted, although the Assembly has wide recommendatory powers as to general policy and long-term objectives, the Council is paramount in the ordinary creation and working of the INMARSAT system. Only in the matter of the creation of additional space segment facilities, the special or primary function of which is to provide radio determination, distress or safety services is the consent of the Assembly necessary (art. 12(1)(c)), and the Convention expressly provides that that consent is not necessary where radio determination, distress or safety facilities are ancillary to public correspondence services (art. 12(1)(c)). The Assembly is concerned with 'external relations', membership and amendment to the Arrangements, which are important matters. But in the day-to-day, and even year-to-year, establishment and operation of maritime (and soon aeronautical) mobile satellite services, it is the Council which is the key organ within INMARSAT.

The Directorate

The Directorate is the executive organ of INMARSAT and functions under a Director-General (art. 9) who is the organization's chief executive and legal representative and is responsible to, and is under the direction of, the Council (art. 16(3)). It has to be said that the impression left by the Convention is that the Director-General of INMARSAT could have the Council breathing down his neck, for its approval is required on many matters, the organization is small, and hence the Directorate may be more amenable to intervention by the Council or interference by a zealous councillor.

The Director-General has a six-year term of office (art. 16(2)). The paramount consideration in his appointment and that of other Directorate personnel is the 'necessity of ensuring the highest standards of integrity, competency and efficiency' (art. 16(6)). He is appointed by the INMARSAT Council from candidates proposed by Parties or by Signatories through Parties, but the appointment by Council is subject to confirmation by the Parties through a special procedure. The appointment is notified to Parties immediately by the UK as Depository of the INMARSAT Convention. Confirmation is automatic unless within 60 days of the Parties being informed, more than one-third of the Parties object in writing. The Director-General may assume his functions during this 60-day period (art. 16.(1)). The Council may remove the Director-General 'on its own authority', reporting the reasons for this action to the Assembly (art. 16(2)).

The Director-General appoints other members of the Directorate, appointments of personnel reporting directly to him being subject to approval by the Council (art. 16(5)). He is responsible for the staffing structure, staffing levels and terms of employment of all employees, consultants and advisers, although subject to Council approval (art. 16(4)).

THE SYSTEM AND SERVICES

It is the purpose of INMARSAT to establish by lease or purchase (art. 6) the space segment required to provide an improved system of maritime communications services for public correspondence, for radio determination, for distress and safety of life, for traffic services and for the efficiency and management of ships (art. 3(1)). In the future, assuming that the Agreements are appropriately amended, such services will also be provided for aircraft as well. The organization seeks to provide a service in all areas where such services are needed (art. 3(2)), needless to say (although it has to be stated in the Convention) for peaceful purposes only (art. 3(3)). To that end it has three satellites in orbit, and leases transponders on other (mostly INTELSAT) satellites.

The INMARSAT system is open for use by ships (and aircraft) of all nations, subject to conditions determined by the Council, which are directed technically to the proper use of the system through establishing requirements as to the nature and characteristics of the terrestrial end of the link,[36] and otherwise have procedural importance. In permitting access to the INMARSAT system, the Council may not discriminate on a basis of nationality. Ships and aircraft of all nations may use the system, subject to the appropriate conditions set by Council, on a global basis (art. 7(1)). In addition, 'structures operating in the marine environment' may also be granted access to the system provided that that use does not detract from the primary use of the system for ships and aircraft (art. 7(2)). Earth stations on land must be within the jurisdiction of a Party and wholly owned by Parties or by entities subject to their jurisdiction (art. 7(3)). The Council is, however, permitted under art. 7(3) to depart from this requirement if it finds it to be in INMARSAT's interests. Acting partly in the spirit of this provision, and also through negotiation, INMARSAT has sponsored an international agreement under which a foreign ship, while within the national waters of Party (for example, in a harbour), may use the INMARSAT system without hindrance.[37] This sensible provision essentially alters the normal requirement that a Party to the ITU Convention exercises jurisdiction over all radio broadcasts within its territory.

PROCUREMENT

As in the case of the other organizations which provide telecommunications by satellite, an important question in the negotiation of the Agreements was the arrangements to be made for the procurement of the necessary satellites and other equipment. In the case of INMARSAT, the presence particularly of the USSR might have been thought to have added to the difficulty. It appears that it has not. Indeed, the USSR has on this, as on other matters, made quite considerable concessions. This may be so that it can thereby remain in touch with Western technology, but whatever the reason it is something to be noted.[38]

Procurement policy is intended to encourage competition on a worldwide basis. As indicated elsewhere in this book in the field of satellite telecommunications there is a distinct danger that a single supplier might become a monopoly supplier. The costs and risk of enterprise are enormous and manufacturers may find it impossible to remain active in the area if they fail to gain regular business, but the inadvertent creation of a monopoly supplier is to be avoided. International organizations cannot be as brusque to overchargers as a national government can be to a national contractor (say in defence). The supply of goods and services for INMARSAT is ordinarily by open international competitive tender. The best combination of quality, price and the most favourable delivery time will normally indicate where a contract is placed, but where different tenders meet that test, the Council is to award the contract so as to encourage worldwide competition (art. 20(1)). Some contracts are, however, exempt from the competition requirement. These are:

1 contracts of an estimated value of under US$ 50,000[39] where award will not prejudice the later application of the general tendering requirement;
2 contracts where procurement is needed urgently to meet an emergency;
3 contracts where there is only a single source or there are so few

sources that open tendering is unreasonable and all such sources are given opportunity to tender;

4   where the contract is for an administrative purpose for which it is neither practicable nor feasible to have open international tender; or

5   contracts where the procurement is for personal services (art. 20(2)(a)–(e)).

INVENTIONS AND TECHNICAL INFORMATION

Concomitant with procurement is the question of the rights and information and the access to, and use of, rights and information which may be gained by the organization or by others in the performance of the work contracted for by the organization. What is done within these rights is sweeping.

By art. 21 of the Convention INMARSAT is required to acquire rights in inventions and technical information arising from work which it performs, or which it contracts to have done for it. It must obtain those rights which are necessary in the common interest of INMARSAT and of its Signatories. In the case of work it has put out to contract the rights are to be acquired on a non-exclusive basis (art. 21(1)). More particularly art. 21(2) provides that the organization, with due regard to its purposes and to general industrial practice, shall ensure that, in cases where there is a significant element of study or research or development, it shall have the right to have disclosed to it without payment all inventions and technical information generated by the work, and the right to disclose such information and inventions to Parties, Signatories and bodies under the jurisdiction of Parties in connection with the space segment or its land and mobile stations. Further, although in the case of work done under contract ownership of the invention or information remains with the contractor (art. 21(3)), Parties, Signatories and bodies under the jurisdiction of Parties may use the information or invention without payment in connection with the INMARSAT space segment or land or mobile stations (art. 21(2)(b)). In addition, INMARSAT is to secure similar rights, on fair and reasonable terms, to inventions and information used to modify or reconstruct INMARSAT equipment (art. 21(4)). Deviation from these principles is permitted only where either a failure to deviate would be detrimental to INMARSAT's interests (art. 21(5)), or in other exceptional circumstances provided that:

1   non-deviation would be detrimental to INMARSAT;
2   Council decides that INMARSAT should be able to ensure patent protection in any country; and

3   the contractor is unable or unwilling to seek patent protection
    within a specified time (art. 21(6)).

In other cases, when INMARSAT has rights to inventions or technical information, it has the right to disclose or have disclosed those
inventions or technical information to any Party or Signatory subject
to reimbursement of any payment it itself makes in the matter (art.
21(7)(a)). It also may make available to a Party on Signatory the right
to disclose inventions or technical information and use or authorize
the use of such material, without payment in the case of use in connection with the INMARSAT system, or on fair and reasonable terms
in other cases (art. 21(7)(b)).

Equally importantly, art. 21 ends with two general provisions. First,
disclosure and use within the organization must be on a non-
discriminatory basis (art. 21(8)). Any information or invention in
which INMARSAT acquires rights is therefore open to all Parties,
their respective Signatories and to persons and bodies under their
jurisdiction. Although it follows that, through INMARSAT, states can
learn what is going on in other states, balancing that, art. 21 does
not wholly tie INMARSAT's hands in the contracts it enters into. If
desirable, INMARSAT may enter into contracts with persons who are
subject to their own domestic laws which regulate and perhaps
restrict the disclosure of technical information (art 21(9)). Using these
two provisions therefore, INMARSAT treads a delicate balance, for
many of its contracts are with companies that are subject to state
control as to what information can be disclosed by them. Notably,
most of the West is subject to the nationally enforced decisions of the
Coordinating Committee on Information (CO-COM) which tries to
impede the passing of 'sensitive' information to the communist and,
sometimes, Third World countries.[40]

INVESTMENT SHARES AND UTILIZATION CHARGES

As in the case of other international telecommunications organizations, much importance can hang on the investment share which a
Signatory holds within INMARSAT, since this determines membership of the Council and also affects voting weight within that body.[41]
The initial investment shares attributable to Parties was laid down in
an Annex to the Operating Agreement (art. V(3)), but that allocation
was made on a negotiated basis and might be altered by further
negotiations.[42] Investment share is now determined on the basis of
utilization of the INMARSAT space segment, each Signatory having a
share equal to its percentage of the total use of the segment by all
Signatories, measured in terms of utilization charges (art. V(1)). There

is a minimum investment share of 0.05 per cent irrespective of actual minimal usage (art. V(8)) and there is one other variation in a special case, discussed later.

Naturally regard had to be paid to the nature of the use within the system. By art. V(2) each use of the system in both directions is halved and is considered as a ship or aircraft part and a land part. The Signatory of the Party in whose territory the signal originates or terminates is credited with the land part, and the Signatory of the Party having jurisdiction over the ship or aircraft is credited with the ship or air part of each utilization (art. V(2)). Structures operating at sea (such as oil platforms) which have been permitted access to the INMARSAT system by the Council (art. 15(c)) are considered as ships for the calculation of utilization (art. V(2)). This splitting of the aircraft – ship and the land utilization base could lead to an awkwardness and possible unfairness if a small state happened to have many aircraft or ships but few land stations. Accordingly, by art. V(2), it is further provided that where, for any Signatory, the ratio of aircraft–ship part to land part utilization is greater than 20:1 the Signatory can apply to the INMARSAT Council and is entitled to be assessed a utilization of the higher of twice its land part utilization or an investment share of 0.1 per cent.

Investment shares are now determined annually in accordance with art. V(5)(a) of the Operating Agreement. Shares are also redetermined on the date of entry into force of the Agreement for a new Signatory, the share attributable to the new Signatory being set by Council until the next 'normal' yearly redetermination (art. V(6)), and on the effective date of withdrawal or expulsion of a Signatory (art. V(5)(b) and (c)). In the cases of alteration because of a change in the membership in the organization, the redetermination is done on the basis of proportionality subject to certain controls. Where membership of the organization is reduced by the departure of a Signatory, a minimum investment share held as a minimum – and not on basis of actual usage – is not increased by the resultant redetermination (art. V(7)). In any redetermination, a share held by a Signatory cannot be increased in one step by more than 50 per cent of the share it first held, nor may it be decreased by more than 50 per cent at one time (art. V(9)). At a redetermination, shares which are unallocated because of the 50 per cent rule are apportioned by the Council among those Signatories wishing to increase their shares, subject to no Signatory increasing its share by more than 50 per cent at a time (art. V(10)). If there still remain unallocated shares, these are distributed among Signatories in proportion to their projected investment shares, subject to the minimum of 0.05 per cent and the 50 per cent increase rule (art. V(11)). Finally, a Signatory may apply to have its investment share reduced, and Council will determine a lesser share if that would not contravene the minimum 0.05 per cent share rule and if other Signatories are willing to take up fully and equitably the proportion of investment share so released (art. V(12)).

The actual calculation of the investment share, taking into account the aircraft–ship part and the land part indicated above, depends on the units of measurement for the various types of utilization of the space segment. These different units are established by Council decision, and charges are set for each of them (art. 19(1)). Signatories are bound to provide such information as the organization requires properly to determine utilization of the space segment (art VII(2)). All Signatories are charged the same rate for each type of utilization (art. 19(2)) although it is competent for the Council to set different rates for non-Signatories permitted to use the system. In that case, non-Signatories are charged the same rate for the same type of use (art. 19(3)). The purpose of the utilization charges is the earning of sufficient revenue from user charges to cover the organization's operating, maintenance and administrative costs, the creation of such operating funds as the Council considers necessary, the amortizing of the investment made by the Signatories, the paying out of sums due to a Signatory which has left the organization, and payment of compensation for the use of capital at a rate set by Council (art. 19(1); art. VIII(1) and (2)).

FINANCE

Much of the above narrative relates to finance, but other matters also relate to the working of the financial arrangements within the organization and between members. In particular, it may be noted that capital contributions are payable to INMARSAT by Signatories in proportion to their investment shares when required by the Council. Repayment of capital is also competent when circumstances so warrant (art. III(1)). (The organization is to keep its holdings of 'surplus' funds to a minimum (art. IX(1)).) Capital contributions fund direct and indirect costs of the design, development, acquisition, construction, and establishment of the space segment (the satellites and their launching) together with the property costs which the organization incurs (art. III(1)). Operating, maintenance and administrative costs are also funded from capital contributions insofar as they are not covered by revenues from the system's operation (art. III)(2)). Alternatively, the Council may arrange overdraft or loan facilities for these operational purposes on a temporary or longer-term basis (art. VIII(3) and art. X). Capital contribution receipts may also be used to settle any liability of the organization to a Signatory (art. XI(4)).

Payments of capital contributions and of utilization charges are to be prompt, and interest may be added at a rate set by Council to sums remaining unpaid (arts. III(3) and VII(4)). Payments in either direction between INMARSAT and a Signatory are made in any freely convertible currency acceptable to the creditor (art. IX(2)). Utilization charges are settled in a manner approximating to the normal accounting for international telecommunication facilities (art. VII(1)). Settlements due between Signatories consequent on the regular determination of investment shares are dealt with by a complex formula and procedure detailed in art. VI and carry interest on unpaid sums at a rate determined by Council. Roughly speaking, the calculation starts with the original acquisition cost with deductions for capitalized return and captialized expenses, accumulated amortization and outstanding loans. The figure is then adjusted for compensation for use of capital and that figure is related to the difference between the old and new investment share. As indicated, payments between INMARSAT and a Signatory are made in any freely convertible currency acceptable to the creditor (art. IX(2)). There is no similar provision as to payments between Signatories, but as most financial adjustments occur between INMARSAT and individual Signatories this may not be important.

When payment of utilization charges has fallen due and has been in default for four months or longer, the Council is required to institute 'any appropriate sanctions' (art. VII(3)). Such sanctions are not further specified but might be thought to include taking the matter up with the Party which designated the Signatory, suspending payments due to the Signatory, suspending the voting privileges of the Signatory or its access to INMARSAT procurement and inventions data, and suspending service to the Signatory. Beyond that, the more fundamental question can arise of whether the Signatory is in default of its obligations. Clearly that question can arise more immediately in the case of a failure timeously to make capital contributions. Such default can raise questions of the continuation of the Signatory's membership of the organization.

SUSPENSION AND TERMINATION OF MEMBERSHIP

Where a Signatory has failed to pay capital contributions within four months of their falling due, the rights of the Signatory under both the Convention and Operating Agreement are automatically suspended (art. 30(3)). If payment is not made within a further three months, or if within that period the Party which designated that Signatory has not been substituted another Signatory (under art. 29(4) which involves the new designee taking on the obligations of its predecessor), after hearing any representations by the Signatory or by the Party the Council 'may decide that the membership of the Signatory has terminated' (art. 30(3)). This curious form of language implies that the Council is less taking a decision than recognizing a situation, and is a form of words which may make Council action easier.

In cases other than that of default in capital contributions, the Council may take note of a failure of a Signatory to comply with any obligation. If the failure is not remedied within three months of the Signatory being informed of the 'taking note' resolution, the Council, taking into account any representations by the Signatory or its designating Party, may suspend the Signatory's rights. If the matter is not resolved within a further three months, and in the light of any representations by the relevant Signatory or Party, Council may recommend to the Assembly that the Signatory's membership be terminated, and the Assembly may so decide (art. 30(2)). In this case, therefore, the Assembly expels the Signatory. The process is much longer, however, than the case of failure in capital contributions, where the failure is very clear and must be swiftly dealt with to ensure that the INMARSAT system is established without delay.

The suspension of a Signatory's rights does not affect its becoming liable for obligations under decisions taken during the period of suspension (art. 30(4)). When the rights are terminated or recognized to have terminated the Operating Agreement immediately ceases to be applicable to the Signatory (art. 30(2) and (3)).

A Party can also be expelled from the organization by decision of the Assembly (art. 30(1)). In this case, the decision may be taken not earlier than one year from the date on which the Directorate receives 'written notice that a Party appears to have failed to comply with any obligation under' the Convention (an extraordinary Assembly may be convened for the purpose). It is not stated from whence that notice may come, but clearly a Council resolution or the award of an arbitration[43] could contain such notice. The Assembly may decide on termination of the membership of a Party after considering representations by the Party, and must find a) that the failure has occurred, and b) that the failure 'impairs the effective operation of' INMARSAT. The second head is a difficult class to which to ascribe content. It may include a direct challenge to the INMARSAT concept through the establishment of a separate system in disregard of INMARSAT's findings on the matter.[44] If there were questions of technical compatibility and, say, orbital location involved, it might be thought that a recalcitrant Party has in effect divorced itself from INMARSAT, while retaining formal membership. Again, a failure of a Party to substitute a new Signatory for one whose membership has been terminated might constitue a default which impairs the organization's effective operation.

Where membership of a Party is terminated, that entails the simultaneous withdrawal of its designated Signatory or itself as Signatory in appropriate cases (art. 30(1)). The Signatory remains liable for capital contributions required prior to the end of its membership, and for its share of liability for acts and omissions of INMARSAT while it was a member (art. 30(1)).

SEPARATE SYSTEMS

The INMARSAT provision for the coordination of systems separate from itself but which are intended to meet the purposes for which INMARSAT exists (that is its equivalent of INTELSAT's art. XIV) are contained in art. 8 of the INMARSAT Convention. Questions of ensuring technical compatibility and of avoiding 'significant economic harm' to the INMARSAT system are involved (art. 8(1)). A Party to the Convention is to notify the organization if it or any person under its jurisdiction intends to create or initiate the use of such a separate system (art. 8(1)), and give all necessary technical information (art. 8(4)). The Council expresses its view as a non-binding recommendation as to technical compatibility, and informs the Assembly of its view on 'economic harm'. Thereafter the Assembly expresses its opinion, also in the form of a non-binding recommendation (art. 8(3)). The Assembly's views must be given within nine months of the original notification, and an extraordinary Assembly may be called for the purpose (art. 8(4)). The exceptions to this requirement of consultation are systems established for national security purposes, and systems created prior to the coming into force of the INMARSAT Convention (art. 8(5)), the latter eliminating the requirement for consulting as to INTELSAT.

The language of art. 8 appears weaker than its INTELSAT equivalent – that the views of the INMARSAT organs are non-binding is made very clear. We await practice in the matter and do not yet know whether the INMARSAT provision will work better or worse than that of INTELSAT. It will also be interesting to see if the INMARSAT view of 'significant economic harm' is at (useful) variance with the INTELSAT development of the concept. This may come in 1989, following the grant in August 1988 by the FCC of a conditional authorization for a radio determination satellite service to the GTE Spacenet Corporation.[45] GTE Spacenet argued that consultation with INMARSAT was not required, but the FCC did not accept its argument and has conditioned its authorization on compliance with the obligations of both art. 8 of the INMARSAT Convention and with art. XIV(e) of the INTELSAT Agreement.[46]

SETTLEMENT OF DISPUTES

Mention has been made of the arbitral settlement of disputes between Parties or between Parties and INMARSAT. The decision of an arbitral tribunal does not affect or prevent a decision of the Assembly that a Party be expelled from the organization (art. 31(1)). Arbitration cannot therefore be used as a final court of appeal against such an Assembly decision.

That case apart, disputes between Parties and between Parties and INMARSAT, between Signatories, between Signatories and INMARSAT and between Signatories and Parties can end in arbitration under art. 31 of the Convention or art. XVI of the Operating Agreement. The processes are very similar to that for INTELSAT and so will not be fully discussed here. Suffice it to say that it is hoped that disputes will be settled by negotiation within one year. Continuing disputes as to the Convention involving Parties or Parties and INMARSAT can be referred to the International Court, to other mechanisms or to arbitration in terms of the Annex to the Convention[47] if the parties to the dispute so agree. Disputes between parties and Signatories as to the Convention may similarly be submitted to arbitration under the Annex. Disputes between Parties and INMARSAT under other agreements between them, or between Signatories and between Signatories and INMARSAT must, if not settled within the year, be submitted for arbitration under the Annex at the request of one party to the dispute.. The reason for this careful distinction is that states are not now normally willing to submit to compulsory arbitral settlement of disputes.

The Annex provides for an arbitral tribunal of three members, with procedures and powers similar to that discussed in relation to the INTELSAT arbitral arrangements. The decision of the arbitral tribunal of three members established in terms of the Annex to the Convention, is binding on all the disputants, and is to be carried out by them in good faith (Annex, art. 11(2)).

THE FUTURE

The future of INMARSAT as an organization seems bright though not entirely tranquil. Clearly it has a satisfactory role in meeting its original objective of providing a maritime service. The shipping market is defined both in terms of ships and in that of shipping and trading companies. The specialist interest of safety of life at sea is also defined, and has recently been given a fuller recognition through the 1986 ITU Maritime Services Conference which dealt with the Global Maritime Distress and Safety System (GDMSS).[48] More difficulty might be encountered if INMARSAT seeks to broaden its activities to other forms of mobile services.

Other mobile services

As indicated above, steps are being taken to amend the INMARSAT Convention and Operating Agreement to permit the organization to provide an aeronautical service. At first sight this would seem an uncontentious matter and an obvious extension of INMARSAT's remit. However, international aviation is largely governed, so far as its contracting parties are concerned, by the Chicago Convention of 1944,[49] part of which sets up the International Civil Aviation Organization (ICAO) which has the major role in matters of air safety, flight procedures and the like. As the ITU is the UN agency dealing with telecommunications, so ICAO is the UN agency entrusted with air navigation matters and the Annexes to the Chicago Convention which it adopts and other procedures and recommended practices established through the organization form the major context in which international civil aviation provides services. Included in the remit of ICAO is aeronautical communications. Art. 28 of the Chicago Convention obliges the Parties, so far as is practicable, to use standard systems of communications using rules and procedures adopted by ICAO. Annex 10 to the Convention adopted by the organization in accordance with art. 37, deals with aeronautical communications and includes the setting-up and operation of a network of fixed stations for the provision of a regular civil aviation telecommunication service which is operated by states. This service is directed towards air safety and traffic control, not public telecommunications services.

INMARSAT could provide a service similar to that arranged through ICAO, could provide navigation facilities for aircraft as it does for ships, and could add the provision of public telecommunications facilities for passengers to a degree not acceptable under the ICAO system as it exists at present. However, many of the Parties to the INMARSAT Convention are also Parties to the prior Chicago Convention, and a conflict of obligation might arise as a matter of law, apart from considerations such as economics, politics and safety. Suffice it to say that, after much discussion and some wrangling, agreement has been reached between INMARSAT and ICAO under which INMARSAT will provide telecommunications services through its satellite system on a non-exclusive basis.[50]

That is not, however, the end of the matter. The US position is unclear. Companies other than COMSAT have sought to enter the air mobile services field in North America, and it is by no means clear that COMSAT will be permitted to engage in an INMARSAT incursion into that form of service.[51] It may be that the USA, either directly through governmental decree or indirectly through FCC action, will seek to separate the question of maritime and air-mobile services and attempt to designate different telecommunications entities for the 'traditional' and the 'new' type of INMARSAT business, COMSAT dealing with maritime service as before and a new Signatory dealing with air mobile services.

If that happens, INMARSAT may find itself in a difficult constitutional position. As the Convention stands, there does not appear to be provision for a Party to designate more than one Signatory to the Operating Agreement. Art. 2(3) of the Convention speaks of a Party designating 'a competent entity . . . which shall sign the Operating Agreement'. That is clear. And the terms of the Operating Agreement imply that there is one Signatory for each Party. For example, in dealing with the division of traffic for the purpose of calculating investment shares, art. V(2) speaks of attributing one half of a utilization of the space segment to 'the Signatory of the Party under whose authority the ship is operating' and the other half to 'the Signatory of the Party in whose territory the traffic originates or terminates'. In short, for the USA to designate two Signatories would be improper unless an appropriate amendment is made to both the Convention and the Operating Agreement. Whether the complexity that would then be required suitably to separate the air and maritime mobile sections of INMARSAT is desirable, or even feasible, is a question which, it is hoped, will not arise.

Of course, the remaining possibility is that, from air-mobile services, INMARSAT will branch out yet further to land-mobile services.[52] Given the immense technical strides that have been made in recent years it is not fanciful to think of lorries and ultimately even cars being provided with aerials capable of connecting with the INMARSAT system. There would be less need for such a system where cellular radio is available on a terrestrial basis, but cellular radio is unlikely to be cost-effective even on a country-wide basis. The UK, for example, is unlikely to be entirely provided with a cellular network, since the highlands and islands of Scotland do not justify the setting up of a terrestrial network sufficient to provide complete coverage. In such cases a satellite system could meet a need. After that, who knows? It was never clear what system Dick Tracey's wrist-radio was channelled through.

NOTES

1   H.H.M. Sondaal, 'The Current Situation in the Field of Maritime Communications Satellites: "INMARSAT" ', *Journal of Space Law*, vol. 8, 1980, pp. 9–39. (Mr

Sondaal was Chairman of the INMARSAT Preparatory Committee (see below));
S. Doyle, 'INMARSAT: The International Maritime Satellite Organization –
Origins and Structure', *Journal of Space Law*, vol. 5, 1977, pp. 45–63. (Mr Doyle was
Chairman of the Economic Assessment Working Group of the Panel of Experts
on Maritime Satellites of the IMCO, 1974–75 (see below)); N. Jasentuliyana, 'The
Establishment of an International Maritime Satellite System', *Annals of Air and
Space Law*, vol. 2, 1977, pp. 323–49. (Mr Jasentuliyana is of the Outer Space
Division of the UN Secretariat, and was a UN observer at the Preparatory
Conference (see below)); P.K. Menon, 'International Maritime Satellite System',
Journal of Maritime Law and Commerce, vol. 8, 1976–77, pp. 95–106; O.J. Haga,
'INMARSAT: An example of Global International Co-operation in the Field of
Telecommunications', *Telecommunication Journal*, vol. 47, 1980, pp. 511–16; O.
Lundberg, 'INMARSAT: the First Year and Next Decade', *Telecommunications
Journal*, vol. 9, 1983, pp. 469–76. (Dr Lundberg is Director-General of
INMARSAT).

2   The Interim INTELSAT Agreement made no specific provision for a maritime
service but simply intends the 'objective of achieving basic global coverage in
the latter part of 1967' (Interim INTELSAT Agreement, art. I(a)(ii)).

3   Convention establishing the Intergovernmental Maritime Consultative Organ-
ization, Geneva, 6 March 1948, 289 UNTS 48; (1958) UKTS No. 54, Cmd. 589; 9
UST 621, TIAS 4044. Convention on the International Maritime Organization,
London 14 November 1975, (1982) UKTS No. 34, Cmnd. 9632.

4   Final Acts of the World Administrative Radio Conference to deal with matters
relating to the Maritime Mobile Service, Geneva 18 September to 4 November
1967, Recommendation No. MAR/3.

5   As to the meaning of 'allocation' see Chapter 8, 'The International Telecommunication Union', subsection 'Radio regulations: allocations', p. 346 and 349.

6   Final Acts of the World Administrative Radio Conference for Space Telecommunications, Geneva 1971. Annex 3 thereof, 'Revision of Article 5 of the Radio
Regulations', which revised allocations within the band 10 kHz to 275 GHz.,
dealt with much more than maritime requirements, but the maritime mobile and
the maritime mobile-satellite services were well represented within the new
allocations. The maritime mobile-satellite frequencies allocated lay in the band
1535–1645 MHz, and are partly shared with aeronautical mobile-satellite
allocations.

7   Outlined in Chapter 1, section, 'Introduction' and Chapter 8, section,
'Introduction'.

8   Cf. International Convention for the Safety of Life at Sea, London 1 November
1974–1 July 1975, (1980) UKTS No. 46, Cmd. 7874; TIAS 9700, which replaces a
succession of predecessors.

9   IMCO Res. A. 305(vii), 23 November 1973.

10 The Sessional Act of the Second Session of the Conference, with Resolutions and
Recommendations made to the final session is printed (1976) 15 ILM 219–248. See
also articles by Sondaal, Doyle, Jasentuliyana, Menon, Haga and Lundberg cited
above in note 1.

11 The 1987 World Administrative Radio Conference for the mobile services
(MOB-87) held in Geneva from 14 September to 17 October 1987, had as a major
item on its agenda the radio requirements of the proposed Global Maritime
Distress and Safety System (GMDSS) which has been developed through the
IMO and which will be introduced in coming years.

12 Special provision made by the 1987 WARC MOB-87 requiring trained personnel
on ships to operate the GDMSS shows how far the displacement of trained
radio and electronic personnel had gone.

13 Cf. R.T. Gallagher, 'The Watch on the Shore', *Ocean Voice*, vol. 7, no. 2, April

1987, pp. 11–13; J.L. Leaf, 'Branch Office Aboard', *Ocean Voice*, vol. 8, no. 2, April 1987, pp. 17–21.

14 Cf. J. Wagland, 'A Question of Weather', *Ocean Voice*, Vol. 5, no. 4, July 1985, pp. 8–11.

15 COMSAT was offering a maritime service (MARISAT) by this time.

16 COMSAT had benefited from its position within the interim INTELSAT consortium.

17 See art. I(l), and cf. art. I(k) of the INTELSAT Agreement.

18 A formal agreement between INTELSAT and the USSR was only arrived at during the First Session of WARC–ORB 85–88 in 1985. On WARC–ORB 85–88 see Chapter 8, section, 'WARC–ORB 1985–88', p. 393 and Chapter 9, p. 410.

19 Ademuni-Odeke, *Protectionism and the Future of International Shipping*, (Dordrecht: Nijhoff, 1984).

20 A note to the text of the investment share Annex printed at (1976) 15 ILM 1075 states that a compromise was required whereby both Poland and Kuwait were allocated initial shares of 1.48 per cent, thus entitling them to individual representation on the Council.

21 See Chapter 8, 'The International Telecommunication Union', subsection, 'Other Participants', p. 331.

22 Convention on the International Maritime Satellite Organization (INMARSAT), London 3 September 1976; (11979) UKTS No. 94, Cmnd. 7722; 31 UST 1, TIAS 9605; (1976) 15 ILM 1051–75. Final Act of the International Conference on the Establishment of an International Maritime Satellite System 1975–1976, IMO, London. *Basic Documents, INMARSAT*, (3rd edn.), (London: INMARSAT, 1986). The Convention is also IMCO Doc. MARSAT/CONF/35 of 2 September 1976.

23 Art. XVII of the Operating Agreement ties the effectiveness of that Agreement to that of the Convention.

24 The list of initial shares goes on: Sweden (2.3 per cent), Denmark (2.1 per cent), Australia (2 per cent), India (2 per cent) and Brazil (1.5 per cent), Kuwait (1.48 per cent), Poland (1.48 per cent) before going below 1 per cent.

25 Assembly, Council and investment shares are discussed below.

26 Amendments to the Agreements to include air mobile and land mobile services within INMARSAT's remit are under way. See n. 50.

27 See also the Protocol on the Privileges and Immunities of the International Maritime Satellite Organization (INMARSAT) printed *Basic Documents, INMARSAT* (3rd edn.), (London: INMARSAT, 1986).

28 Headquarters Agreement between the Government of the United Kingdom of Great Britain and Northern Ireland and the International Maritime Satellite Organisation, London 25 February 1980, (1980) UKTS No. 44, Cmnd. 7917. The version printed in *Basic Documents, INMARSAT*, (3rd edn.), (London: INMARSAT, 1986) also contains the Exchange of Letters between the UK and the INMARSAT Director General as to the interpretation of the Agreement.

29 Note the possibility of difficulty with the COMSAT designation implicit in US moves on airmobile services indicated at the end of the chapter on COMSAT, p. 63 above.

30 Statute on the All-Union Maritime Satellite Communications Association 'Morsviaz' sputnik', confirmed by the Minister of the Maritime Fleet of the USSR, 29 April 1976, (1982) 20 ILM 1365–70.

31 In December 1988 INMARSAT had 52 members. The 22 member Council was made up as follows: of the eighteen seats filled by those 'largest by investment share' twelve were held by single members. These (with investment share taken from *Ocean Voice*, vol. 8, no. 4, October 1988) were: Australia (1.29719), Canada (1.49840), Denmark (1.88014), France (3.41418), West Germany (1.83138), Greece (2.68666), Italy (1.55367), Japan (9.47408), Norway (13.98511), Singapore (2.69342),

UK (15.14069) and the USA (27.47075). In addition six groups were represented on the basis of their cumulative investment shares: Brazil (1.80597), Portugal (0.13888) and Peru (0.05000), a total of 1.99485; The Netherlands (2.19766) and Belgium (0.49427), a total of 2.69193; Saudi Arabia (1.06774), Kuwait (0.89753) and Oman (0.05000), a total of 2.01527; Spain (1.99787) and Colombia (0.05000), a total of 2.04787; Sweden (0.74534) and Finland (0.34688), a total of 1.09212; and the USSR (3.29473) and Bulgaria (0.24098), a total of 3.53571. Of these, Brazil, Kuwait, the Netherlands, Saudi Arabia, Spain and the USSR would have been entitled to separate representation on the basis of 'largest share holders'. Twenty-six members were represented either individually or as part of a group.

In addition, as outlined in the text below, four seats on the Council are elected from among those not otherwise represented to as to provide geographic representation. These seats are held in December 1988 by China (0.30453), Chile (0.05000), Gabon (0.06229) and Poland (0.36636). Thirty members therefore represented on the Council either individually or as part of a group.

The total investment shares represented on Council was 97.0866 per cent, the remainder being distributed between the other 22 members: that seems a satisfactory representation ratio.

32   In 1987 there were three such groups: The Netherlands and Belgium with an aggregate investment share of 2.82726 per cent and a voting participation of 2.99856 per cent; Saudi Arabia, Kuwait and Oman with an aggregate investment share of 1.51700 per cent and a voting participation of 1.68830; and Sweden and Finland with an aggregate investment share of 0.82668 and a voting participation of 0.99799 per cent. See 'Representation and Voting Participation on the Council', DG, Council/26/17, 2 March 1987. As explained below, the voting participation varies from the investment share because of redistribution of the extent by which the US share would otherwise exceed 25 per cent.

33   In 1987 the four 'special' seats were held by China, Bulgaria, Gabon and Chile.

34   Bound with *Basic Documents, INMARSAT*, (3rd edn.) op. cit.

35   In the redistribution of 1987, 3.59735 per cent otherwise 'belonging to' the USA was redistributed equally to the other 21 members of the INMARSAT Council for voting purposes: 'Representation and Voting Participation on the Council', DG, Council/26/17, 2 March 1987.

36   Earth station is inappropriate in speaking of ships or aircraft.

37   International Agreement for the Use of INMARSAT Ship Earth Stations within the Territorial Sea Ports, London 16 October 1985, (1987) UK Misc. No., Cm 149; INMARSAT 1986. Twenty ratifications are required and the Agreement is not yet in force (December 1988).

38   Cf. H.H.M. Sondaal, 'The Current Situation in the Field of Maritime Communication Satellites: "INMARSAT"', op. cit., pp. 9–39 at p. 39.

39   The limit may be revised by the Council to take account of currency fluctuations (art. 20(2)(a)).

40   This is a matter which affects not only procurement of equipment but also the launch opportunities for INMARSAT. Naturally the satellites have to be designed to be capable of launch by a variety of launch vehicles, and the USSR's Proton rocket is a possible launcher. However, the military in the West may have reservations about using a Russian rocket for INMARSAT launches. I am not aware whether INMARSAT has a similar arrangement to that of INTELSAT in respect of the requirements of secrecy imposed on US contractors.

41   See above, section, 'The Council', p. 223.

42   Annex: Investment Shares Prior to the First Determination on the Basis of Utilization, paras. (b)–(f).

43   See following section 'Settlement of Disputes', p. 237.

44 See below, Section 'Separate Systems', p. 236.
45 See 'In the Matter of GTE Spacenet Corporation; For Modification of Authoriza-
    tion for the GSTAR III Domestic Fixed-Satellite; and Geostar Positioning
    Corporation; Application for Blanket License for User Terminals in the Radio
    determination Satellite Service', FCC File No. 829–DSS–MP/LA; File Nos. 974–
    DSE–P/L–88; CSG–88–023–P/1, adopted 12 August 1988, released 15 August
    1988.
46 The INTELSAT coordination was accomplished in October 1988: see INTELSAT
    Assembly of Parties, Record of Decisions of the Thirteenth (Extraordinary) Meet-
    ing, Washington DC, 11–13 October 1988, AP-13-3, paras 15 and 16. The
    INMARSAT coordination is likely to be settled in January 1989.
47 Annex: Procedures for the Settlement of Disputes referred to in Article 31 of the
    Convention and Article XVI of the Operating Agreement, annexed to the
    INMARSAT Convention.
48 See Chapter 8, The International Telecommunication Union, Section 'Other De-
    velopments prior to WARC–GRB 1985–88', p. 381. See now also 'Agreement of
    Cooperation between the International Civil Aviation Organisation (ICAO) and
    INMARSAT', INMARSAT Council/30/44, DG, 8 July 1988.
49 Convention on International Civil Aviation, Chicago, 7 December 1944, 15 UNTS
    295; (1953) 8 UKTS, Cmd. 8742; TIAS 1591, 61 Stat. 1180.
50 See for a history of the matter, J.-L. Magdelenat, 'INMARSAT and Satellites for
    Air Navigation Service', *Air Law*, vol. 12, 1987, pp. 266–81. The amendments are
    printed (1988) 21 ILM 691–694.
51 See the discussion of this problem, Chapter 2, above, at note 122, p. 63.
52 See the paper by the Director-General of INMARSAT, O. Lundberg, 'Mobile
    Satellite Communications in the 1990s: An International Overview', *Telecom-
    munication Journal*, vol. 54, 1987, pp. 611–18.

6 Europe: A Regional Response

Apart from the national development of telecommunications services which have some spill-over into international traffic, and apart from the major global telecommunications systems of INTELSAT and INMARSAT two other telecommunications networks were likely to develop – one global and one regional. The global system, INTER-SPUTNIK, developed within the Communist bloc, cannot be said to be a major network in terms of its membership, nor, it is rumoured, in terms of its traffic. It will be considered later in these pages. The regional development is the telecommunications satellite system which we know as EUTELSAT, the European Telecommunications Satellite Organization.

Europe is one of the major centres of international and national telecommunications traffic. Intercontinental traffic was always likely to be carried by INTELSAT, but intracontinental traffic was a different matter. The nascent space industries of the region, the local telecommunications providers and the governments all saw satellite telecommunications as something which might usefully be approached on a regional basis. Telecommunications was, therefore, a matter which was to be fitted within the general developments in space which occurred in Europe. Therefore we deal first with the emergence of the European Space Agency (ESA) before turning to the specific response in the field of telecommunications which led, through the European Conference on Posts and Telecommunications (CEPT) and through ESA, to EUTELSAT.

THE EUROPEAN SPACE AGENCY (ESA)

Introduction

While it is clearly necessary to treat EUTELSAT separately, that organization fits into a more general European effort in space matters which must also be dealt with. The principal avenue through which that European effort is cooperatively directed is the European Space Agency (ESA).

Space research and its technological applications are areas where success normally comes only through the investment of vast quantities of money and the efforts of many minds. It is expensive in terms of cash, time and skill. These constraints have required cooperation amongst European nations, but it was a cooperation not easily obtained and which even now still has its difficulties.

In the immediate aftermath of the Second World War various European countries tried to keep a foothold in the developments in technology which had been accelerated by the war. There were both civilian and military aspects to this effort, although naturally in the light of the then recent history the military took precedence. Rocketry was an attractive technology which, if it had not proved, had at least indicated its potential. At that time, countries wanted to have their own technologies based in their own industries under their own control, in order to be militarily secure: they were not willing to be dependent on purchase even from an ally. Britain, for example, sought through the Blue Streak project to develop a military rocket using a technology which had also application to space launchers. Of course, such state concerns were not limited to what we would now term space matters. The UK and other countries tried to maintain their position in nuclear developments as well as military rockets and the like. However, in many of these fields of endeavour it was soon obvious that the USSR and the USA had established a commanding lead.

In the 1950s it became clear that, if Europe was to remain in touch with the progress being made elsewhere, some European cooperation would be necessary in space research and development. No European country could afford separately to finance endeavours across the broad front of space activity. No single European country could supply the numbers of experts and specialists needed to match those available for such enterprises within the USA or USSR. The supply of 'brains' was spread over a great number of projects in each country, thus reducing their impact on any one area. However, it seemed possible that the countries of Europe could make significant contributions by combining their resources, and reap corresponding rewards through new applications of new technology. One early result was the European Organization for Nuclear Research (CERN), whose programme, centring on the huge nuclear accelerator outside Geneva, has done much to further the understanding of particle physics and the four basic forces. Curiously though, it may be noted that CERN was fostered into being through the United Nations Educational, Scientific and Cultural Organization (UNESCO), which in the 1950s was seeking to encourage regional developments in the sciences.

Broad-scale collaboration on space research and technology within Europe was slow to develop. Only in the 1960s were the first European space organizations created – the European Space Research Organization (ESRO) and the European Launcher Development Organization (ELDO). To them must be added the European Conference of Space Telecommunications (CETS), set up as a sub-group of the European Conference of Posts and Telecommunications (CEPT) discussed below in connection with the establishment of EUTELSAT. The members of these various bodies also came together to form an important extra-legal discussion forum, the European Space Conference (ESC) which first came into existence in 1966. Although it became permanent in 1967, it never had formal legal personality and operated as an arena for the discussion of space matters of interest to its members, sometimes at a ministerial level and sometimes at levels below that. All members of ESRO, ELDO and the CETS were invited to join in the ESC, and this body was instrumental in the later development of ESA as will be seen below. Through its coordination of much of the European space programme throughout the 1960s and 1970s the ESC was of major importance, although it never left much legal trace of its existence.

The European Space Research Organization (ESRO)

The European Space Research Organization (ESRO) is an example of how, occasionally, the efforts of individuals acting outside normal diplomatic channels can have major effects. It is said that the success of the European Organization for Nuclear Research (CERN) provided a spur, but whatever the truth of that, it is clear that Professor Amaldi of Italy, Professor Auger of France and Sir Harry Massie of the UK were instrumental in bringing about ESRO. Following discussions which they arranged, the Council of Europe recommended, on 24 September 1960, the setting-up of a European agency to develop and build a space vehicle and to promote the peaceful uses of outer space (Council of Europe Recommendation 251). In December 1960 a formal intergovernmental meeting was held at Meyrin near Geneva, the headquarters of CERN, and set up a European Preparatory Commission on Space Research (COPERS) with 12 participating countries (Austria, Belgium, Denmark, France, West Germany, Italy, the Netherlands, Norway, Spain, Sweden, Switzerland and the UK).[1] COPERS both helped form the eventual organization through considering possible constitutional structures, and was also important in working out the initial programme for it.

In the COPERS discussions it soon became obvious that the requirements of space vehicle research and development and those of pure space research were different, and it was thought that they would probably best be served by separate organizations. In addition, certain countries which were keen to engage in cooperative space research felt that launcher development had too many military overtones. Accordingly it was decided that separate organizations would be the best solution to the difficulties – a decision which led in due course to the formation of ESRO and the European Launcher Development Organization (ELDO) dealt with below. The Convention for the Establishment of a European Space Research Organizatiron (ESRO) was signed at Paris on 14 June 1962 and came into force on 20 March 1964.[2] Ten countries – Belgium, Denmark, France, West Germany, Italy, The Netherlands, Spain, Sweden, Switzerland and the UK (that is, the membership of COPERS less Austria and Norway, which were later accorded observer status) – signed and ratified the Convention.

COPERS, the Preparatory Commission, continued in existence until ESRO came into being in 1964. As indicated above, it was important in providing a forum for consideration of the constitutional structures of ESRO, and also in working out the programme for the organization's initial years.

The purpose of ESRO was 'to provide for, and to promote collaboration among European States in space research and technology, exclusively for peaceful purposes' (art. 2). The programme indicated in art. 5 was extensive, including the design and construction of sounding rockets, procurement of launch facilities and vehicles, research and the dissemination of information. To that end, art. 7 laid down that the organization would provide for the launching of sounding rockets, small satellites in near-earth orbit and small space probes, and large satellites and large space probes. A European Space Technology Centre with a research laboratory was to be established (art. 6(a) and (b)), and a Data Centre and tracking, telemetry and command facilities (art. 6(d)). Sounding rocket facilities were also to be established and operated (art. 6(c)).

Other important elements in the ESRO Convention included a requirement that 'the scientific results of experiments carried out with the assistance of the Organization shall be published or otherwise made generally available' (art. 3(1)), and that 'subject to patent rights', the technical results of activities were similarly to be made available (art. 3(2)). Members were to facilitate the exchange of all scientific and technical information, but the obligation of interchange of non-ESRO information was subject to saving clauses for security and military reasons (art. 3(3)). The exchange of personnel was also to be facilitated (art. 4).

Organizationally, ESRO consisted of a Council (art. 10(1)) composed of not more than two representatives per member, plus their advisers, which met at least twice a year (art. 10(2)), and a Director-General with appropriate staff (art. 10(1)). As well as having extensive supervisory powers, the Council determined policy in scientific, technical and administrative matters, and approved programmes and annual work plans (art. 10(4)). Most importantly, every three years it was responsible for determining the level of resources to be available to the organization for the next triennium (art. 10(4)(c)). In this matter the Council required the unanimous decision of all member states. In other matters, various levels of voting were required, with each member having one vote (art. 10(5)(a)) and there was provision for loss of vote if in financial default (art. 10(5)(b)). The organization's finances were set out in a separate financial protocol, a main element of which was that no member was to be liable for more than one quarter of the budget. (The UK had that level of liability.)

It has to be said that ESRO was reasonably successful while it existed, and attained a number of satellite flights. As it developed, it set up planning and administrative headquarters at Paris (art. 1(3)), a Space Research and Technology Centre at Noordwijk in Holland, and had its Data Centre and its Space Operations Centre at Darmstadt in West Germany. Although the initial Convention indicated a commitment to space research, the exploitation of space technology was not long left off the ESRO agenda. Thus in 1966 the European Space Conference assigned applications programmes to ESRO, and Resolution 1 of ESRO of July 1970 required ESRO to undertake a communications satellite programme which would produce a service for the European Broadcasting Union (EBU) and the European Conference on Posts and Telecommunications (CEPT). However, this produced constitutional problems, requiring a revision of the Convention. Until that was executed, use was made of art. 8 of the Convention under which the organization had power, by a two-thirds majority, to make available its assistance to projects being engaged in by its members which fell outside the formal ESRO programme.[3] Acting under that kind of legal arrangement and the general ESRO umbrella in 1971, Belgium, France, Germany, Italy, Sweden, Switzerland, the UK, USA and Canada launched the AEROSAT programme;[4] the European states listed started a meterological programme; and the same participants also began an experimental communications satellite programme intended to meet the terms of Resolution 1 of 1970 mentioned above. Later, in 1973, a maritime satellite programme (MAROTS) was added as a fourth applications programme.[5]

The European Launcher Development Organization ELDO

The European Launcher Development Organization (ELDO) grew from the UK's abandonment of its 'Blue Streak' military rocket development project. Work on Blue Streak had been carried out throughout the 1950s, but in 1960 it was clear that no significant return was to be had from that investment. At the same time France had been developing a smaller rocket for small satellites, 'Coralie', and it was thought that the two projects could be combined to produce a heavy launch vehicle. Accordingly the two governments arranged a series of conferences and consultations with other European countries. As we have seen, in autumn 1960 the Council of Europe recommended the setting-up of a European agency to develop and build a space vehicle and to promote the peaceful uses of outer space. The European Preparatory Commission on Space Research (COPERS) discussed in detail whether to keep space research and launcher development within the compass of one organization, but eventually decided in separate organizations for the two significantly different enterprises. Accordingly, further meetings about launcher development were held in 1961 and the result was a Convention for the Establishing of a European Organization for the Development and Construction of Space Vehicle Launchers signed on 29 March 1962, which entered into force on 29 February 1964.[6] Six European countries originally signed the treaty (Belgium, France, West Germany, Italy, The Netherlands, and the UK), along with Australia, whose participation was sought in order to gain access to the Australian facilities at Woomera for the testing and launching of the rockets. It will be seen that this was a membership significantly less than that of ESRO, and reflects the divergent interests in space research and launcher development.

Experience with the eventual ELDO rocket was not happy. In order both to make use of existing expertise and to spread the development work, various ELDO members had charge over the separate elements. The rocket had three stages, a Blue Streak base, a Coralie second stage and third stage, 'Astris', of German origin. Italy built the first satellites. Holland did the telemetry, and Belgium was responsible for the command and guidance control systems. By 1972 there had been no successful launches, and when in that year NASA offered either to make available launch vehicles for European use, or to launch European satellites through NASA facilities, ELDO's justification was considerably reduced and, as will be seen below, it was merged into ESA when that came into being. Even so, the picture is not one of entire failure. ELDO's technology has contributed to ESA and the semi-private French development of the Ariane launcher, which is presently a major competitor in the international satellite-launching business. It has also, at one remove, resulted in the development of the Littleo project, the 'Little Launcher for Low Earth Orbit'.[7]

Organizationally, ELDO consisted of a Council (art. 14) and a Secretary-General with appropriate staff (art. 15). The Council was composed of two delegates from each member state (art. 14) and was responsible for policy-making. The Convention also provided for members to have access to information gained through ELDO work (art. 8), for the distribution and placing of contracts (art. 6) and for the commercial exploitation of inventions and developments made in the course of the work of the organization (art. 10). The financial burden of the organization was distributed among the Parties to the Agreement, roughly on the basis of means and interest (art. 18 and Financial Protocol to the Convention). Elements to be found in later International agreements can therefore be seen to have been foreshadowed in this relatively early space agreement, showing how fundamental these matters are. Information, a policy which distributes procurement contracts along members, and a commercial return on investment are basic to all applied space technologies, except perhaps in the military sphere. In addition, one other provision of the ELDO Convention should be specially noted. Under art. 4, states party to the Convention were bound to the initial programme set out in art. 16, but programmes later decided upon by the ELDO Council were optional (art. 4(3)), access to them being conditional on acceptance of a share of financial responsibility. Although the obligation to participate was expressed in art. 4(3), that paragraph also provided that a member could formally declare itself 'not interested' in a further programme.[8]

The European Space Agency (ESA)

Genesis

The European Space Conference (ESC) was mentioned earlier as having been an arena for international discussion on developments within Europe in space research and technology. It was an institution which did not require to have legal personality, serving merely as a forum for the official and unofficial interchange of views and ideas. As such it was useful.

The initiative for the establishment of the ESC came from ELDO. That organization, concerned as we have seen with the launch of space satellites, found that it was dealing both with ESRO and with the European Conference on Satellite Telecommunications (CETS). In addition, proposals were also coming to ELDO (often through CETS) from Eurospace – a private organization which represented different elements of the European space industry – and certain countries, notably Germany and France, were announcing their own projects. Some forum for the harmonization of policies was therefore clearly required. Accordingly, in July 1966, the ELDO Council adopted a resolution proposing that the Committee of Alternatives (the civil service subordinates of the government ministers forming the Council) should examine the question and suggest ways through which such harmonization could be attained. Previous experience made it obvious that a simple Conference mechanism, without formal legal existence, would be suitable for that purpose. Progress was swift. The first meeting of the ESC was convened in Paris in December 1966. That Conference agreed that the coordination of the European effort was desirable and could be attained through its mechanism. At its second meeting in Rome in July 1967, the ESC was therefore constituted as a permanent body with legal personality, its decisions to be binding on all its members, but to be taken by unanimous resolution. Thereafter the ESC did good work in bringing some coherence to the space acitivites of the several European states and organizations. Its work in the area of satellite telecommunications is referred to in the section on EUTELSAT.

For our purposes in this chapter an important meeting of the ESC was that held at Brussels in July 1970. One item of discussion was a NASA offer of participation in the US space programme to follow on the Apollo missions, which included the Shuttle programme. That proposal took most attention at the time, but also on the agenda was the first consideration of the merging of ESRO and ELDO.[9] Discussion of a merger was inconclusive, but it showed that the possibility of a return to a single organization with both spacecraft and space launch capabilities was once more on the cards. Experience was indicating that that decision taken in the early 1960s to separate the competences into ELDO and ESRO was proving unfortunate, if not wrong.

Time strengthened the force of argument. By 1972, ELDO's launch programme was known to be in serious difficulties. By then also, negotiations had progressed on a further NASA offer to provide launch facilities and also separately to include ESRO experiments on certain NASA launches. At a meeting in December 1972, therefore, the ESC took the decision that a new organization should be established merging ESRO and ELDO and, for many purposes, in effect the ESC itself. However, as noted by one commentator,[10] the major matter which preoccupied the participants was the settling of the future space programmes rather than the drafting of the constitution of the new body. Broad agreement on certain space applications programmes which had been begun in ESRO was achieved in December 1971, with later legal results.[11] In December 1972 and June 1973 further agreement was reached on the MAROTS programme,[12] on SPACELAB (participation in the US Shuttle programme), and on Ariane (the ESA heavy launcher). In passing, it may be noted that these programmes represented a tilt in the general European programme to one weighted in the direction of space applications. Purely scientific investigations were not excluded, as for example in SPACELAB, but the new emphasis on applications shows that arguments in favour of Europeans consciously developing a space industry and of obtaining a satisfactory return on investment were having an effect. The earlier ideal of ESRO as a purely space research organization had, as we have seen, been diverted towards applications programmes by the interests of governments and industry. That diversion was now formalized, and, some would say, legitimized: but pure research has arguably been hindered. Such decisions do, however, have to be made in a high-cost, high-risk area of endeavour.

The ESA Convention

The Convention

The Convention for the Establishment of a European Space Agency was opened for signature at Paris on 30 May 1975.[13] Its history thereafter was legally unusual because, although the Parties to it immediately behaved as if it were in force, the ESA Convention did not come into force until more than five years later on 30 October 1980. By its art. XXI.1 the Convention was to enter into force when ratified by named parties, these being the members of ESRO and/or ELDO, and the ESRO and ELDO Conventions were to cease to have legal effect on the coming into force of the ESA Convention (art. XXI.2). Clearly it was, however, undesirable that there should be uncertainty as to the coming into force of the Convention because of the differing timetables each state might have for taking the appropriate action in terms of its own constitution. It was also undesirable that the organizsations being replaced should limp on until that time. Accordingly, and unusually in international law, by its Resolution No. 1 the Conference of Plenipotentiaries at which the Convention was signed resolved that as from the day following the signature of the Final Act of the Conference (which included signature of the Convention) the ESRO and ELDO Councils should meet jointly, and that 'in the application of the Conventions for the establishment of ESRO and ELDO the provisions of the Convention for the Establishment of a European Space Agency should be taken into account to the greatest possible extent'. The effect of compliance with this recommendation was, as the resolution itself acknowledged and intended, to permit the immediate functioning of ESA on a *de facto* basis. Therefore until the ESA Convention did come into force on 30 October 1980, the legal basis of what was done by 'ESA' in the intervening five years and five months remained the ESRO and ELDO Conventions.

I believe that this device permitting the new institution to operate was a legal novelty, although it does have affinities with the interim organizations of INTELSAT, INMARSAT and EUTELSAT. The difference in law between ESA history and that of the telecommunications organizations lies in two respects. First, the ESA practice was predicated upon the technical continuation in force of the two predecessor conventions and on the continued existence of two organizations which were each possessed of full legal personality (art. 14 ESRO Convention; art. 20 ELDO Convention). Second, while the arrangements for the interim telecommunications organizations differ from the definitive arrangements, the interim arrangements expired coincidentally with the new systems being brought into operation. In ESA the transition period while the Convention was signed but not yet in force was used by the two organizations which were to cease to exist to establish and operate the new systems.

Purposes

As laid down in art. II of the ESA Convention, the purpose of ESA is to provide for and to promote cooperation amongst European states in space research and technology. Such cooperation is for peaceful purposes only and covers both pure scientific research and the use of discoveries made both in science and in technology. Operational space applications systems are expressly mentioned. To that end, ESA determines and implements a long-term European space policy both through recommendations to its members and through coordinating the policies of the separate members (art. II(a)). It also implements space programmes (art. II(b)). It is the mechanism through which the member states' space programmes are coordinated and through which national programmes are progressively integrated as completely as possible into the overall European programme (art. II(c)). In this, the development of applications satellites is expressly indicated (art. II(c)). Finally, although not the least importantly, ESA elaborates and implements industrial policy affecting its programme and also recommends what is called a 'coherent industrial policy' to its members (art. II(d)).

In order to achieve these purposes ESA has legal personality (art. XV.1), and it, its staff and the members' representatives have full legal capacity and the privileges and immunities provided for in a separate Annex (Annex I) to the Convention (art. XV.2). It has the capacity to enter into agreements necessary for the accomplishment of its purposes, and to hold land and other facilities which may be necessary (art. XV.3, and by implication from *inter alia* art. VI). It also itself engages in programmes and in that connection carries on the differentiation made in the ELDO Convention between the mandatory and optional activities of the organization (see below).

The ESA Council

The organs of ESA are its Council and the Director-General with his staff. The Council comprises representatives of the member states (art. XI.1) and may meet either at delegate or at ministerial level (art. XI.2). Meetings are held as and when required (art. XV.2). When the Council meets at ministerial level it elects a chairman for that meeting, who also convenes the next ministerial meeting (art. XI.4). In other cases, the chairman of meetings is the chairman of the Council who, along with two vice-chairmen, is elected for a two-year term by the Council itself, and may be re-elected for one further year. As set out in art. XI.3(a) the chairman directs the Council proceedings, deals with minutes, informs members of proposals on any optional programme and coordinates the subordinate organs of the Agency. He liaises with Members on policy matters affecting ESA and attempts to harmonize views. He also advises the Director-General between meetings and is informed by him as necessary. In the performance of his duties the chairman is assisted by a Bureau, which serves the same function as the secretariat in such organizations as INTELSAT and INMARSAT (art. XI.3(b)).

Quorum and voting   The quorum of the Council is composed of representatives from a majority of all member states (art. XI.6(c)). Each member has one vote (art. XI.6(a)), but no member may vote on an optional programme in which it is not a participant (art. XI.6(a)). Unless the Convention provides otherwise, the majority required for a decision by the Council is a simple majority of representatives present and voting (art. XI.6(d)). However, in determining whether any vote meets its required level, no account is taken of a state which has no vote (art. XI.6(e)). This situation could occur where a state has no interest in an optional programme (art. XI.6(a), cf. above), or where it loses its vote because financial default. Arrears equal to the current year's worth of contributions causes a Member to lose its voting rights (art. XI.6(b)). Where the default is in relation to a particular programme, the loss of voting rights is restricted to questions relating to that programme (art. XI.6(b)). However, a two-thirds majority of all Members of the Council may decide that any default is due to circumstances beyond the defaulter's control. In this case, the defaulting member may retain its voting rights notwithstanding the default (art. XI.6(b)).

The Council is responsible for the approval of the Agency's mandatory activities and programme (see below). This it does by a majority but, once adopted, a decision on these activities can be changed only by a two-thirds majority of all Members (art. XI.5(a)(i)). However, by art. XI.5(a)(ii) and (iii) the Agency works on a five-year financing basis, and decisions as to the level of funding for mandatory programmes have to be unanimous. There is, therefore, some additional room for control or influence on Agency mandatory activities by a dissentient Member. Decisions on the acceptance of optional programmes is by a majority (art. XI.5(c)(i)).

Activities

In respect of its own activities the ESA Convention picks up the differentiation made in the ELDO Conventions between the organization's mandatory and optional programmes. Art. V of the ESA Convention states that the 'activities of the Agency shall include mandatory activities, in which all Member States participate apart from those that formally declare themselves not interested in participating therein'. As in the case of ELDO, therefore, the default position is that all Members participate in all activities, and must separately contract out of optional activities in which they do not wish to be involved. There is a voting effect in that case (see above), and to be effective the decision to opt out must be communicated to the Agency within three months of the Council's decision on the particular activity (Annex III, art. I.2).

Mandatory activities   ESA's functions are more wide-ranging in respect of its mandatory activities than for the optional activities. These functions include basic matters such as documentation and study, technological research and 'education' (art. V.1(a)(i)). The collection and dissemination of relevant information, the duty of drawing attention to gaps and duplications, and the provision of advice and assistance towards the harmonization of both international and the national programmes of its members are also part of ESA's duties (art. V.1(a)(iii)). It is also required to maintain regular contacts with the users of space techniques and to stay informed as to the requirements of users (art. V.1(a)(iv)). Buried in the list of duties in respect of mandatory activities, the Agency is to 'ensure the elaboration and execution of a scientific programme including satellites and other space systems' (art. V.1(ii)): a requirement which sums its central endeavour.

Optional programmes   In relation to its optional activities, the Agency is to ensure that programmes decided upon are carried out. The kind of programmes to be engaged upon are indicated in art. V.1(b) and include the placing in orbit and control of satellites and other space systems, and the operation of launch facilities and space transport systems. Naturally the design, development, construction and all other necessary elements of such space programmes are also included. This catalogue would seem potentially to encompass every space activity, but it remains possible for member states to engage upon work outwith the 'Optional Programme' category; such, work is, however, also brought within the influence of the Agency by the requirements of the internationalization of space programmes.

The procedural requirements for an optional programme are laid down in Annex III. A proposal is communicated by the chairman of the Council to all members for examination (Annex III, art. I.1). The matter is then decided upon by the Council, on a two-thirds majority (art. XI.5(a)(i)), and, as noted, Members then have three months to intimate that they do not wish to participate (Annex III, art. I.2: Convention art. V). Thereafter, the participating states execute a Declaration setting out the phases of the programme, the conditions under which it is to be carried out (including an indicative financial envelope), the scale of contributions to it, and the duration and amount of the first binding financial commitment (Annex III, art. I.2). In the absence of unanimity among the participants, the financial arrangements are similar to those for the Agency itself, being based upon average national income for the last three years for which statistics are available (revisable by the Council triennially or as it sees fit), subject to a maximum of 25 per cent of the programme cost, and subject to a minimum for each participant equivalent to 25 per cent of its assessed contribution to ESA (Annex III, art. I.2(c), Convention art. XIII.2).

The Declaration is not, however, final. There are points in each programme at which participants may withdraw, or the programme itself may be discontinued. As to the first, participants have a period stipulated within the Declaration itself in which further to consider its position, and a participant may withdraw before the Declaration becomes binding (Annex III, art. I.3). There is a further opportunity to withdraw where the programme has a project definition phase. At the end of such a phase, if the reassessed cost of the programme is more than 20 per cent higher than the originally indicated financial envelope, any participant may withdraw (Annex III, art. III.1), and those that remain may rearrange the programme accordingly. A similar facility exists at each stage, but only where the indicative financial envelope (which may have been modified by the Council is in the time to take account of price fluctuations (Annex III, art. III.3) has been overrun by more than 20 per cent (Annex III, art. III.4)). Naturally also, the withdrawal of a member from ESA entails its withdrawal from any optional programmes in which it may be concerned (Annex III, art. V).

An optional programme may be discontinued by the decision of a two-thirds majority of participants liable for at least two-thirds of the contributions to the programme (Annex III, art. VI.1). On a happier note, however, the programme may also terminate upon its completion in terms of the rules set down in the relevant Declaration (Annex III, art. 2).

During the currency of an optional programme, it is for the Council to take decisions on matters of budget and implementation in accordance with the agreed rules and details set out in the applicable Declaration (Annex III, art. II). Decisions on the start of new phases of a programme ordinarily require a two-thirds majority of the participants but, failing that, participants wishing to carry on with a project may determine how to do this among themselves. Council will then take any necessary measures (Annex III, art. 2).

Cooperation

As indicated, the purpose of ESA is the eventual implementation of a European space programme (art. II, particularly II(a)). Naturally, that objective has not been wholly achieved in the limited time which has elapsed since the *de facto* implementation of the Convention in 1975. However, there are provisions within the Convention which will tend towards its attainment through the internationalization of the national space programmes.

Art. I of Annex IV states that the object of the internationalization of national programmes is that each member state makes available for participation by other members any new civil space project which it intends to undertake either by itself or with other members. Therefore, the member beginning a project has to notify the Director-General before embarking on the project's definition phase (Annex IV, art. I(a)). Timing and the content of the proposals for participation 'should make it possible for other Member States to undertake a significant share of the work involved', and any restrictions, together with an explanation of them, have to be furnished (Annex IV, art. I(b)). In addition, the state initiating the project is to explain its proposals for its management (Annex IV, art. I(c)). All such information is disseminated to other members by the Agency. Therefore other member states must indicate their willingness or their wish to participate in the proposed project and the initiating state is to 'use its best endeavours to accommodate all reasonable responses' subject to agreement on, finance, work-sharing and the like (Annex IV, art. I(d)). The inability of an initiating State to obtain the desired level of participation by other members does not mean that the state can then go ahead with the project outside the framework of the Agency. Although the particular provision on that matter (Annex IV, art. I(d)), is carefully couched so as not to give offence, it is clear that that provision also means that national programmes cannot be kept out-side the purview of ESA by their being originally unrealistically framed. The Agency can also act as a mechanism through which proposals can be revised. After agreement on participation, if the project is to proceed as an optional programme of the Agency, the procedures of Annex III are begun (Annex IV, art. I(d)).[14] In any event, however, the Agency retains an interest in the matter through the Convention's other provisions which require the dissemination of information amongst members.

The effect of these provisions is to reduce the undue duplication of industrial investment and work within Europe as each state active in the space business attempts to make itself self-supporting. That it has worked can best be seen positively rather than negatively. Thus, while it can indeed be said that there has not been a growth of national space vehicle launcher technology and programmes, it is easier to point to the specialization that has occurred in certain areas of space industry. France and West Germany have concentrated on launchers, notably through the establishment of Arianespace – technically a commercial company but one in which the French Government has a majority shareholding, while the UK and Italy have given their attention to satellite technology.

Nonetheless, it should not be thought that the effect of ESA has been to stifle other development outside its ambit. Annex IV, art. II specifically deals with cooperation by members with non-members, allowing for it, but attempting to harmonize such developments also with the European effort. Members are to 'use their best endeavours' to ensure that bilateral and multilateral space projects with non-members do not prejudice ESA's scientific, economic or industrial objectives (Annex IV, art. II). So far as possible without prejudicing such projects, members negotiating such 'outside' projects are to in-form ESA of them (Annex IV, art. II(a)), and it is even contemplated that such projects could be drawn within the scope of ESA's own participative processes which have just been outlined (Annex IV, art. II(b)).

Facilities

Under arts. VI and IX of the ESA Convention, the Agency and the member states undertake various obligations as to facilities and ser-vices which are also directed, first, towards cooperation, and then towards the eventual integration of the European space programmes. Under art. VI, the Agency has the power and duty to establish and maintain facilities and establishments needed for its tasks, and also may enter into special arrangements for carrying out certain parts of the Agency's activities, including the contracting out to member states. Conversely, the Agency can take on the management of national facilities (art. VI.1(b)).

As noted, art. V lists the activities of the Agency, and these include the provision of the various operational facilities and services contemplated in art. V.2. Clearly, those first on the list are those that the Agency requires for the carrying out of its own activities, but it is important to note that art. V.2(c) indicates that the Agency can provide facilities at the request of users as well as those needed for ESA programmes themselves. While art. V.2 stipulates that these latter specially supplied facilities and services are to be paid for by those that request them, nonetheless it is clear that the result can be the development of a standing facility or service which can then be bought on a user basis by those requiring it. Such a development is aided by the obligation accepted by both members and the Agency to 'endeavour to make the best use of their existing facilities and available services as a first priority, and to rationalise them' (art. VI.2). Accordingly, art. VI.2 continues that 'they shall not set up new facilities or services without having first examined the possibility of using the existing means'. In the area of launchers and other space transport systems, the Agency is bound in terms of art. VIII to make use of launchers with which it or its members have a substantial involvement, provided that they are suitable. Under art. IX (which expands the limited reference in art. V.2(c)) the Agency can make its facilities available to members for their own programmes, at the cost to the state or states concerned.

By art. V.3 members are required 'in good time' to inform the Agency of projects relating to new space programmes. Thereafter the Agency facilitates consultations among the members; we have already seen the provisions of Annex IV directed towards the internationalization of such programmes.[15] Obviously national facilities and services are initially most likely to be used, but as time passes it will make more sense to develop national facilities in order to play a role within the European programme as well as for any remanent national programmes. It may also make sense for the Agency itself to provide facilities, as it has the power to do in terms of its Convention.

Resolution No. 7 of the Final Act of the Paris Conference which adopted the ESA Convention underlines the point. This, while recognizing the 'need to give preference to the use of the potential and facilities developed by the European Space Agency or belonging to it, and also the need to avoid setting up redundant facilities in Europe', it nonetheless invites the Agency 'when it has need to make use of the potential and facilities of the Member States, provided that there exists an economic case for so doing'. Costs would, of course, have to be agreed, but the Resolution indicates Parties' determination not to dissipate their efforts by duplication, to the detriment of a European stake in the space business.

The effect of such provisions is both to enhance the attractiveness of cooperation through encouragement, and to stimulate it through more direct legal obligation. Interstate cooperation becomes attractive both financially and otherwise. Members know that by rationalizing their programmes and facilities or services they do not lose through giving up a national facility, for they have access to the Agency's facilities, or through it to other national capabilities. The difficulty, of course, is that the Agency, or the other members, may not place as high a priority on a particular development or experiment or whatever, as does one member state. If it has completely given up its national capability, a state may therefore have to wait until a time or cost-slot is available for a purpose which is dear to it, or which it considers to be in its national interest. The fear of such an eventuality has resulted in some national capabilities being retained, although in European terms they are redundant. Nonetheless, that is how the real world operates. It is untidy, but it works.

Industrial aspects

Art. II of the ESA Convention gives the overarching purpose of the Agency as being 'to provide for and to promote', for peaceful purposes only, cooperation among the European states in space matters. The Article goes on to list various methods by which that provision and promotion should occur, the last, contained in art. II(*d*) being the elaboration and implementation of 'the industrial policy appropriate to [the Agency's] programme' and through recommending 'a coherent industrial policy to member states.

That industrial policy is further elaborated in art. VII, as expanded by Annex V to the Convention. The policy is particularly designed first to meet the requirements of the European space programme and the coordinated national space programmes in a cost-effective manner second, the worldwide competitiveness of European industry through maintaining and developing technology, and through encouraging rationalization. Here the aim is to develop an industrial structure appropriate to the market, making primary use of members' existing industrial potential. Third, the policy is to ensure the equitable participation of all members in the implementation of a space programme and its consequent technology, having due regard to their financial stake. As far as possible, preference in the execution of those programmes undertaken by the Agency is to be given to industry located in member states, and they are to be given the maximum possible participation in technological procurement. Last, the Agency is to use free competitive bidding in all cases where this is not incompatible with the other stated aims of the industrial policy. This listing of objectives of the industrial policy can be added to by a unanimous decision of the ESA Council, but no such decision has been made.

Annex V sets up procedures designed to implement the requirements of art. VII. In particular, mention should be made of certain provisions. Thus, the Director-General is required to act in accordance with art. VII, Annex V and Council directives (Annex V, art. I 1). The Council is required to keep under review the industrial potential and industrial structure in order to be able to monitor and, where necessary, adapt the Agency's industrial policy. In particular, it considers the general structure of industry and industrial groupings, the degree of specialization which is desirable and how to achieve it, the coordination of relevant national industrial policies, the interaction with relevant industrial policies of other international bodies (which would include the telecommunications organizations), the relationship between production capacity and potential markets, and the organization of contacts with industry (Annex V, art. I 1). However, although rationalization is one of the objectives which the Agency has in view in the adoption of industrial policy, member states still retain one control. Any decision taken on industrial policy grounds which has the effect of excluding a particular firm or organization of a member state from competing from Agency contracts requires the agreement of that member state (Annex V, art. VI).

Naturally the question of placing of contracts enters largely into the implementation of the ESA industrial policy. Art. II of Annex V therefore states the requirement of member preference in appropriate cases, and allows the Council to determine the extent of any derogation from that requirement. Whether or not a potential contractor belongs to a Member is determined by having regard to the location of its registered office, of its decision-making and research centres, and the territory where the work is to be carried out. In cases where doubt remains the Council itself determines the matter (Annex V, art. II 3).

Prior to inviting tenders for contracts above a financial limit defined in subsidiary rules, or in cases where the Director-General considers the policy rules and any additional guidelines set by the Council do not adequately cover the case concerned, the Director-General submits to the Council proposals as to the procurement policy to be followed (Annex V, art. III 1 and 2). Thereafter Agency contracts are awarded directly by the Director-General, except when the evaluation of the tenders received indicate an award which would transgress the Council's indicators (Annex C, art. III 3(a)), or where the Council has decided for specific reasons to review the matter before the contract is awarded (Annex V, art. III 3(b)). Furthermore, the Director-General regularly reports to the Council both on the contracts previously let and on planned contracts, so that the Council may monitor the Agency's implementation industrial policy, thereby allowing it to modify that implementation as it considers necessary to attain the objectives of the Convention in such matters (Annex V, art. III 3).

In order to secure the proper distribution of Agency contracts, Annex V art. IV gives general rules to secure geographic spread. The indicator used is called the 'overall return coefficient', which is the ratio between a state's percentage share of the total value of all the contracts placed by the agency among all member states and the state's total percentage contributions. In making this calculation, no account is taken of special (optional) projects undertaken under art. 8 of the ESRO Convention if the Arrangement under which the project is undertaken so stipulates, or of optional projects under art. V 1(b) of the ESA Convention, provided that all the states originally participating in the particular project so agree (Annex V, art. IV 1). Weighting factors are also applied to contracts on the basis of their technological interest, thus allowing for example, the attendant development of expertise to be taken into account (Annex V, art. IV 2). Return coefficients are computed quarterly, and formal reviews take place every three years (Annex V, art. IV 4 and 5). Separate assessments of coefficients of return are made for different categories of contracts, particularly for advanced research and development contracts and project-related technology contracts. These assessments are discussed by the Director-General with the Council at regular intervals in order to monitor and redress any imbalances (Annex V, art. IV 7). Ideally, the return coefficient will be 1 (that is, a state receives in contract value 100 per cent of its contributions to the Agency) (Annex V, art. IV 3). However, it was recognized that this was a counsel of perfection, and the first triennium coefficient returns were set at a lower limit of 0.8 (that is, a state would receive a minimum of 80 per cent of its contributions), with the Council there-after having the power to adjust the minimum coefficient of return, although never to less than 0.8 (Annex V, art. IV 6). Where the trien-nial review indicates that a member state has a return of less than the stipulated level, the Director-General is required to propose to the Council methods to redress that imbalance within one year, always, however, within the overall rules for the placing of contracts (Annex V, art. V 1). Only if the situation persists after that year can action be taken outside the normal guidelines for the placing of contracts (Annex V, art. V 2).

Such then are the ways in which the ESA Convention and Arrange-ments seek to deal with the vexed question of procurement, the dis-tribution of contracts and the harmonization and/or rationalization of the space industries within the membership of the Convention. Since the Agency's purpose lies in the development of the industries and not in the provision of commercial services, the question of the distribution of contracts is subject to parameters and criteria somewhat different from those attaching to similar decisions taken by the telecommunications organizations. In these latter, a primary aim of many members is the provision of a satisfactory service at economic cost, and questions of procurement distribution rank accordingly. In ESA the member states have come together to develop their own potential within the space business, and, although cost-effectiveness is an important factor in the elaboration of industrial policy (art. VII 1 (*a*)), it appears to rank more lowly it might first be thought. Member states have other concerns. While ESA provides a very useful forum for cooperative endeavour, and has achieved a good deal, national interests and the protection of domestic industries has, to a degree, hobbled the development of a true pan-European industrial policy for space. Whether the result has been good or bad is unproven. Certainly some members of ESA have been able to use the technical skills developed under ESA (and its predecessors) in bidding for contracts by other agencies, (notably Arianespace, and the British satellite producers), but of the nature of things we just do not know what would have happened under other circumstances. That said, it is also true that ESA has provided an umbrella under which European space industries have made progress, and which has led to a useful independence of the US space programme and contractors. For example, the abrupt limitation of civilian access to the space shuttle, imposed by President Reagan in the aftermath of the Challenger tragedy of February 1986, would have crippled an industry wholly reliant on the shuttle as a launcher. The ESA-related Ariane avenue, (though not without its problems) provides a solution, less subject to possible political intervention by governments over which Europe has little influence. The Chinese and Japanese launch facilities, to say nothing of the Russian option, are now there, but carry the same element of risk as the United States has been proved to have. It is better that Europe has the Ariane option – an option which it must be said is to the particular credit of France and West Germany. Since its decision to downgrade the Blue Streak project, the UK has not figured greatly in launch vehicle technology, although the announcement of the Little Launcher for Low Earth Orbit (Littleo) may rectify that situation.[16] In the UK, government and industry has preferred to concentrate on satellites, and immense strides have been taken in all forms of satellite work, ranging from remote sensing to telecommunications.

However, it has to be said that within the ESA format the UK has not been particularly helpful or cooperative. Notably there was the decision in August 1987 not to increase the UK budgetary contribution for space research available for ESA developments and monitored through the British National Space Centre. The charitable interpretation of the UK government's action is that the decision not to increase the £110 million allocated to such matters from government funding was taken in ignorance. Space research is expensive, and the Concorde and TSR-2 projects have demonstrated the ease with which prestige projects can overrun budgets. But the procedures for the allocation of procurement contracts within ESA indicated above do mean that, without taking a substantial investment share in the appropriate programmes, the UK will find that its contractors do not get contracts for hardware which will enable UK industry to retain its lead in techniques and skills. Despite pronouncements on price and quality, the requirement of cost-effectiveness ranks behind that of returning to member states a proportion of contracts which will relate to their 'subscription' for each project. Thus, in the aftermath of the UK refusal in August 1987 to increase its budgetary contribution, various British manufacturers lost ESA contracts and perhaps thereby leadership in various areas. The main argument from the UK government seems to be that, if the returns on space research are as commercially attractive as they are said to be, then private industry should step in and meet the bill for research and development, commercial operations and return not being the function of government. But that argument can be overstated to the point at which the interests of the UK as a whole are neglected and prejudiced. Other countries do not seem to make that mistake. They, perhaps, fall off the other side of the tightrope. But Ariane and similar projects, despite their problems and forays down expensive blind alleys, cannot be said to have been mistakes. Furthermore, being 'in' the institution of ESA, or participant in a project, the UK could exercise an adequate control over finance while still reaping commensurate benefit. The danger now facing the UK is that it is left outside major scientific and technological development and becomes merely a client, buying a service which it itself does not know how to provide, with all the dangers attendant upon that position.[17] Perhaps the device of impeachment should be disinterred.

THE EUROPEAN TELECOMMUNICATIONS SATELLITE ORGANIZATION (EUTELSTAT)

Introduction

It is not surprising that a regional satellite telecommunications organization with its own satellite system has been established to serve the

European area. The positions taken by the European nations at the negotiations of both the interim and definitive arrangements for INTELSAT and in the development of INMARSAT indicated that Europe was likely to seek some regional development within which it could play a primary role, and cater for its own particular interests.

Like INMARSAT and INTELSAT, the European Telecommunications Satellite Organization (EUTELSAT) was first established on a provisional basis pending the negotiation of definitive arrangements once the organization had built up some expertise and the problems of its activities had been discovered and in measure, solved. In the case of EUTELSAT, interim arrangements were entered into in 1977 and 1978. The definitive arrangements were opened for signature at Paris on 15 July 1982 and came into force on 1 September 1985 following some delay in both signatures and ratifications.

The concept of EUTELSAT has many sources. As indicated interim EUTELSAT came into being towards the end of the 1970s. It follows that the European participation in the worldwide telecommunications structures such as INTELSAT together with the European developments in other satellite and space endeavours contributed to the decision to proceed with a new regional organization. Some of that detail is presented earlier in this chapter in dealing with ESA and some in the chapter on INTELSAT. Here it will suffice first to recapitulate certain elements.

After the Second World War the various European nations were relatively slow to re-enter what would now be classified as space studies. There were other pressing problems of the reconstitution of economies to be dealt with. Once such studies again became possible, the Western European nations soon appreciated that progress in space matters would be expensive and might be more efficiently attained by collaborative effort. No single country could afford to maintain a full scientific programme across the different fields of space science inquiry. However collaboration was difficult to establish. Which country should concentrate on which areas of endeavour? For various reasons (mainly military) some countries were unwilling to drop out of certain areas. At length in the 1950s such organizations as ELDO and ESRO were founded, and operated with limited success. The reconstitution of these bodies into ESA in 1975 was a useful advance.[18] That is not, however, to diminish work done by ESA's predecessor organizations, and particular mention must be made of the European Communications Satellite (ECS) programme which ESRO began in 1971. That programme was a clear spur towards the development of the European regional space telecommunications arrangements. A further spur came through the meeting of post and telecommunications administrations within Europe.

The European Conference of Postal and Telecommunications Administrations (CEPT)

An important development in the provision and administration of telecommunications took place in Europe on 18 June 1959 when at Montreux, Switzerland, a conference of European postal and telecommunications administrations adopted an Arrangement constituting the European Conference of Postal and Telecommunications Administrations (CEPT).[19] That step owed its origin to an initiative of 1951 in the Consultative Assembly of the Council of Europe which had proposed a European Postal Union (a step which is within the contemplation of the Constitution of the Universal Postal Union).[20] Other proposals for a regional Postal Union were made in subsequent years as the idea was further refined, including a recommendation in 1958 by the then six members of the European Economic Community (EEC) that they establish a European Community of Posts and Telecommunications. However, in the following year the broader CEPT Agreement was adopted by the Montreux Conference, and the EEC proposal was dropped having been overtaken by events.

CEPT is principally devoted to the coordination of the administration of postal and telecommunication services within Europe. Membership is open to European postal and telecommunications administrations which are members of the ITU or the UPU as appropriate (art. 3(1)). As the Conference is merely an arrangement between technical bodies, there is no need for states to be involved and all that is needed after signature is the 'confirmation' of its signature by the agency involved (art. 3(2)). Although consultative, the Conference does have permanent organs including a Plenary Assembly meeting annually (art. 6(1)) and a Bureau. Below the Assembly level and reporting to it, the Conference has a Postal Committee and a Telecommunications Committee, dealing with these functions (art. 5(3)). There is no permanent secretariat, however, and annually the responsibility for the Conference and any necessary implementations of decisions during the ensuing year rotates to the Member Administration which has chaired the previous Conference (art. 7(2)). Two-thirds of the members form the quorum of the Conference, and each member administration has one vote. Members may represent themselves and one other member. Decisions are by simple majority (art. 8(2)), except where they concern a revision of the Arrangement itself which requires a two-thirds majority (art. 12(3)). Decisions as to the working of the Conference bind members. All other decisions have only status as recommendations (art. 8(3)).

Interim EUTELSAT

If we now return more particularly to telecommunications we find that CEPT was concerned specifically with satellite telecommunications from 1960 when a special working group of its Telecommunications Committee was set up to deal with the question. Thereafter, as indicated in the discussion of the interim INTELSAT arrangements, CEPT and its offspring, the European Conference on Satellite Communications (CETS) (established for the purpose), were important in helping form and influence the European participation in the negotiation of Interim INTELSAT. Indeed the European stance had a major effect on the eventual shape of the interim organization. However, and again as noted above, CETS was not so successful in promoting a unified European approach to the negotiations of the definitive INTELSAT arrangements and was terminated following upon their adoption in 1973. CEPT, however, continues to play its part in matters of European postal and terrestrial telecommunications.

The 13-year existence of CETS was not, however, solely devoted to questions of INTELSAT. Thus, and although possibly in part, as a bargaining counter in the initial INTELSAT negotiations, CETS produced and detailed 5-year programme for the development of a European telecommunications satellite through a specially constituted Space Technology Committee. In 1964 it established a Planning Committee to reassess these plans with the purpose of fleshing them out with regard both to the interim and permanent INTELSAT arrangements. Two years later it requested ESRO to study television broadcasting satellites. It also pressed for studies by its own planning staff and by CEPT on the financial aspect of telecommunications satellites.[21]

Apart from CEPT and its CETS, other organizations had also become involved in satellite studies. In the later 1960s Eurospace, an association of members of the European space industry, began economic studies some of which dealt with telecommunications requirements and opportunities. As a result in 1967 ESRO was approached to incorporate into its studies the specific requirements of the European Broadcasting Union for broadcasting satellite facilities. In addition, France and West Germany put forward the 'Symphonie' project on a bilateral basis.

As indicated earlier in this chapter, by this time the problems of the separate functioning of ESRO, ELDO and, indeed, the CEPT with its subdivision CETS were proving awkward. The solution found was the establishment of the European Space Conference (ESC) in December 1966 to act as a coordinating body, although without legal status.[22] In 1969 the ESC formed a group composed of representatives of CEPT, CETS, EBU, ELDO, ESRO and the ESC itself to consider how

the needs of CEPT and EBU might be met by a European telecommunications satellite system. In short, even the needs of
broadcasting as well as telecommunications were being brought
together. The group recommended in 1970 that the states members of
the various space organizations should commit themselves to a programme of satellites for a variety of space applications including
telecommunications and broadcasting, that studies be embarked on
and that a single organization should be established in order to carry
out that programme. For our purposes it is enough here to note that
from 1970 the ESC was active in promoting studies and work on a
variety of space projects. A major thrust of these activities were directed towards protecting the European stake in an independent
launch facility, which has proved to be a wise step. But telecommunications have only provided an element of the ESC's work. That,
and the various negotiations and compromises which proved necessary, were the forcing house through which ESA was developed.

When ESA started to operate *de facto* in 1975, telecommunications
was an area which the ESA Council was very willing to encourage.
EUTELSAT was the result. There were many reasons for this.
Telecommunications was clearly an area of cooperation in space
technology which had many advantages. As we have seen, the very
negotiation of the interim INTELSAT arrangements had been
instrumental in welding together the European telecommunications
agencies in order to present a united front to the US negotiators. The
schemes drawn up through CETS showed the feasibility of a European system. There was a large telecommunications market to be
served. The Europeans wanted actively to be involved in the
development of space telecommunications for many reasons, not all
of which were fully satisfied in the eventual result of the INTELSAT
negotiations. Many state members of ESA wanted national companies to be kept in touch with developments both in telecommunications technologies and in the satellite technologies. The technological
spin-offs, the development of national space industries and the allied
question of remaining abreast with developments which had military
and security as well as technological aspects had already led to a
variety of developments in Europe. Launcher development had been
forwarded by ELDO. It was seen as useful to continue an independent launcher facility which would free from dependence on the
USA, and the prospect of telecommunications launches added to the
viability of the launcher programme. A telecommunications facility
therefore had the advantage of justifying other endeavours within
the ESA jurisdiction. Finally, and undergirding many of these reasons, telecommunications was seen as an area of space activity in
which governments and other funding bodies might see some return
for their investment.

For all these reasons the development of a European telecommunications satellite programme was seen as desirable and attainable. Couple that to the perceived demand for telecommunications within the European region of the globe and the case for proceeding with a regional organization was considerable. One cannot say, however, that the case was wholly compelling merely on pure telecommunications grounds, since INTELSAT could have provided the service which was required. It has been suggested to me that INTELSAT was willing to permit the European development because it was taking the view that its function was to serve the world, and to cope with Europe would have been a diversion of resources required elsewhere. I doubt it. A major element was certainly the will of a group of fairly major INTELSAT members to develop their own industries, competence and expertise and to have, for regional purposes at least, the option of an alternative to INTELSAT service. It should be remembered that the European space industry was not particularly strong and as a result had not gained INTELSAT satellite contracts in the face of competition from US firms. Without the creation of a 'domestic' market for its skills the European space industrial complex would not have been able to break into the INTELSAT market. But this is unascertainable on the record. Let us leave it that the precise balance of reasons which led each participant to proceed with its involvement in EUTELSAT is unique to each country.

When it began *de facto* operation in 1975 ESA took on board the European Communications Satellite programme (ECS) which ESRO has begun in 1971. That programme and others were carried forward particularly through the Orbital Test Satellites (OTS), OTS–2 being directed to test a variety of telecommunications purposes including the provision of aeronautical, maritime and fixed services.[23] The results obtained were favourable. In early 1977 therefore the ESA Council determined that the ECS programme should be operated on a commercial basis, and that a new international organization separate from itself should be established for the purpose. CEPT members were willing to take that step on the understanding that, as in the case of INTELSAT, an interim agreement would be adopted, to be replaced in due course by a permanent constitution once the major problems of the organization had been discovered. In addition, it was thought best that the new organization should take on board all communications matters, and therefore that it should have responsibility for the provision of space segments for both a European communications system by fixed service (the ECS system) and for a maritime mobile communications system using the MAROTS technology which ESA had already decided to persist with. Even so, it cannot be said that the PTTs were eager. The interest of the European space industry mentioned above was not a major factor for PTTs.

They were doubtful of the profitability of a system meeting a need which they considered might be equally and less expensively met by terrestrial system development. No thought had really been given to the provision of satellite television links as a major revenue earner, nor was there any guarantee that the European Broadcasting Union would, in fact, contract to use the planned system. EUTELSAT was originally being conceived as a telecommunications system largely for message traffic. In the case of the UK, it was government pressure and financial encouragement given in order to strengthen the space industry which sweetened the pill for the British Telecom, and other governments seem similarly to have persuaded their PTTs – a task made easier when the PTT is part of government. And there was another pressure on the PTTs. As matters stood they were monopolists in telecommunications. If they did not participate in the new proposal there was a possibility that ESA itself might go ahead and offer a telecommunications service using the ECS system, thereby rivalling the PTT monopoly. The PTTs therefore entered interim EUTELSAT with some reservations, herded by self-interest and extraneous factors rather than what had been their normal run of commercial decision. If it worked, it worked: if not they would be able to withdraw without too much damage.

A Constitutive Agreement in the French language only creating the interim EUTELSAT arrangements was signed at Paris on 13 May 1977 and entered into force on 30 June 1977.[24] A Supplementary Agreement (Annex A1 to the constitutive Agreement) relating to the space segment for the fixed service (the ECS Agreement) was signed at Paris on 10 March 1978 and entered into force on 14 September 1978.[25] A further Supplementary Agreement (Annex A2) relating to the maritime mobile service (MAROTS) was signed at London on 7 July 1977, entered into force on 22 October 1977, and was amended as from 29 November 1978.[26] Sixteen countries were members of interim EUTELSAT when it commenced operation.[27]

The interim arrangements

Membership of interim EUTELSAT was open initially to any member of CEPT or a recognized private operating agency (an ITU classification) duly authorized by its CEPT member administration, provided that each country might be represented by one Signatory Party only (Agreement art. 13(a)). The Agreement was open for signature until 15 September 1977. Thereafter qualified administrations or private operating agencies might accede to the Agreements on conditions determined by Interim EUTELSAT's Assembly of Signatory Parties.

The organization (like that of interim INTELSAT) was one without legal personality. It was a joint-venture composed of states and telecommunications entities designed to take further a joint project, and to be superseded by a formal international legal persona when that step was required or justified. Accordingly, it was necessary that some legal moves were taken to provide enough legal 'being' for these interim purposes. Art. 10(b) therefore appointed the UK to act as the legal representative of interim EUTELSAT in matters exclusively connected with the MAROTS space segment, and France as legal representative for all other purposes. These two countries were administrations mandated to deal with the matters entrusted to them, and the several members of the organization bound themselves to joint and several liability to the extent of their financial shares in the organization for all commitments duly entered into by the mandated administrations on behalf of interim EUTELSAT and directed towards the establishment operation and maintenance of the space segments concerned (art. 10(b) and Annex B. Annex B contains the terms of the Mandate.).

Structurally, interim EUTELSAT consisted of an Assembly of Signatory Parties, two Councils (the ECS Council in charge of the ECS space segment and the MAROTS Council in charge of the MAROTS space segment), and the permanent General Secretariat under the direction of a Secretary-General (Agreement, art. 4). Of this last, little requires to be said: its function was to undergird the activities of the other parts of the organization.

The Assembly of Signatory Parties which met at least once a year (art. 5(g)) was composed of all Signatory Parties to the Agreement, but a Signatory might delegate another Signatory to represent it, subject to no Signatory representing more than two other Parties (Agreement, art. 5(a)). In addition, any other member administration of CEPT or a recognized private operating agency duly authorized by its CEPT member administration could attend the Assembly with observer status (art. 5(b)). Although it was intended so far as possible to take decisions unanimously (art. 5(e)), in the Assembly each member had one vote (art. 5(c)), and the usual rules as to matters of substance and of procedure applied (art. 5(e) and (f)).

The Assembly's functions included the expression of views on general policy; representation of interim EUTELSAT to external bodies; approval of budget, the defining of principles applicable in, and the approval of, procurement contracts; the definition of principles on inventions and intellectual property; approval of annual reports and consideration of any proposed amendment of the Agreement (art. 6).

The function of each Council was of more practical importance, they being analogous in power and responsibility to the Interim Communications Satellite Committee of Interim INTELSAT. After the entry into force of each Supplementary Agreement each Council was composed of one representative of each Signatory to that Supplementary Agreement (Agreement, art. 7). In relation respectively to the ECS fixed service and the maritime mobile service, the appropriate Council dealt with questions of establishment, maintenance and operation and utilization of the ECS or MAROTS satellite system. A member might itself sign one or more of the Supplementary Agreements or designate a telecommunications entity to sign the Agreement (Agreement, art. 3(a) and (b)). As in the case of the optional programmes within the ESRO/ELDO/ESA regimes, members of interim EUTELSAT were free to choose whether to participate in particular programmes, and to set the extent of their involvement in financial terms. Such decision were not simply financial; they affected both the influence and the right of system use available to each country. Once the Supplementary Agreements were in force, each member of each Council was to have a weighted vote corresponding to the financial share of the Signatory he represented (Interim EUTELSAT Agreement, art. 7 (e)(2)). Further, in the ECS Supplementary Agreement, it was provided that the right to use the fixed service facility was open to members who had entered the Supplementary Agreement in proportion to their financial share of the project, although use in excess of that right could be arranged subject to payment therefor (Interim EUTELSAT, ECS Supplementary Agreement art. 2(b)). Attachment 2 to the ECS Supplementary Agreement allocated the initial financial shares amongst the various members as follows: France and the UK 15 per cent each; Italy 10.5 per cent; Germany 9.9 per cent, The Netherlands and Sweden 5 per cent each; Spain 4.25 per cent; Switzerland 4 per cent; Belgium 3.5 per cent; Denmark 3 per cent; Portugal 2.8 per cent; Finland 2.5 per cent; Norway 2.3 per cent; Austria 1.2 per cent; Turkey 0.85 per cent; and Luxembourg 0.2 per cent. The remaining 15 per cent was reserved as unallocated, with a note that it might be taken up either by other European countries agreeing to accept an initial financial share and/or by the listed countries agreeing to increase their financial allocation.[28]

For the space segment which each Council dealt with, that Council was the principal organ of interim EUTELSAT (art. 8). Within the terms of the Constitutive Agreement, each Council had all powers necessary to fulfil its remit (art. 8). Specific enumeration of particular powers included those of approving budgets, establishing conditions of use of the space segment, approval of all earth stations, specification of units of utilization, establishing charges for utilization, and allotting space segment capacity (art. 8). The Councils were to meet as necessary, but in any case at least twice a year (art. 7(i)).

In fact, however, the MAROTS Council provided to be unnecessary. In the case of both the ECS and the MAROTS system, interim

EUTELSAT was to secure the service envisaged through acquiring, or acquiring the use of, a space segment. The organization had power 'to conclude the necessary agreements to that end, particularly with the European Space Agency' (Agreement art. 2(a)), and such an agreement was signed with ESA on 25 May 1979. However, in both the maritime and fixed service instances, interim EUTELSAT was formed in the knowledge that there were other international organizations active in providing such services. The Preamble to the Interim EUTELSAT Agreement narrates the intention of the Signatory Parties to abide by their existing obligations in regard to INTELSAT and INMARSAT. INTELSAT proved to be no real problem, but, before the MAROTS Council of EUTELSAT had properly established itself, it became clear that INMARSAT was interested in acquiring at least a portion of its space segment through the use of the MAROTS/MARECS system.[29] With the consent of interim EUTELSAT, ESA undertook directly to supply the MARECS space segment to INMARSAT, and interim EUTELSAT therefore no longer required to have a MAROTS Council.[30] By contrast, however, the ECS Council carried out its tasks and was responsible for laying the foundations upon which the permanent EUTELSAT organization has built. Although we will review the achievements of interim EUTELSAT as part of what EUTELSAT itself has done, suffice it here to say that interim EUTELSAT created a satellite system for Europe and, through contracts with such organizations as Eurovision and the European Broadcasting Union, has provided a system which is of utility to broadcasting as well as to the more orthodox requirements of a public international telecommunications service.

The definitive arrangements: negotiations

Article 20 of the Agreement on interim EUTELSAT, required that, at least six months before the ECS space segment was planned to become operational, the Assembly of Signatory Parties was required to make recommendations to the Signatory Parties on the definitive arrangements for the EUTELSAT organization (art. 20(a)). The new organization would, of course, be a much more important and binding arrangement through the involvement of governments as members. It was no longer to be a club of PTTs. Definitive EUTELSAT was to have purposes consistent with that of its interim predecessor (art. 20(b)(1)), all CEPT members were to be eligible to join as were other European administrations with the approval of the new Assembly of Signatory Parties (art. 20(b)(2)), and investments already made were to be safeguarded (art. 20(b)(3)). The Assembly was also to ensure that the proposed new agreements dealt with the legal personality of the new organization, the determination and adjustment of financial shares, the conditions for return on investment, and the conditions on which space segment utilization charges were to be determined (art. 20(b)) – a list which indicates the preoccupations of the negotiators of the Interim Agreements. Provision was also to be made for the transfer to EUTELSAT from member administrations of functions which they had been mandated to carry out by the interim organization and procedures were to be established to regulate relationships with ESA (ibid.).[31]

Within three months of its submission, the report of the Assembly of Signatory Parties was to be examined by an international conference of the Signatory Parties at which European telecommunications administrations not members of the interim arrangements were to have observer status (art. 20(c)). The Signatory Parties further bound themselves to endeavour to create the new definitive organization 'as soon as possible' so that it could enter into force no later than the operational date of the ECS space segment (art. 20(c)). However, and wisely (or realistically) the interim documents remained in force and interim EUTELSAT remained responsible for the operation and maintenance of the space segment until the Definitive Agreements were effective in law (art. 20(d)).

Preparatory work on the definitive arrangements began within EUTELSAT in 1980, the matters indicated in art. 20 which are laid out above being a list of principal areas on which agreement had to be reached. Eventually early in 1982 the Twelfth Meeting of the Assembly of Parties of interim EUTELSAT convened in extraordinary session to adopt a report with recommendations and a draft set of agreements was circulated to the by then 20 Signatory Parties,[32] and other interested telecommunications administrations. Thereafter France convened the required international conference to deal with unresolved matters, and, it was hoped, with the adoption of the definitive arrangements. This conference met in Paris from 3–14 May 1982,[33] and the definitive arrangements for EUTELSAT which it adopted were opened for signature at Paris on 15 July 1982. At the time, of the 20 interim EUTELSAT members, four were bearing half the cost of the ECS programme.[34]

The definitive arrangements

The pattern of the definitive arrangements for EUTELSAT conforms to that pioneered by INTELSAT and INMARSAT. There is an intergovernmental agreement which sets up the constitution of the organization and deals with other necessary matters which are the responsibility of states. In the case of EUTELSAT this agreement is referred to as the Convention (art. I(a)). There is also an Operating Agreement between telecommunications entities.[35] No reservation can be made to either the Convention or the Operating Agreement (art. XXI(d)), thus avoiding any problem arising from different versions of a multilateral agreement applying between different Parties to it. The life of the Operating Agreement is tied to that of the Convention (art. 23(c)).

The definitive arrangements for EUTELSAT were opened for signature at Paris on 15 July 1982, in English and French texts, both being equally authentic (Docquet and art. XXI).[36] In terms of art. XXII of the EUTELSAT Convention, the Convention, and hence the Operating Agreement, was to come into force 60 days after Parties who (or whose Signatories to the ECS Agreement) held two-thirds of the financial shares under the ECS Agreement had completed their constitutional processes for making the Convention binding on them, and the appropriate arrangements had been made in respect of the Operating Agreement (art. XXII (a)). This was not, however, the sole condition. The exercise was subject to time limits. The Convention was to come into force not earlier than eight, nor more than 18 months from the agreements opening for signature on 15 July 1982. For various reasons, not least problems of getting the appropriate ratification through the French political process,[37] that timetable slipped. Accordingly, a Protocol was signed on 15 December 1983 by members of interim EUTELSAT (including the Holy See, Liechtenstein, Monaco and San Marino, which had joined in the period from 1982), extending the time-limit to 36 months.[38] Even so, the new deadline also nearly killed off the organization prior to its birth, the appropriate level being reached only some two weeks before the expiry of the extended time-limit. In fact the definitive arrangements for EUTELSAT came into effect, replacing the interim arrangements, on 1 September 1985.

Membership

EUTELSAT is established by art. II(a) of the Convention. Each Party signing the Convention was bound to designate as a Signatory to the Operating Agreement a single telecommunications entity which, subject to its jurisdiction or it itself, was bound to sign the Operating Agreement (art. II(b)). It is competent for a government to sign the Operating Agreement in the case where a state Party to the Convention is itself responsible for the operating of telecommunications services within the country of which it is the government. 'Signatory' of the Operating Agreement can therefore mean either a state or a telecommunications entity as appropriate (arts. I(f)). The application of the Operating Agreement to a Signatory is contingent upon the membership of the state Party to the Convention under whose jurisdiction it functions (art. 23(a)), although the Party cannot become a member of the Convention until the appropriate signature has been attached to the Operating Agreement (art. XXI(c)). Thereafter the two instruments come into force simultaneously for Party and Signatory.

Any state whose telecommunications administration or recognized private operating agency either has or had the right to become a Signatory Party to the Provisional Agreement could become a member by signature not subject to further procedure, or by signature followed by completion of the appropriate internal procedure for that state to become bound at international law by its signature (usually by ratification of the agreement), or by accession. The Convention was open for signature until it came into force, and thereafter accession remained possible for states which qualified as potential members of interim EUTELSAT (art. XXI(b)) but only for two years (art. XXIII(a)).[39] Accession is now open to such formerly 'qualified' states and to any other 'European state' which is a member of the ITU. However, accession is not entirely at the will of the acceding Party: it is now subject to controls which show that EUTELSAT is conscious of its regional character and that it is a small and balanced organization (art. XXIII(b)(i) and (ii)).[40]

A state now wishing to accede to the Convention must apply to the Director-General in writing giving various data which the Board of Signatories may require, including information on the proposed use it will make of the EUTELSAT space segment. The Board examines this information from the technical, operational and financial viewpoints to ascertain the compatability of the application with the interests of EUTELSAT and the Signatories within the scope of the organization, and makes a recommendation to the Assembly of Parties (art. XXIII(d)). Within six months of the Board deciding it has received all the required data for performance of its role, and if necessary at an extraordinary meeting, the Assembly makes its decision on the application, taking the recommendation of the Board into account. The Assembly's decision is by secret vote (an interesting provision to find *in gremio* of the Convention) and counts as a decision on a matter of substance, therefore requiring two-thirds of the Parties present, or represented and voting (art. XXIII(e)). If the state is permitted to accede, a protocol annexed to the instrument of accession will deal with any necessary matters (art. XXIII(f)). Financial adjustments will also be required.[41]

Legal existence

The definitive organization, EUTELSAT, has legal personality (art. IV(a)) and the full capacity necessary for the exercise of its functions and the achievement of its purposes (art. IV(b)). Without prejudice to that generality, and those powers which it would be implied to have by international law in order to carry out its purposes,[42] under the Convention EUTELSAT specifically has powers:

1 to contract;
2 to acquire, lease, hold and dispose of moveable and immoveable property;
3 to be party to legal proceedings; and
4 to conclude agreements with states or international organizations (art. IV(b) (i) to (iv)).

In accordance with art. XVII of the Convention, a Headquarters Agreement was entered into between EUTELSAT and France on 15 November 1985, and a Protocol on Privileges and Immunities was signed at Paris on 13 February 1987.[43] Once this Protocol is in force for the members of the organization, EUTELSAT will, within the terms of international law, have as full a legal existence as any other international organization.

Purposes

Practice has produced an interesting divergence between the language of the Convention and Operating Agreement and the actual way in which the system has been developed. A category of 'primary services' unknown to the legal agreements has been developed more accurately to reflect the organization's actual main revenue-earning operations. But first we must deal with the 'law'.

Since the expected members of EUTELSAT were also Signatories to the INTELSAT and INMARSAT Agreements, it was necessary that the scope of the activities of EUTELSAT should be formulated so as not to breach the responsibilities of Parties and Signatories under those prior agreements. The Preamble to the Convention therefore narrates that the Parties wish to establish the European system 'without prejudice to any rights and obligations of the States which are parties to the' INTELSAT and INMARSAT Agreements. Further the scope of EUTELSAT's activities are set out in Art. III and in the main contain seriatim references to European activities.

According to the Convention, the main purpose of EUTELSAT is the design, development, construction, establishment, operation and maintenance of the space segment of the European telecommunications satellite system or systems. The prime objective within that main purpose is the provision of the space segment required for international public telecommunications services in Europe (art. III(a)).[44] Such language clearly indicates the organization's regional thrust of and its intention to provide primarily for the public service. The 'space segment' involved is the set of telecommunications satellites needed for the job, together with all necessary tracking, telemetering, command, control, monitoring and related facilities, and the equipment required for the operational support of the satellite system (art. I(g)). The term 'public telecommunications services' is similarly widely drawn and refers to fixed or mobile telecommunications services which can be provided by satellite and which are available to the public. The defining paragraph in this case (art. I(k)) goes on to give the examples of telephony, telegraphy, telex, facsimile, data transmission, videotex, the transmission of radio or television programmes between approved earth stations having access to the EUTELSAT system for further transmission to the public, and multi-service transmissions. The provision of leased circuits to be used for any of these services is also included (art. I(k)). The listing is, however, only a set of examples. The ruling language of art. I(k), defining 'public telecommunications services' as 'fixed or mobile tele-communications services which can be provided by satellite and which are available to the public', is sufficiently open-ended to allow EUTELSAT to take advantage of future developments.[45]

While EUTELSAT's principal stated objective is the provision of international public telecommunications services within Europe, the drafters of the Convention were careful to ensure that the organization can also provide facilities for domestic services and provide other (perhaps non-European?) international services. But there is a priority of provision. The space segment is available for domestic services on the same basis as for international services in the case where areas under the jurisdiction of a Party are either separated by the territory of another Party, or by the high seas (art. III(b)). Other domestic or international services can also be supplied by EU-TELSAT provided that their provision does not impair its ability to achieve the primary objective of providing international public tele-communications services in Europe (art. III(c)). The hierarchy of provision therefore runs: top priority, international services or domestic services between areas of a Party separated by the territory of another Party or the high seas; then, other domestic or international services. The provision of a link between Aberdeen and Marseilles would therefore rank above one from Aberdeen to London, or one from Paris to Marseilles. A link between Madrid and Tenerife would rank equally with an international service.

But the restriction of EUTELSAT to the provision of public tele-communication services would have been both short-sighted and economically unjustifiable. In any satellite system it is likely that the space segment will contain capacity in excess of that required for the expected demand for public telecommunication services, even if only to provide for emergency as well as some redundancy to cope with crises such as one satellite going out of commission earlier than its expected life. That excess capacity ought to be used so that the Parties and Signatories can obtain the maximum possible return on their investment. Accordingly, by other clauses of art. III of the Convention, EUTELSAT is empowered to provide within Europe specialized telecommunications services, and is able to provide hardware which is not connected with the EUTELSAT space segment.

'Specialized telecommunications services' are defined as 'telecommunications services which can be provided by satellite' other than the public telecommunications services dealt with in art. I(k) (art. I(l)). Examples of such specialized services are listed, but only as examples. The list runs 'radio navigation services, broadcasting satellite services, space research services, and remote sensing of earth resources' (art. I(l)). The sole absolute exclusion from these services is that EUTELSAT may not provide such services for military purposes (art. III(e)). Two other conditions are, however, imposed by the Convention. First, the provision of specialized services must not be to the detriment of the public telecommunications services (art. III(e)(ii)). Second, the arrangements under which the specialized services are to be provided must be acceptable from both a technical and an economic point of view (art. III(e)(ii)). Subject to compliance with these conditions, and the negotiation of appropriate terms and conditions, EUTELSAT may, on request, provide any of these specialized services within the operational capacity of the system as it exists for the time being. This potential market is something which is taken into account when the space segment is being designed and when future developments of the system are planned.

Apart from such use of the EUTELSAT system itself to provide services, art. III(f) opens the possibility of the organization providing satellites and associated equipment entirely separate from the EUTELSAT space segment. On request, and subject to the negotiation of appropriate terms and conditions, EUTELSAT can provide what would amount to a separate space segment:

1 for domestic public telecommunications services;
2 for international public telecommunciations services;
3 for specialized services other than for military purposes (art. III(f)(i)–(iii)).

Of course, the efficiency and economic operation of EUTELSAT's own space activities must not be unfavourably affected by such ancillary activities (art. III(f) proviso), language which immediately evokes memories of INTELSAT's problems with 'significant economic harm' in relation to separate systems. EUTELSAT may itself finance the provision of such separate facilities and services, but only by the unanimous decision of the Board of Signatories (art. V(f)). In the absence of such unanimity, the satellites and equipment required are to be financed by those requesting the provision, on terms and conditions to be set by the Board of Signatories 'with a view to covering at least all relevant costs borne by EUTELSAT' (art. V(f)). Those costs are not to form part of the capital requirements of EUTELSAT itself (art. V(f)) and hence will not be part of the liabilities of Parties and Signatories. Nor will the satellite system and ancillary equipment form part of the EUTELSAT space segment (art. V(f)). EUTELSAT can therefore operate as a contractor for the establishment of other space systems, but its own telecommunications operations are kept quite separate from those activities, and in the last analysis such secondary activities give way before the organization's primary objective – the European system (art. III(a)).

However, the entire preceding explanation depends upon a non-initiate's reading of the text of the Convention. In fact EUTELSAT, without sacrificing its commitment to public telecommunications services has also established as a working concept within EUTELSAT the notion of 'primary services'. To the 'prime objective' of the telecommunications facility has been added the provision of television and entertainment transmission as well as international business services, and it is to these that the organization looks economically to justify the system. In short, perhaps the drafters of the Convention incorrectly defined the importance of the various services to the economic viability of the organization. No matter, the Convention is such that it need not be modified to include this practical development, and EUTELSAT can easily carry on lawfully expanding its 'primary services' including both telecommunications, and the other services from which it now draws revenue, and which are often secured for their users by being provided on a non-pre-emptible basis.

Structure

The structure of EUTELSAT is tripartite. By art. VI(a) of the EUTELSAT Convention, the organs of EUTELSAT are the Assembly of Parties, the Board of Signatories, and an Executive Organ with a Director-General at its head. The reduction in structure by comparison with INTELSAT is explained by the fact that the European organization is far smaller. EUTELSAT exists to provide services for a comparatively small geographic area and involves a limited number of states and telecommunications entities. It has therefore been found possible, and indeed desirable, that all Signatories should have the option of participation in decision-making at the Board level. As a result in EUTELSAT those responsibilities and functions of the Meeting of Signatories and of the Board of Governors in INTELSAT have been fused and given to the Board of Signatories.

As with the other satellite organizations, each organ is to act within its powers (*intra vires*), and is prohibited from acting in such a way as to harm the *intra vires* exercise of the powers of any other organ (art. VI(b)). It follows that the allocation of powers and responsibilities between the organs is important, although as a matter of prudence there is also the catch-all provision that the Assembly of Parties exercises any function necessary for EUTELSAT's purpose that is not otherwise allocated to another organ (art. IX(b)).

The Assembly of Parties

As in the case of INMARSAT and INTELSAT, the nominally highest echelon of the tripartite structure of EUTELSAT is the Assembly of Parties, a gathering of the representatives of states. The Assembly is composed of all the Parties to the Convention (art. VII(a)). It is, however, competent for a Party to be represented by another Party at meetings of the Assembly, although no Party may represent more than two other Parties in addition to itself (art. VII(b)). This provision would appear designed to allow such countries as Liechtenstein, Monaco and San Marino to become Signatories and yet to have their interests represented by larger neighbours who traditionally act in such a manner.

Ordinary meetings of the Assembly of Parties are held every two years, unless the Assembly itself decides otherwise (art. VII(c)). Extraordinary meetings are competent, and are held either at the request of the Board of Signatories or at the request of one or more Parties supported by at least one-third of the Parties (art. VII(d)). The request for an extraordinary meeting must state the purpose for which it is sought (art. VII(d)). One's first assumption would be that an extraordinary meeting would deal only with the matter for which it is called, but INTELSAT practice indicates that, while an extraordinary meeting is called for a purpose, once it is agreed to hold an extraordinary meeting, other matters may also be put on the agenda. The costs of meetings of the Assembly are an administrative cost of the organization, although each Party bears its own costs of representation (art. VII(e)).

The quorum for the Assembly of Parties is composed of the representatives of a simple numerical majority of the Parties, provided that not less than one-third of all Parties are present (art. VIII(d)). Each Party has a single vote in the Assembly, Parties abstaining from voting being considered as not voting (art. VIII(a)). As is usually the case in similar organizations, particular care is taken over the decision-making of the Assembly in matters of substance; procedural matters are less stressed and require only a simple majority. In the case of EUTELSAT, a decision on a matter of substance is taken by the affirmative vote of at least two-thirds of the Parties present, or represented and voting (art. VIII(b)). A Party representative of Parties not present may vote separately for each Party that it represents (art. VIII(b)). Apart from these rules contained in the Convention, the Assembly adopts its own Rules of Procedure, which include the provisions for the election of its chairman and officers, for the convening of meetings, for representation and accreditation, and for the actual voting procedures to be used at meetings (art. VIII(e)).

The functions of the Assembly of Parties are laid out in art. IX of the Convention. There is an all-inclusive statement at the start of art. IX(a): 'The Assembly of Parties, which may concern itself with any aspect of EUTELSAT which affects the interests of the Parties, shall have the following functions: . . .' Such formulation makes it clear that the Assembly can in fact deal with anything of relevance to the organization, and that the functions thereafter listed are not exhaustive of its powers. Whether such a formulation will in the future be regretted by the Board of Signatories, or by specific interests within Europe remains to be seen. It could provide a justification for what may be considered interference by governments in the work of a service-providing and operating organization. On the other hand, in the small geographic area of Europe, and with its high commitment to, and penetration by, telecommuncations, it is understandable that governments would wish to have an open-ended right to raise matters of concern.

Technically, states did not have as extensive a right to raise any matter they wished under the interim arrangements. Art. 6 of the Interim Agreement simply listed the functions of the Assembly of Signatory Parties, the only generality mentioned being the duty to 'deal with all general questions of interest to both the ECS and MAROTS space segments . . .' That formula does not equate with the language of the present art. IX(a)), which clearly reflects political reality and, to an extent, past practice.

In the performance of its functions, the Assembly is to take into account relevant recommendations from the Board of Signatories (art. IX(c)). What this exactly means could, of course, be a matter of dispute. It seems likely, however, that, as in the experience of other satellite organizations, the Assembly, composed of representatives of governments, is likely to give due weight to the recommendations of the organs responsible for the operation of the system in technical matters. The interaction of technical, financial and political elements could, however, present future difficulties. Indeed, such difficulties could be particularly accented within the smaller organization dealing with a smaller area and membership. Governments may well take a closer (though not necessarily better-informed) interest in EUTELSAT than in, say, INTELSAT, and may be concerned to make greater use of their powers within the Assembly of Parties of an organization which is, in a sense, on their doorstep and in which they have a proportionately greater voice.

The Assembly of Parties is to exercise any function, contained in the Convention, which is necessary for achieving EUTELSAT's purpose and which is not expressly attributed to another organ under the Convention (art. IX(b)). Apart from that somewhat indefinite but useful formulation, the functions are listed as belonging to the Assembly of Parties are wide-ranging, but, naturally, concern overall policy and matters which are properly the province of states as members of an international organization. The functions stipulated under art. IX(a) are:

1 the consideration of general policy and EUTELSAT's long-term objectives together with the making of recommendations on such as appropriate;
2 recommending to the Board measures to prevent EUTELSAT's activities 'from conflicting with obligations laid on its Parties by any multilateral treaty acceded to by a simple majority of' them. This cumbersome language, means that obligations under the INTELSAT, INMARSAT and ITU Conventions and Agreements may well take precedence over members' EUTELSAT obligations;
3 the Assembly of Parties authorizes, through general rules or specific decisions, the utilization of the EUTELSAT space segment and the provision of any of the separate telecommunications service systems countenanced by art. III(f);
4 the Assembly deals with the Board's recommendations and expresses views on reports made to it;
5 it expresses views on proposed satellite systems to be established within the EUTELSAT area of operation;[46]
6 it decides on formal relations between the organization and states and with other international organizations;
7 the Assembly has a role to play in the withdrawal of a Party from the Organization;
8 it is involved in the procedures for the amendment of the Convention or the Operating Agreement;
9 it has a role in the accession of new Parties to the Convention (art. IX(a)(i)–(x)).

The Board of Signatories

EUTELSAT's Board of Signatories is composed of Board members, each of whom represent one Signatory (art. X(a)). However, as in the case of the Assembly of Parties, a Board member may also represent up to two Signatories additional to his own national Signatory (art. X(b)). As in the case of INTELSAT, therefore, a Signatory may not in fact be represented on the Board of Signatories but, unlike the INTELSAT position, every Signatory can be represented if it so wishes, either through its own representative or through a member representing another Signatory.

Like the other organizations we have considered, decision-making within the central governing organ, the Board, is keyed to investment participation in the project. In EUTELSAT the voting participation of each Board member is equal to the investment share held by the Signatory represented, subject to various modifications in certain instances (art. XI(a) and (e)). Recalculations of voting participation take effect along with the redetermination of an investment share (art. XI(e)). The initial investment share was fixed by Annex B to the Operating Agreement (arts. XI(b) and 6(c)), and will be replaced by an investment share based upon utilization of the EUTELSAT space segment in due course (art. XI(b), (d) and (e)).[47] As we will see, it is possible for a Signatory voluntarily to accept an increased investment share as well. However, and with a limited exception, by art. XI(c) of the Convention no Signatory (and hence no Board member acting for a single Signatory) can have a voting participation of more than 20 per cent of the total voting participation in EUTELSAT. The limited exception referred to is that if EUTELSAT decides to provide services other than the primary ones of international public telecommunication services and domestic services between 'detached' parts of a Party, the cost of such provision is not to devolve on Parties which are not directly interested in the extended service before it enters into operational use (art. 4(d)). The cost of such 'special' provisions is to be borne by the Signatory or Signatories interested in the extended service, and their investment participation (and hence their voting participation) can rise by up to a limit of 5 per cent irrespective of the normal general 20 per cent maximum voting weight. Where the ordinary limit of 20 per cent, or the special 25 per cent maximum is exceeded, the excess investment participation is distributed equally among the other Signatories for voting purposes (art. XI(c)).

The Board of Signatories meets as necessary and at least three times per year (art. XI(j)). The quorum for a Board meeting is composed of either representatives of a simple majority of the Signatories, provided that the voting participation which they represent is more than two-thirds of the total voting participation of Signatories with voting rights, or representatives of the total number of Signatories with voting rights less three, regardless of their voting participation (art. XI(f)).

The Board is bound to try to take decisions unanimously (art. XI(g)), but, recognizing that this may not always be attainable, the Convention makes specific provision for certain types of decision.

Decisions on procedural matters are taken by a simple majority of Board members present and voting (art. XI(g)(iv)): each Board Member has one vote (art. XI(g)(iv)) irrespective of the number of Signatories he represents (art. XI(g)(v)). It is the Board's responsibility to adopt its own Rules of Procedure, dealing *inter alia* with such matters as the election of its chairman and officers, the convening of meetings, representation and the accrediting of representatives, and the actual voting procedures to be used (art. XI(h)).

In substantive decisions, a Board Member may vote separately for each Signatory he represents (art. XI(g)(v)). Apart from certain financial questions, decisions on substantive matters are taken by a special majority, made up either by members representing at least four Signatories having at least two-thirds of the total voting participation of Signatories with voting rights (art. XI(g)(i)), or by an affirmative vote of the total number of Signatories present or represented less three, regardless of the voting participation involved (art. XI(g)(i)). In certain financial matters the majorities required are raised. Decisions on adjustment of the capital ceiling of the organization in order to meet its basic purposes of the provision of public international telecommunications services and the limited and special domestic services require a simple majority of Signatories present and voting, provided that the majority holds at least two-thirds of the total voting participation (art. XI(g)(ii)). Decisions on adjustment of the capital ceiling in order to meet objectives other than these basic purposes (the extended services) require an affirmative vote of two-thirds of the Signatories present and voting, representing at least two-thirds of the total voting participation (art. XI(g)(iii)).

The Board of Signatories is responsible for the design, development, construction, operation and maintenance of the EUTELSAT space segment (art. XII(a)). It may acquire property by purchase or lease for for the purpose (art. XII(a)). It also has the responsibility for the conduct of any other activities which EUTELSAT is authorized to undertake (art. XII(a)). In its activities, the Board is to take 'due account' of the views of the Assembly of Parties. It is uncertain whether, in the smaller organization, this means that the Assembly of Parties has a stronger voice in the workings of EUTELSAT than does the equivalent body in other organizations. Opinions vary.

In order to carry out these responsibilities the Board is assigned a non-inclusive list of functions by art. XII(b), which range from planning (art. XII(b)(i)), powers concerning matters of procurement (art. XII(b)(ii) and extensive financial powers (art. XII(b)(v) and (x)), to the designation of an arbitrator if EUTELSAT becomes involved in arbitral proceedings in which it is required so to do (art. XII(b)(xxi)). It also has functions in the approval of earth stations (art. XII(b)(vi) and (vii)), in the allotment of space segment capacity (art. XII(b)(viii)) and affording access by other entities to the system (art. XII(b)(ix)), in the supervision of the spectrum and orbital use to ensure compliance with ITU requirements (art. XII(b)(xi)), and in coordination of the EUTELSAT system with other systems (art. XII(b)(xiii) and (xiv)). In addition, of course, it has considerable functions in matters of the organization's constitution and constitutional working (art. XII(b)(xv)–(xxiv)).

The Executive Organ

The Executive Organ of EUTELSAT is headed by a Director-General, who is appointed by the Board of Signatories (arts. XIII(a) and XII(b)(xvi)) subject to confirmation by the Parties. His appointment is automatically confirmed unless more than one-third object in writing to the Depository of the Convention (France) within 60 days of its notifying Parties of the appointment (art. XIII(a)). The Director-General normally serves for a 6-year period although the Board may determine otherwise (art. XIII(b)). The Board may remove the Director-General 'for cause' prior to the end of his period of appointment, reporting to the Assembly on the reasons for their action (art. XIII(c)).

The Director-General is the chief executive and legal representative of EUTELSAT. He acts on the direction of the Board, and is responsible to it for performance of all the functions of the Executive Organ (art. XIII(d)). The Board approves the structure and staffing of the Executive Organ, together with the conditions of employment of consultants and other advisers (art. XIII(e)). The Director-General makes staff appointments but is subject to the Board's approval of senior officers reporting directly to himself (art. XIII(f)). As in the case of INTELSAT and INMARSAT, the paramount consideration in the appointment of the Director-General and other staff of the Executive Organ is 'the need to ensure the highest standards of integrity, competence and efficiency' (art. XIII(h)), and to underline that, art. XIII(i) stipulates that the holders of these appointments 'shall refrain from any action incompatible with their responsibilities to EUTELSAT'.

Financial and other matters

Investment shares

The financial principles on which EUTELSAT operates are contained in art. V. The organization operates 'on a sound economic and financial basis having regard to accepted commercial principles' (art. V(b)), an explicit statement borrowed from the INMARSAT Convention. (art. 5(3) thereof), but not found in the INTELSAT Agreement.

Each Signatory is responsible for financing EUTELSAT (art. V(a)) on proportion to its investment share (art. V(c)). An initial investment share was allocated to each Signatory under Annex B to the Operating Agreement as amended to take account of later accessions to the organizations (art. 6). The timing of the replacement of the initial investment shares by shares based on utilization was left deliberately uncertain. It was to occur seven years from the date on which the first EUTELSAT satellite was positioned and working (art. 6(d)(iii)) or at a date not earlier than four years after the date of operation of the first working EUTELSAT satellite[48] provided that at least 10 Signatories had been accessing the space segment, and EUTELSAT utilizatiron revenues 'during a six-month period have been greater than the revenues that would have derived from the utilization by Signatories for the same period of the Space Segment capacity required to establish 5000 telephone circuits using digital speech interpolation' (art 6(d)(ii)). These conditions were met in 1987, and the new investment shares were first set in October of that year, calculated in percentages on the basis of utilization of the system in terms of art. 6.[49] A nil use of the system still attracts a minimum investment share of 0.05 per cent (arts. V(c) and 6(g)). Recalculation of shares now occurs on:

1 the effective date of withdrawal of a Signatory (art. 6(e)(iii);
2 on the accession of a new Signatory (art. 6(e)(ii)); or
3 on 1 March each year provided that the total utilization charges payable to EUTELSAT in respect of the preceding six months are not more than 20 per cent lower than the sums payable for the 6-month period running from 18 to 12 months prior to that 1 March (art. 6(e)(i)).

It is possible for a Signatory to apply to have its investment share reduced from its mathematical total, and this is done provided that other Signatories take up the portion the Signatory wishes to be relieved of (art. 6(h)). There have been no problems so far in such adjustments.

The provisions limiting the timing of the introduction of utilization set by use, and on the recalculation of shares for cases other than the increase or loss of membership illustrate starkly the commercial element present in the EUTELSAT endeavour. The earlier-mentioned reservations of the PTTs at the inception of interim EUTELSAT have been largely dispelled, in large measure through the non-telecommunications development of the system. It will be interesting to see how the PTTs react to the changes that occur in future years, although at present it has to be said the financial position is healthy. All capacity on the EUTELSAT system is at present committed, and the next satellite in the series, the F–5, due for launch in late 1988, has also been fully 'booked'.

Utilization charges

The utilization units and the charges which users of the EUTELSAT system pay in terms of art. V(e) are set by the Board of Signatories (art. 8(a)) and paid under arrangements established by the Board. Charges 'have the objective of earning sufficient revenues to cover operating, maintenance and administrative costs' as well as providing working funds, the amortizing of investment and compensation to Signatories for use of capital (art. 8(a)). The charge for a particular type of utilization is intended to cover all expenditures related to that use (art. 8(a)) and the rates for each type of utilization are identical for all users within the jurisdiction of Parties to the Convention (art. V(e)(i)). Users outside the jurisdiction of Parties may be charged a rate for each type of utilization different from that operative 'within' the territory of EUTELSAT members, but 'outside users' of the system all pay the same rate for each type of utilization (art. V(e)(ii)).

Procurement and rights

Given that the obtaining of expertise for national space industries was a motive behind the establishment of EUTELSAT, it comes as no surprise that 'due consideration to the general and industrial interests of the Parties' is an element in the awarding of contracts (art. XIV(c)). However, that consideration is secondary to the 'widest possible competition in the supply of goods and services' (art. XIV(a)) secured ordinarily through the use of open international tender procedures (art. XIV(b)). Only when these processes have resulted in a comparable combination of quality, price, delivery and undefined other 'criteria of relevance to EUTELSAT' is the 'general and industrial interests' of Parties weighed up.

Broadly speaking, the EUTELSAT arrangements follow those of INTELSAT with provision for Board approval of requests for proposals, invitations to tender and for contracts of over 150,000

European Currency Units (ECUS) (art. 17(b)). Smaller contracts, cases of urgency, single-source and limited-source supply contracts, local procurement of administrative natures and personal service contracts are not subjected to the requirement of an international tender (art. 17(d)).

Again, the INTELSAT form is followed in regard to the rights which EUTELSAT seeks to acquire in intellectual property. Reference is made to the observance of 'generally accepted industrial practices'.

Coordination with other systems

Naturally the problem of 'separate systems' is one which had to be dealt with by the EUTELSAT Agreements, even though EUTELSAT itself was coming into being in a world where INTELSAT and INMARSAT were already operational. As we have seen, by art. IX(a)(ii) the Assembly of Parties can recommend to the Board of Signatories measures to prevent EUTELSAT's activities conflicting with the international obligations of at least a simple majority of its members under any general multilateral agreement. In addition, EUTELSAT's own coordination requirements do not apply to INTEL-SAT or INMARSAT developments (art. XVI(c)(i)). The INTELSAT and INMARSAT Agreements thus have a measure of protection or priority over the EUTELSAT Convention, as the Law of Treaties would dictate. Indeed, the EUTELSAT system had to be coordinated with the INTELSAT system through the mechanisms of art. XIV(d) of the INTELSAT Agreement as we have seen elsewhere. In that series of coordinations INTELSAT gave great attention to the question of 'significant economic harm' to its system, and the plans for the expanded EUTELSAT system coordinated in 1982 was applicable only until 1988, INTELSAT reserving the question of the use of the 11/14 GHz frequencies thereafter, and with an implicit threat that INTELSAT might expand its European services to the extent that EUTELSAT might be damaged economically. This remains a possibility, but the appropriate coordination has now been extended for a further period and it may be that other INTELSAT competitors will serve to shelter EUTELSAT for long enough to get its services well established and the European market developed to the extent that INTELSAT competes but does not damage the regional venture. Certainly INTELSAT will continue to carry intercontinental traffic. EUTELSAT Signatories are required, however, to 'endeavour to route a reasonable portion of their traffic' between them by the EUTELSAT space segment in their separate traffic agreements (art. 2(b)), which may be a pointer for the future.

Coordination with INMARSAT must also be mentioned, although, as yet, EUTELSAT has not entered the maritime mobile market and is not likely to do so other than providing channels for INMARSAT to use. EUTELSAT is not providing a maritime service to end-users; hence coordination with INMARSAT under art. VIII of the INMARSAT Convention has not yet arisen.

The other possibility is that a EUTELSAT Party or Signatory, or a person or company within its jurisdiction, may intend to set up a system separate from that of EUTELSAT to provide international public or domestic telecommunications within the EUTELSAT service area. In such cases, the new system has to be coordinated with the EUTELSAT system in terms of art. XVI of the Convention. The headings of concern are similar to, but not identical with, the INTELSAT requirements. Technical compatibility of the two systems and the efficient use of the radio spectrum and orbital space is assured in terms of guidelines adopted by the Board and intended to be considered by the appropriate Party or Signatory (art. XVI(b)). There is nothing in the Convention similar to the 'prejudice to direct links' category of concern indicated in art. XIV(d) of the INTELSAT Agreement. Of course, economic harm has to be dealt with. Information must be supplied by the appropriate Party or Signatory to the Board of Signatories which considers whether there is likely to be significant economic harm caused to EUTELSAT by the proposed system (art. XVI(a)). The Board submits a report and conclusions to the Assembly of Parties which gives its views within six months of the start of the coordination procedure, if necessary by a special meeting (art. XVI(a)). It is to be noted that the Assembly expresses its views (art. XVI(a) and IX(a)(v)). Neither the mode of expression nor its weight is indicated in the Convention. As no coordinations have yet been undertaken under art. XVI, it is not yet known whether EUTELSAT considers 'significant' and 'economic harm' in the same terms as INTELSAT.

Disputes

Disputes arising from the EUTELSAT Convention or Operating Agreement are referred to arbitration under their respective art. XX or art. 20, although the referral of a dispute between a Party and a Signatory is subject to their agreement. An arbitration tribunal of three members is constituted and operates under Annex B to the Convention, which is much less detailed (or complex) than the equivalent provision in the INTELSAT Agreement. One interesting provision is that a member of the tribunal may not abstain from voting on the final award to be made (Annex B, para. 6).

Sanctions

The failure of a Party or a Signatory to comply with an obligation carries the same range of sanctions as in INTELSAT. The Assembly of Parties, either on its own initiative or having had notice about the failure, may deem a Party to have withdrawn from the organization (art. XVIII (b)(i)). The deemed withdrawal implies the withdrawal of its accredited Signatory as well (art. XVIII(b)(i)). The rights of a Signatory are automatically suspended three months after failure to pay a capital contribution after it becomes due, and it may be deemed to have withdrawn three months after than the suspension (art. XVIII(b)(iii)). In other cases of default a Signatory's rights are automatically suspended three months after it fails to comply with a resolution of the Board of Signatories notified to it by the Executive Organ calling on it to remedy the failure. It retains its obligations and liabilities during this period (art. XVIII(ii)(A)). Thereafter the Board of Signatories, after due process, may deem the Signatory to have withdrawn from the organization (art. XVIII(ii)(B)). In both cases of deemed withdrawal of a Signatory the relevant Party then assumes the responsibilities of the Signatory until it either nominates another Signatory or itself withdraws from the organization (art. III(b)(ii)(B) and (iii)(B)).

Withdrawal

Withdrawal from the organization is competent for either a Party or a Signatory either involuntarily as just described, or voluntarily under art. XVIII(a)). The withdrawal of a Party involves the withdrawal of its Signatory. Where a Signatory withdraws, the relevant Party assumes its responsibilities until it nominates another Signatory or itself withdraws from the organization. Voluntary withdrawal takes effect three months after the receipt of notification by the Depository for the Convention or, in the case of a Signatory, the Director-General. Financial adjustments are then made between the organization and the withdrawers, and between the remaining members of EUTELSAT.

The future

The future of EUTELSAT seems reasonably secure. The traffic generated in telecommunications within Europe is such that a satellite system is likely to have a proportion of the market adequate to sustain it. In addition, EUTELSAT has been active in providing services other than telecommunications (for example, television), and in allowing its members to acquire non-pre-emptible rights in transponders for domestic purposes – thus, to a degree, tying EUTELSAT integrally to the telecommunications structure of individual countries. These steps, coupled with the continued interest in the regional effort as affording a niche within space technology for domestic industries mean that countries would be loath to see the organization founder.

That said, one awaits with interest to see whether INTELSAT will seek to use the opportunity it reserved to itself in the coordination of the expanded EUTELSAT system to develop a stronger presence in European telecommunications. Much would depend on how such steps were taken. On one set of facts there might well be room for competition between EUTELSAT and INTELSAT in providing intra-European links particularly with the development of the cohort of INTELSAT business services. It does not seem likely, however, that such limited competition will cause a problem, although it might benefit the consumer or end-user.

Problems would really arise only if INTELSAT finds itself nudged into aggressively seeking revenue from Europe as a major telecommunications arena because of loss of revenues on other routes and in other areas to the US authorized separate systems. If, for these reasons, INTELSAT were to refuse an acceptable coordination under art. XIV(d) with EUTELSAT in order to gain an entry to the European market, trouble would ensue. The European members of INTELSAT might well become disenchanted with the global system and consider their interest served better by a strengthened EUTELSAT than a weak INTELSAT. In that case EUTELSAT itself might even seek a share of the transatlantic traffic and consider altering its coverage, either by relocating or even merely by tilting or redirecting its satellites so as to permit a transatlantic service. Of course, such steps would imply a loose adherence to the obligations of both the INTELSAT and EUTELSAT Agreements. It is, however, a general principle in civilized legal systems that one cannot complain of another doing what one has done oneself. There is a matter of personal bar or estoppel.

But, in such a case, the world would have lost much, and the notion that outer space is to be used solely for the benefit of all would have been called in question.

NOTES

1   Agreement setting up a Preparatory Commission to study the possibilities of European Collaboration in the Field of Space Research, (1961) UKTS No. 60, Cmnd. 1425; 10 *European Yearbook*, pp. 1,111–15. Various extensions of the Com-

mission were agreed as necessary by Protocol; First Protocol 1962–3 Cmnd. 1898 and 2091; Second Protocol 1962–63 Cmnd. 2122; Third Protocol, 1963–64 Cmnd. 2173; Fourth Protocol 1963–64 Cmnd. 2350.

2   Convention for the Establishment of a European Space Research Organisation (ESRO), Paris, 14 June 1962 (1964) UKTS No. 56, Cmnd. 2489; 528 UNTS 33; 10 *European Yearbook* p. 1,115. The Final Act of the Conference, with Convention, Financial Protocols and Resolutions addressed to the future Council of the Organization was published as (1961–2) Cmnd. 1840.

3   This is not quite the same as the so-called 'optional programmes' available in ELDO which members might join, but were not compelled to do so. See below.

4   Aerosat, an aeronautical satellite system, foundered in 1977 when the US budget appropriation for the project was cancelled domestically.

5   This programme, which transmuted into the MARECS programme has been swallowed by the INMARSAT developments, but was most helpful to that organization in its early days. Originally, interim EUTELSAT was conceived as having two objectives, the creation of the European fixed service (ECS) and a European maritime service using a derivative of the OTS satellite (MAROTS/ MARECS). See the discussion of interim EUTELSAT, below.

6   'Convention for the Establishing of a European Organisation for the Development and Construction of Space Vehicle Launchers' (1964) UKTS No. 30, Cmnd. 2391; 507 UNTS 177; *European Yearbook*, p. 1153.

7   *The Times*, 12 March 1988, p. 1.

8   This is a device adopted in other European cooperative agreements, not only on space matters. However, to give a 'space' example: under the EUTELSAT Convention, facilities separate from that of the EUTELSAT space segment are financed by EUTELSAT only by the unanimous decision of the Board of Signatories (art. V(f), EUTELSAT Convention).

9   Such a proposal had been made by an ESC Working Group on Institutions in 1968, but the 1970 meeting was first to air the matter at Conference level.

10   R.F. von Preushen, 'The European Space Agency' *International and Comparative Law Quarterly* 1978, pp. 46–60 at p. 48.

11   These programmes were METEOSAT (a meteorological satellite); TELECOM (a telecommunications satellite): 'Arrangement . . . concerning the Execution of a Communication Satellite Programme'. (1976) UKTS No. 12, Cmnd. 6414; and AEROSAT (an air traffic satellite, which was to have involved also the USA and Canada).

12   A maritime satellite, not unconnected with proposals for what became INMARSAT: 'Arrangement . . . concerning the Execution of a Maritime Satellite Programme', (1976) UKTS No. 53, 1976 Cmnd. 6528.

13   Convention for the Establishment of a European Space Agency With Final Act and Resolutions, 1975 Cmnd. 6272; (1981) UKTS No. 30, Cmnd. 8200; (1975) 14 ILM 855–908; 23 *European Yearbook*, p. 825. M-G. Bourely, 'Le nouveau cadre de la coopération spatiale européene. L'Agence spatiale européene', *Revue française de droit aérien*, 29, (1975) pp. 233–67; K.J. Madders, 'European Space Agency', *Encyclopedia of Public International Law*, vol. 6, (Max Planck Institute for Comparative Law: Amsterdam: North Holland, 1983), pp. 203–6; R.F. von Preushen, op. cit.

14   See section 'Optional programmes' above, p. 255.

15   See section 'Cooperation' above, p. 257.

16   See *The Times*, 12 March 1988, p. 1.

17   Cf. *United Kingdom Space Policy;* vol. 1, 'Report'; vol. 2, 'Minutes of Evidence', (London: HMSO, House of Lords Select Committee on Science and Technology, 1987) (HL 41–I and II), esp. Report at paras. 5.10–19 and 5.330–117. Cf. also *House*

of Lords' Select Committee on Science and Technology: UK Space Policy, Government Response, (London: HMSO, 1988).

18   The European Space Agency (ESA) technically came into being in 1980 following upon the coming into force of the Convention for its establishment which had been signed in May 1975. However, by agreement the several Parties to the ELDO and ESRO Conventions had acted since 1975 as if the ESA Convention was effective. See section on ESA in this chapter.

19   'Arrangement Constituting the European Conference of Postal and Telecom-munciations Administrations (CEPT)', (1959) 7 *European Yearbook,* pp. 639–59: A.H. Robertson, *European Integration* (3rd edn.), (London: Stevens 1973), pp. 366–70. L. Weber, 'European Conference of Posts and Telecommunications Admin-istrations', *Encyclopedia of Public International Law,* vol. 6 (Max Planck Institute for Comparative Law, Amsterdam: North Holland, 1983) pp. 147–8; R. Rutschi, 'The European Conference of Postal and Telecommunications Administrations', *Tele-communications Journal,* vol. 34, 1967, pp. 481–3.

20   Restricted unions are permitted by art. 8 of the Constitution of the Universal Postal Union; cf. George A. Codding, *The Universal Postal Union. Coordinator of the International Mails,* (New York: New York University Press, 1964) pp. 229–34.

21   N. Simmons, 'A Review of the Work of the Technical Planning Staff of CETS', *Spaceflight,* no. 10, 1968, p. 87; N.M. Matte, *Aerospace Law: Telecommunications Satellites,* (Toronto and London: Butterworths, 1982) pp. 54–5.

22   See the development of ESA, first section of this chapter.

23   The OTS was launched by NASA. OTS–1 suffered a launch failure.

24   'Agreement on the Constitution of a Provisional European Telecommunications Satellite Organisation "Interim EUTELSAT"', available from interim Eutelsat itself: printed in *Space Law: Selected Basic Documents* (2nd edn.), Committee on Commerce, Science and Transportation, US Senate, 1978, 95th Congress 2nd Sess., pp. 469–85. Extracts from the Agreement are printed in unofficial transla-tion as Appendix IX to N.M. Matte, *Aerospace Law: Telecommunications Satellites,* op. cit., pp. 312–21.

25   Available annexed to the constitutive Agreement, and as Appendix X to N.M. Matte, *Aerospace Law: Telecommunications Satellites,* op. cit., pp. 322–5.

26   Available annexed to the constitutive Agreement, and (unamended) as Appen-dix XI to N.M. Matte, *Aerospace Law: Telecommunications Satellites,* op. cit., pp. 326–8.

27   Austria, Belgium, Denmark, West Germany, Finland, France, Italy, Luxemburg, The Netherlands, Norway, Portugal, Spain, Sweden, Switzerland, Turkey, the United Kingdom. Intriguingly Yugoslavia adhered to the interim arrangements at a later stage.

28   Art. 12 of the Interim EUTELSAT Agreement and arts. 3 and 4 of the ECS Supplementary Agreement dealt with financial matters.

29   MAROTS is the Maritime Orbital Test Satellite (from the OTS series). MARECS is the Maritime European Communications Satellite, the result of the use of the ECS series (the successor of the OTS series) for maritime service.

30   Amendment No. 1 to the Agreement which is dated 15 October 1979 altered art. 21(e) to provide for the termination of the Interim EUTELSAT Agreement if ESA decided not to provide 'either' of the two space segments specified in the Agree-ment. The previous language terminated the Agreement if ESA decided not to provide 'one' of these segments. The result was that the 'failure' of the MAROTS project did not affect continuance of the interim EUTELSAT organization.

31   In fact, because, as described above, interim EUTELSAT had not proceeded with the MAROTS project, only France held a mandate under the interim arrange-ments. It had dealt with all matters other than the maritime project: see above.

32   Austria, Belgium, Cyprus, Denmark, Finland, France, Greece, Ireland, Italy, Luxemburg, The Netherlands, Norway, Portugal, Spain, Sweden, Switzerland, Turkey, the United Kingdom, West Germany and Yugoslavia.

33   Apart from members, Liechtenstein, Monaco and St Marino attended.

34   France and UK 16.4 per cent each, Italy 11.48 per cent and West Germany 10.82 per cent.

35   In common with other similar pairs of documents, the Articles of the Convention are given Roman numbers, and those of the Operating Agreement arabic numerals.

36   Convention and Operating Agreement of the European Telecommunications Satellite Organisation (EUTELSAT), Paris, 15 July 1982, (1983) Misc No. 25, Cmnd. 9069.

37   There were technical problems, not objections in principle to the treaty.

38   'Protocol Amending the Convention establishing the European Telecommunications Satellite Organisation' (EUTELSAT) Paris, 15 December 1983, (1984) Misc No. 4, Cmnd. 9154; Amendment No. 1 to the EUTELSAT Convention, Intergovernmental Conference, Paris 15 December 1983, (printed with the Convention as available from EUTELSAT).

39   Every member of interim EUTELSAT joined within the time-limits.

40   'European states' (a term undefined in the Convention) not qualified to be members as not having been members of interim EUTELSAT could use this accession procedure only once the Convention had entered into force. Membership is, of course, separate from becoming a user of the system. Non-member users include Hungary, Poland and the USSR.

41   See section, 'Financial and Other Matters' below, p. 287.

42   *Reparation for Injuries Suffered in the Service of the United Nations,* International Court of Justice, Advisory Opinion 1949, 1949 ICJ Rep 174. See also *American Journal of International Law,* vol. 43, 1949, p. 589.

43   'Protocol on the Privileges and Immunities of the European Telecommunications Satellite Organisation (EUTELSAT), Paris 13 February 1987', (1988) Misc No. 3, Cmnd. 305.

44   See below as to the possible mistake in defining this as the 'prime objective'.

45   Specialized services are considered below.

46   See subsection 'Coordination with other systems', p. 289 below.

47   See subsection 'Investment Shares', p. 287 below.

48   The EUTELSAT system began operation in October 1983.

49   The UK had a 24 per cent share on this basis, although it dropped to 17.5 per cent from 1 March 1988 at the annual review, as to which see below.

7   INTERSPUTNIK and ARABSAT: Two other responses

There are two other organizations which are major responses to the challenges and opportunities of space telecommunications. The first, INTERSPUTNIK is the creation of the USSR and the communist bloc, in part offering an alternative global organization to INTELSAT. The second, ARABSAT, is, like EUTELSAT, a regional response although the 'region' concerned is unusual being 'long and thin' and its organizing principle is more that of a common cultural and religious heritage than geographic. ARABSAT is also akin to EUTELSAT in that its members are also INTELSAT members, and hence the establishment of the ARABSAT system was subject to coordination and consultation with INTELSAT.

INTERSPUTNIK

As indicated in Chapter 1, proposals for the exploration and use of space by satellite were made within the planning stages of the International Geophysical Year 1957. These, although also bearing fruit in other ways, lead directly to the present systems of satellite telecommunications. Most of the initiatives originated from the USA. However, the USSR astonished the world by the launch of Sputnik 1 on 4 October 1957 and, since then, the USSR has been a major participant in manned and unmanned space exploration. It has been less active in the field of satellite telecommunications although it has not neglected this field.[1]

Ideally, one would have wanted to see a single global satellite telecommunications system established on the lines of the general tenor of the UN Resolutions on the matter. That was unduly hopeful, particularly in the political climate of the time and given the interconnection of communications and state and military needs. Notwithstanding, the participation of the USSR in the embryonic

INTELSAT was sought, a delegation of the US negotiators of the interim arrangements meeting Soviet Union representatives in Geneva in 1964. However, the USSR refused to join the proposed organization, or even to attend the Washington Conference in autumn 1964 at which the Interim Arrangements for INTELSAT were settled.[2]

There appear to have been a variety of factors in this decision. The negotiations of the interim INTELSAT were sufficiently advanced to worry the USSR. It was already clear that INTELSAT would seek to link financial commitment and voting power, and would key both to the actual or projected usage of the system by the members of the organization. The USSR share of international telecommunications traffic was small and therefore, under the most likely arrangements, the USSR would not have had a significant say within the organization. In contrast to the USA, the UK and the other major users of international telecommunications which were assured of seats on the Board of Governors under a 'usage' basis, the USSR would have had at best a 'shared' Governor on the Board of Governors and weighed but small in the affairs of interim INTELSAT.[3] Again, perhaps the timing of the proposals precluded the USSR from taking any other stance. In 1963–64 the Cuba crisis was very recent, and it was unrealistic to expect the USSR meekly to join an organization in which the USA was the dominant partner.[4] Finally, an unnamed commentator suggested to me that the USSR had convertible currency problems in the early 1960s and was unlikely to join an organization from which it would derive little benefit, but to which it might be called on disproportionately to contribute in scarce currency. Whatever the truth of weight of these factors, if there were to be a single global system, the USSR certainly favoured a structural system representative of states in which all states would have had an equal vote, although the financial arrangements it preferred were obscure. This was not acceptable to the major telecommunications nations. To satisfy them, usage, share of cost and weight of vote had all to be interrelated.

Consequently, the USSR and the communist bloc stayed out of interim INTELSAT and accordingly had to present some alternative if they were to remain active in international satellite telecommunications. Factors other than determination to provide an alternative to INTELSAT also played a part in the decision. As in other cases of similar geography, satellites could readily meet the telecommunications needs of the Soviet Union itself. Indeed, the size and extent of the country and the inhospitable nature of much of its terrain make satellites especially attractive as a means of providing communications links. Like Canada, the major advantages of a satellite system were present in the specialized requirements of the single country.

In January 1964 the launching of two satellites in the Electron series

showed that the USSR had a commitment to satellite telecommunication, and this was underlined in 1965 and 1966 by the launch of the
Molyniya series. That series (the latter of which had also a weather
observation function) provided a high-level but non-geostationary
system with satellite apogees high above the northern hemisphere in
order to provide maximum visibility ground stations in the communist countries to be served – most communist countries being in
the northern hemisphere. Thereafter telecommunications formed an
important part of the Soviet space programme.

In 1967 a meeting of experts from Bulgaria, Cuba, Czechoslovakia,
East Germany, Hungary, Mongolia, Romania and the Soviet Union
drew up a programme for international cooperation in the exploration and use of outer space, which narrated the United Nations
principles as amongst its objectives. That programme evolved in 1976
into the Intercosmos programme.[5] In the meantime, however, one of
its main projects – that relating to telecommunications – had produced a separate organization, INTERSPUTNIK.

The 1967 group of experts had indicated that it wished to see developed a satellite telecommunication system that would be open to
all countries which wanted to take part, as the relevant UN Resolutions had envisaged.[6] A major step was taken at the first United
Nations Conference on the Peaceful Uses of Outer Space, Vienna 1968
where the Intercosmos group put forward a draft Agreement on an
international satellite telecommunications organization and system;
but other countries preferred the INTELSAT alternative.[7]

Notwithstanding, the nine negotiating countries of Intercosmos
signed the 'Agreement on the Establishment of the "INTERSPUT-
NIK" International System and Organization of Space Communications' at Moscow on 15 November 1971.[8] Ratification by six of the
Signatories was required for the Agreement to come into force (art.
21). This was easily met and the INTERSPUTNIK Agreement came
into force on 12 July 1972. As at December 1986 the organization had
14 members: Afghanistan, Bulgaria, Cuba, the German Democratic
Republic, Hungary, the Korean People's Democratic Republic, Laos,
Mongolia, Poland, Roumania, the USSR, Vietnam, and the People's
Democratic Republic of Yemen. In addition, the Syrian Arab Republic
had deposited an Instrument of Accession.[9]

INTERSPUTNIK organization

The Preamble of the INTERSPUTNIK Agreement narrates the recognition by the contracting Parties of the 'need to contribute to the
strengthening and development of comprehensive economic, scientific, technical, cultural and other relations by communications as

well as by radio and television broadcasting via satellites', and the advantages of cooperative effort in research and the development of international satellite communications. Accordingly, 'in the interests of the development of international cooperation based on respect for the sovereignty and independence of nations, equality and non-interference in the internal affairs as well as mutual assistance and mutual benefit', and in the light of the UN Resolution 1721 (XVI),[10] and the Outer Space Treaty, the Parties agree that there shall be an international system of communication by satellite (art. 1.1), and the establishment of INTERSPUTNIK to ensure cooperation and coordination of efforts to set up the system (art. 1.2). INTERSPUTNIK is an open organization (art. 2.1) in the sense that there are no restrictions on membership – for example, by requiring membership of the ITU[11] – although the organization itself is only an intergovernmental organization, only governments being members of it (art. 2.2).[12] In practical terms, of course, the operational responsibilities of each Party are dealt with by the appropriate ministry or whatever arrangements it makes for its telecommunications interests, but this is without altering the interstate nature of INTERSPUTNIK itself.

Membership

Membership of INTERSPUTNIK is open to the initial Contracting Parties, and the Agreement may be acceded to by the government of any state which did not sign the Agreement, provided that acceding government sends to the INTERSPUTNIK Board a statement that it 'shares the goals and principals of the activities of the Organization and assumes the obligations under (the) Agreement' (art. 22.1). Instruments of ratification or accession take effect on the date of their deposit (art. 23) with the Depository (art. 22.2), which is the USSR (art. 20).[13]

Suspension from membership

A member which fails to meet its financial obligations under the Agreement within one year may have its rights of membership partly or completely suspended by the Board (art. 15.8).

Withdrawal from membership

Withdrawal from membership of the organization is competent and takes place by written notice to the Depository government. The withdrawal is effective at the end of the financial year following that in which the denunciation takes place (art. 17). Financial consequences are considered below in 'Financial arrangements'.

Dissolution

The INTERSPUTNIK Agreement may be terminated by the agreement of all the Contracting Parties (art. 18.1) and this will result in the dissolution of the organization (art. 18.1). The precise procedure for dissolution will be a matter for the Board (art. 18.1). However, art. 18.2 provides that, on the dissolution of the organization, its assets will be realized and members paid monetary compensation proportionate to their capital expenditure in establishing the system, with due regard to 'physical and moral depreciation' of the fixed assets. Available current assets, less those required to defray the extant obligations of the organization, will be divided among members in proportion to their monetary contributions as at the date of dissolution (art. 18.2).

Structure

INTERSPUTNIK consists of a governing body, the Board, and an executive body, the Directorate.

The Board

The Board comprises one representative from each member of the organization (art. 12.1), each having a single vote in decision-taking (art. 12.2). The Board meets at least annually, but extraordinary sessions may be held at the request of any member or the Director-General provided that at least one-third of the members agree to the holding of such a session (art. 12.3). Meetings are held according to the constituent treaty 'as a rule, at the seat of the Organisation' (art. 12.4), but information supplied by INTERSPUTNIK indicates that the Board has made use of its power also contained in art. 12.4 to meet elsewhere on the invitation of members, and has adopted a policy of meeting in the different member countries in annual rotation. The chairmanship of the Board rotates annually on an alphabetical Russian basis (art. 12.5).

The powers and functions of the Board cover all matters dealt with under the Agreement. Particularly enumerated in art. 12.6 are powers to examine and approve measures for establishing and operating the INTERSPUTNIK system, the approval of plans for future development, and for the deploying of satellites, the determining of specifications for the satellites and associated earth stations, the allocation of channels among members, and procedures and conditions of use of the system. The Board also elects the Director-General and his deputies and that of the Auditing Commission (which deals with financial matters). It supervises the work of both the Directorate and the Auditing Commission, approves the budget and annual reports from both, deals with the fixing of contributions, adjusts contribution shares, sets rates for system use, and deals with amendments to the Agreement.

The Board tries to act unanimously, failing which a two-thirds majority vote is required (art. 12.7). Board decisions do not bind members which do not vote for them, provided that they express their reservations in writing. Such a reservation may be withdrawn by a member which changes its mind (art. 12.7).

The Directorate

The Director-General heads up the Directorate and is the chief executive of the organization representing it in all matters relating to the satellite system as well as to other states and international organizations (art. 13.2). As indicated, the Director-General and his deputy are elected by the Board (art. 12.6.8) and he is responsible to it (art. 13.3). Both must be nationals of two different members of the organization. They serve for four years and, ordinarily, the deputy is elected for one term only (art. 13.5). The Director-General deals with the implementation of Board decisions, negotiates the setting-up of the satellite system, concludes necessary agreements, deals with budgets, planning, and makes reports to as well as servicing the Board (art. 13.4). The Directorate is responsible for the working of the system. The staff must be nationals of member states of the organization and are appointed with 'due ragard for their professional qualifications and the equitable geographical distribution' (art. 13.6).

Financial arrangements

It is the Board's duty to 'act within the resources determined by the Contracting Parties' (art. 12.8). Financial matters are overseen by an Auditing Commission of three members each from different member states (art. 14.1) elected by the Board (art. 12.6.9) and who hold no other office in INTERSPUTNIK (art. 14.1). The Commission reports annually to the Board (art. 14.3) and has access to all necessary documentation (art. 14.2).

The organization is funded partly through income generated from system use (as to which see below) and by a Statutory Fund established from 1 January 1983 by a Special Protocol of 26 November 1982 (not seen) (art. 15.1).[14]

Contributions to the Statutory Fund are proportional to the use made of the satellite system by each member (art. 15.2) but the apportionment of any increases later found necessary in the Fund are related both to use and to the consent of members (art. 15.3). The share of members is adjusted on the accession of new members or withdrawal of old. As indicated above, failure to pay contributions can ultimately result in a complete suspension of a member's rights (art. 15.8). According to data from INTERSPUTNIK, a member joining the organization makes a compulsory minimum payment of 1 per cent of the Statutory Fund, payable either entirely in convertible currency, or 90 per cent in transferable roubles and the remainder in convertible currency.

The Statutory Fund finances the setting up and operating of the system to the extent that income fails to meet these requirements (art. 15.1). Profits are shared in proportion to contribution, or may either be added to the Statutory Fund or put into special funds (art. 15.9).

Use of the system is determined by the Board, distributing channels according to need (art. 16.2). Capacity in excess of need may be allocated to other users by lease (art. 16.2). Payment for use of the system is at rates set by the Board (arts. 16.3 and 12.6.17), which are calculated in gold francs (art. 16.3).[15]

Summary

The structure and financial arrangements of INTERSPUTNIK are more akin to that of INMARSAT rather than INTELSAT as one might expect, given that the organization is smaller and more restricted in its scope. At present it leases and operates two Stationar satellites, Stationar–4 positioned at 14°W over the Atlantic and Stationar–13 at 80°E over the Indian Ocean with a network of earth stations throughout its members.

Structurally the main difference from INTELSAT stems from the principle that members are states, each having an equal vote in decision-making and that all members have seats on the Board. Operational entities have no role to play within the organization, although obviously they could play a part within the operation of the system. However, this is unimportant as I believe that telecommunications services within all members of INTERSPUTNIK are a matter of government responsibility for both regulation and operation.

I have no data on the volume of traffic carried in the INTERSPUTNIK system, nor as to the actual financial working of the organization. That it conducts its finances on the basis of convertible currencies is a matter for notice, not because INTELSAT is different – it is not – but because it underlines the known economic difficulty of international settlements within most of the countries represented in the organization.

The last point to be made in this brief account is to draw attention to the Agreement between INTERSPUTNIK and INTELSAT which was reached in 1985.[16] Most of the members of INTERSPUTNIK also use the INTELSAT system on a service basis without becoming members. The official move towards interrelationships between the two non-specialized global systems marks progress towards the single global system which was the original hope of the UN Resolutions of the early 1960s.

ARABSAT

The establishment and operation of telecommunications within the area of northern Africa and over into Arabia and the Persian Gulf have always been difficult. Problems of natural topographic obstacles were added to by the presence of different colonial regimes amid some independent states, and latterly, following upon decolonization, by continuing political differences among the states in the area. But there are also common features. Most of the states in the region are in a general sense 'Arab' and a high proportion of their nationals are of the Moslem faith. Within these commonalities a sufficient regional base has been discovered on which to found a satellite communications system.

The original impetus towards what is now the ARABSAT system was an ITU survey funded by the UN Development Programme which recommended steps to meet the national and international telecommunications needs of the Middle East and North Africa areas. Discussions between the affected countries was also facilitated through the Arab Information Institute, the Arab States Broadcasting Union and the Arab League, with UNESCO also playing a role.[17] In the later stages of discussions, and once the organization was established in February 1977, COMSAT General acted as consultants, that Corporation having much useful experience through its involvement with INTELSAT and INMARSAT. However, the prime contractor for the ARABSAT system is the French company, Aerospatiale.

The ARABSAT system was conceived as providing both public telecommunications facilities and direct broadcast services to the entire area between the Atlantic seaboard at Mauritania to the Persian Gulf, and there have been subsequent suggestions that an eastward extension to other Moslem countries further east as far as Malaysia might be feasible. In its present form, providing telecommunications services from the Atlantic to the Gulf, ARABSAT has been approved by all the Arab members of the Arab Telecommunication Union. ARABSAT is an independent organization within the Arab League (ARABSAT Agreement, art. 2) and is intended to provide communications, information, culture, education and any other service that can be provided by it and 'towards the fulfilment of the Arab League Charter' (Preamble).

The ARABSAT Agreement

The constituent document of ARABSAT (properly the Arab Space Communications Organization or the Arab Corporation for Space Communications) is its Charter, the 'Agreement of the Arab Corporation for Space Communications' signed by 21 Parties on 14 April 1976.[18] The original Parties were Algeria, Bahrain, Democratic Yemen, Egypt, Iraq, Jordan, Kuwait, Lebanon, Libya, Mauritania, Morocco, Oman, Palestine, Qatar, Saudi Arabia, Somalia, Sudan, Syria, Tunisia, United Arab Emirates and Yemen. Djibouti has subsequently joined the organization. As the Parties to the Agreement are stated by the Preamble to the Charter to be governments, the presence of Palestine is interesting. Members of the organization are states (art. 1(c)), and there is no provision for membership equivalent to that of the Signatories of INTELSAT or INMARSAT. This reflects the arrangements for telecommunications within the constitution of the member states.

The Agreement was subject to ratification and entered into force 60 days after ratifications were deposited from at least seven members whose combined holding of shares in the organization amounted to at least 60 per cent of the organization's capital (arts. 20 and 22(1)). That requirement was quickly fulfilled, the organization coming into being in February 1977. Its headquarters is in Riyadh, Saudi Arabia, with branches in other Arab countries (art. 4(2)).

No reservation is permitted to the Agreement in the instrument of ratification or accession (art. 21). Amendment of the Agreement is competent in terms of art. 18 and is effected through a proposal by one or more members submitted to the Director-General and supported by one-third of the members or through a proposal by the Board of Directors (art. 18(1)).[19] The proposal is considered by the next ordinary meeting of the General Assembly[20] or by an extraordinary meeting of that Assembly where the proposal has been circulated 90 days prior to its meeting (art. 18(2)). The proposed amendment may be adopted by a two-thirds vote of members of the General Assembly (art. 18(3)), and takes effect on receipt by the Secretary-General of the Arab League (art. 20 and 22(2)) of instruments of ratification of the decision from one-third of the members of the organization holding at least 60 per cent of the organization's capital (art. 18(4)).

The organization has legal personality including the right to enter into contracts, to own property and funds and to take part in all legal proceedings (art. 2). It has the privileges and immunities normal under the Arab League Resolution on such matters – Resolution No. 575 of 10 May 1953 (art. 16).

Membership and withdrawal

Membership of ARABSAT is open to Arab states which are members of the Arab League (reflecting its position within that League) (art. 2) and which contribute to the capital of the organization (art. 4(1)). Original membership was available through ratification of the Agreement and thereafter through accession deposited with the Arab League (art. 20). Withdrawal from the organization is competent and is effected by giving notice to the Secretary-General of the Arab League (art. 17(1)). The withdrawal takes effect one year from the date of submission of the letter of notice but may be withdrawn within that period (art. 17(2)). A withdrawing member remains liable for obligations contracted by the organization up to the effective date of its withdrawal (art. 17(3)), and accounts are then settled and financial shares among the remaining members readjusted (art. 17(4) and (5): see below).

Structure

ARABSAT has a tripartite structure, a General Assembly, a Board of Directors and an Executive Organ headed by a Director-General.

General Assembly

The General Assembly consists of the Ministers of Communications of all member states, or their duly authorized representatives (art. 10(1)). In addition, observers from various Arab organizations are invited. The Agreement specifies that observers will be invited from the Arab League itself, the Arab States Broadcasting Union, the Arab League Telecommunications Union and the Arab League Educational, Scientific and Cultural Organization (art. 10(7)). In addition, the Assembly may approve the attendance of other observers from other organizations that share ARABSAT's objective (art. 10(7)).

Chairmanship of the Assembly rotates annually among members in alphabetical order (art. 10(2)). The Assembly meets annually at its headquarters or at one of its branches or in a member state (art. 10(3)). It may also meet extraordinarily at the request of the Board of Directors or on the requisition of one or more members supported by one-third of the membership for a particular purpose stated in the requisition. In this case, three months' notice of meeting is given (art.

10(4)). A majority of members form a quorum (art. 10(5)). Each member has one vote (art. 10(1)). Procedural questions are dealt with by simple majority (art. 10(6)). Substantive questions are decided by a two-thirds majority of those present (art. 10(6)). Whether a matter is one of substance is decided by a simple majority of those present and voting. In the event of a tie, the side having the Chairman's vote wins (art. 10(6)).

The Assembly's functions are specified in art. 11 and include defining general policy, planning, the setting of rates for space segment usage, adopting general specifications for earth stations, dealing with suspensions and withdrawal, financial allocations and adjustments, investment shares, the ratification of budgets, deciding on representation on and electing to the Board of Directors, dealing with amendments to the Agreement, and settling its own internal rules. Notable is power to delegate some of its authority to the Board of Directors (art. 11(21)).

Board of Directors

As in the other organizations, the real power lies with its Board of Directors. The Board of ARABSAT consists of nine representatives from separate member states (art. 12(1)). Like other such institutions, the Board has a core membership representing those states having the largest financial share in the undertaking and an element representing the other Parties to the Agreement. In the case of ARABSAT, five of the nine members of the Board are the five largest holders of capital investment. These five hold seats as long as they qualify in terms of capital share. In the event of there being an equality of shares, the General Assembly elects the required number (art. 12(1)(a)).[21] The remaining four seats are held by members elected by the General Assembly from among those not entitled to reserved seats. The 'unreserved' seats are occupied on a rotating basis for two years, and a holding member cannot be re-elected immediately (art. 12(1)(b)).

Meetings of the Board have a quorum of seven members, and are postponed for two weeks if the quorum is not present. If the quorum is still not in attendance, the Director-General must call a meeting of the General Assembly for one month thereafter and it then meets to exercise the powers of the Board (art. 12(5)). Decisions are taken by a simple majority of Board members (art. 12(7)), each member having one vote (art. 12(6)). Board meetings may be open or closed and, if open, may be attended by observers from organizations thus entitled (art. 12(9)). The Board has power to invite the attendance of other 'necessary' persons (art. 12(9)).

The Board of Directors executes the policy of ARABSAT, as indicated by the Assembly, to provide and maintain the ARABSAT space segment (art. 13). In particular, it arranges questions of design and implementation and, subject to the Assembly's approval, sets rates, defines standards for earth station access to the space segment, approves applications for access, distributes space segment capacity and deals with accounts and budget. It appoints the Director-General and his deputy (an interim post) and considers reports, ideas and recommendations coming from him.

Executive Organ

The Executive Organ is headed by the Director-General. As just noted he is appointed by the Board (art. 13(15)), is under a 3-year renewable contract (art. 15(1)), but is dismissable by the Board 'for cause' (art. 15(3)). The Director-General represents the organization in law and is responsible to the Board (art. 15(2)). Under his chairmanship, the Executive Organ deals with the administration of the organization. Its senior personnel are appointed by the Director-General, subject to Board approval (art. 13(16)). Appointees to the Executive Organ, including that of technical and administrative officials, are chosen on the basis of high capability and efficiency and with regard, if possible, to geographic distribution (art. 14(2)).

Finance

The initial capital of ARABSAT was US$ 100 million, divided into 1,000 equal shares (art. 5). That figure may be increased on the proposal of the Board with the approval of the Assembly (art. 5). Article 6(1) set the original capital shares in accordance with a financial appendix to the Agreement showing the initial shares as ranging from 26.6 per cent for Saudi Arabia to 0.2 per cent for Mauritania and Palestine.[22] The minimum participation is one share – 0.1 per cent of the organization's capital (art. 6(4)). Shares are reassessed on the accession or withdrawal of a member (art. 6(5)). Two years after the coming into operation of the space segment ARABSAT was to consider the basing of shares on usage of the system, and the reduction of the participation of members without ground stations to one share (art. 6(2)). Members may apply to have their share reduced, and the General Assembly must decide on the distribution of unwanted shares (art. 6(3)).

Five per cent of each member's initial capital share was payable on its ratification of the Agreement and the rest in accordance with calls by the Board of Directors (art. 7(1) and (2)). Settlement was required within 60 days with interest on arrears of 1 per cent per month (art. 7(3)).[23] A member in arrears for more than one year can be suspended from membership by the Assembly (art. 11(11)) and (implicitly) may be deemed to have withdrawn by the Assembly (art. 11(9)). Profits are shared among members in proportion to their investment (art. 8).

Disputes

Disputes among members, or between a member and the organization, are dealt with by the General Assembly whose decision is effective 90 days after its issue (arts. 11(7) and 19).

Other Systems

It is the duty of the General Assembly to arrange that relations between ARABSAT and other organizations are in accordance with international agreements (art. 11(6)). This is weaker than the language in the EUTELSAT Convention, for example. There is no requirement that members shall coordinate their use of other systems with ARAB-SAT. However, as the members of ARABSAT are also INTELSAT members, coordination was required with that organization. That is more fully treated in relation to INTELSAT's development of its concept of 'significant economic harm' and the problems of traffic not otherwise available to INTELSAT and traffic diversion from INTELSAT. Suffice it to say that the Arab interest in the EUTELSAT and PALAPA coordinations reflected their own future arguments. ARABSAT was duly coordinated with INTELSAT by the Fifth Assembly of INTELSAT Parties in 1980 on the basis of the proposals submitted to INTELSAT by the Arab group. Material changes in the system or material extensions of the system beyond 1990 will require further coordination.[24]

Summary

The ARABSAT system, established through the French company Aerospatiale, consists of two in-orbit satellites (at 19° East and 26° East) and one on-ground spare. Commercial operation began on 22 August 1985. I have no information as to its traffic loading or the financial results. It remains possible therefore that INTELSAT will find itself significantly affected by a successful ARABSAT, or that the question of 'cumulative diversion' will require INTELSAT to reconsider its relationship with ARABSAT. However, if the ARABSAT system is established and working to the satisfaction of its members it seems likely that anything other than compromise can be the result.

NOTES

1   An early analysis is *Soviet Space Programs: Organisation, Plans, Goals, and International Implications*, Staff Report prepared for the Use of the Committee on Aeronautical and Space Sciences, US Senate, ed. Glen P. Wilson, 87th Cong. 2nd Sess., 31 May 1962.
2   N. Mateesco Matte, *Aerospace Law* (Toronto: Carswell; London: Sweet and Maxwell, 1969) p. 197.
3   Cf. the role the USSR has under a similarly based voting division within INMARSAT where its usage of the system is greater. In 1987 its usage-based participation was 2.34204 per cent.
4   The USA started off in interim INTELSAT with 61 per cent.
5   See V.S. Vereshchein, 'Agreement on Co-operation in the Exploration and Use of Outer Space for Peaceful Purposes (INTERCOSMOS)', N. Jasentulyana and R. Lee (eds), *Manual on Space Law*, vol. 1 (New York, Oceana, 1979–82) pp. 415–26; Text of Agreement vol. 2, p. 253.
6   It is not clear that interim INTELSAT did not meet this ideal, but clearly there was a view that what was considered to be the lack of full equality of states within interim INTELSAT was an obstacle.
7   'Draft Agreement on the Establishing of an International Communications System using Artificial Earth Satellites', UNGA A/AC.105/46 9 August 1968, (1968) ILM 1365–75; *Telecommunication Journal*, vol. 35, 1968, pp. 508–10.
8   862 UNTS 3; see also Y.M. Kolossov, 'International System and Organisation of Space Communication (INTERSPUTNIK)', N. Jasentulyana and R. Lee (eds), *Manual on Space Law*, op. cit., pp. 401–14, with text of the Agreement in vol. 2 at p. 159. It is printed in *Space Law*, op. cit., n. 18 below at p. 385 and is available from INTERSPUTNIK.
9   Information provided by INTERSPUTNIK.
10 The language of Res. 1721 (XVI) is also referred to in the Preamble to the US Communications Satellite Act of 1962.
11 Cf. INTELSAT Agreement, art. XIX(a), (c).
12 Cf. INTELSAT and INMARSAT, where operational entities are part of the organization through the relative Operating Agreement.
13 See section 'Financial arrangements', p. 301 below.
14 Earlier finance was agreed on the recommendation of the Board on an ad hoc basis: art. 15.4.
15 INTERSPUTNIK informs me that the 1986 rates were: for telephony and telegraphy – each voice channel 1750+2500 gold francs per month on the satellite and each earth station dependent on type of modulation and transmission: for video and associated audio transmission; a satellite use charge of 40.6 gold francs per minute; and an earth station charge of 996 gold francs for the initial 10 minutes plus 28.2 gold francs for each additional minute. Long-term leases, night transmissions and INTERVISION services are discounted. Radio was charged at two gold francs per 10 KHz channel and three gold francs per 15 KHz per minute for the satellite and each earth station. The 1987 rates were not likely to be different.

16 Memorandum of Understanding between INTELSAT and the Ministry of Posts and Telecommunications of the USSR, Attachment 1 to INTELSAT Doc. BG–64–69, 27 August 1985. See also Chapter 4 above, p. 153.

17 S. Gorove, 'The Arab Corporation for Space Communications (ARABSAT)', N. Jasentulyana and R. Lee (eds), *Manual on Space Law*, op. cit., vol. 1, pp. 467–79, with text of Agreement at vol. 2, pp. 345–60.

18 A translation by the Gulf Public Relations Translation Service of the relevant decree of the Amir of Bahrain (Amiri Decree No. 25/1976), Issue 1185 of the *Official Gazette of the State of Bahrain* (22 July 1976) and containing the text of the Agreement is reprinted in *Space Law: Selected Basic Documents* (2nd edn.), Committee on Commerce, Science and Transportation, US Senate, 95th Cong. 2nd Sess., December 1978. The Agreement itself, but omitting the financial appendix, is reprinted from the same source in N. Jasentulyana and R. Lee (eds.), *Manual on Space Law* op. cit., vol. 2, pp. 345–60. Another translation is printed in E.M.M. Abdallah, 'The Arab Satellite', 1977 44 *Telecommunication Journal*, vol. 44, 1977, pp. 422–6. There is a problem of terminology in these translations. In this text (as ARABSAT itself does) I use the more familiar terms of Director-General, Executive Organ, and General Assembly instead of General Manager, Executive Body or Department, and General Body.

19 See section 'Structure', p. 305 below.

20 See section 'General Assembly', p. 305 below.

21 No translated text is satisfactory here, the superficial impression being given that the Assembly elects 'if more than five members are equal', or that it must be 'impossible to choose five members . . . owing to equality of participation', opening the five seats to election from the number. Presumably what is intended is that the larger holders (if any) get seats by right and that the Assembly selects between members having an equal share which would otherwise entitle them to a reserved seat.

22 The figures were Saudi Arabia 26.6 per cent, Libya 18.2 per cent, Egypt 10.4 per cent, Kuwait 8.2 per cent, United Arab Emirates 6.6 per cent, Lebanon 6.3 per cent, Qatar 5 per cent, Bahrein 4 per cent, Jordon 3.3 per cent, Iraq 2.6 per cent, Sudan 2.1 per cent, Syria 1.7 per cent, Algeria 0.9 per cent, Yemen (Arab Rep.) 0.7 per cent, Oman, Yemen (People's Republic) and Tunisia all 0.6 per cent, Morocco 0.5 per cent, Somalia 0.3 per cent, and Mauritania and Palestine 0.2 per cent each. As noted, Djibouti later joined and I do not have figures following the consequent adjustment.

23 An interesting provision to see in a document with strong Islamic associations.

24 AP–5–5 and relative 'Report of the Board of Governors to the Assembly of Parties pursuant to Article XIV(d) concerning Coordination of the Arab Communications Satellite System (ARABSAT)', AP–5–8, BG–41–51, 14 March 1980. See also pp. 175–77 above.

8 The International Telecommunication Union (ITU)

INTRODUCTION

The International Telecommunication Union serves three major functions in regard to satellite communications. First it has a regulatory function, dealing with the use of the radio spectrum. Second, it has a role in rate-setting for telecommunications and the setting of international equipment standards. Third, it has a role in the use of the geostationary orbit by particular satellites.

International agreement on the regulation of the use made of the radio spectrum is necessary. Turn on a radio on a winter's evening anywhere in Europe and tune to a station other than the most powerful lying in the 520–1600 KHz waveband (medium wave), and one becomes immediately conscious that the radio spectrum is crowded. Several weaker stations will cross-interfere, sometimes one and sometimes another becoming briefly dominant, and listening to such a babel becomes difficult. This, in measure, is the problem facing all satellite communications.

The requirement of radio spectrum space for communication with satellites is basic to all space activities. Without adequate radio links there can be no guidance, little tracking, no telemetry or systems command, no contact with astronauts, and no reception of scientific data. These considerations are entirely separate from requirement of spectrum space for a telecommunications link by satellite, but must be considered as part and parcel of the problem which space presents. In addition, immediately on the first satellite being in orbit it was found that its signals were interfering with ground-based communications links.[1]

Apart from such 'adjustable' matters, the needs of the emergent science of radio astronomy must be met. Interference with the observations of radio astronomers by man-made radio signals would at worst destroy their work, and even limited interference with signals from faint radio sources by earth-based stations seriously inhibits the science. The frequencies which the astronomers use are set by physical laws and lie outside our control. They cannot be changed by international agreement. The protection of scientific inquiry therefore requires that the frequencies important in radio astronomy are not used by man for other purposes.

The major agency through which the use of the radio spectrum is arranged is the International Telecommunication Union (ITU). It is the second oldest international organization still in operation,[2] and despite (or because of) its long history has proved itself adaptable to cope with the new demands upon its skills. Some of that history will explain why this should be so. However, adaptability has its pitfalls and difficulties.

The ITU is one of the international organizations which do valuable work while remaining in relative obscurity. It is mentioned in passing in most books on public international law, but little has been written on it. This is unfortunate because it is the culmination of many years and much experience. On the other hand, it is arguable that the effectiveness of an international organization is in inverse proportion to its visibility, and one can only view with concern recent attempts to introduce into the deliberations of the ITU the posturing and politicking which so bedevils other organizations. The ITU may not survive such an abuse of its forums.

The obscurity of the ITU does have certain other awkward effects, notably through the importance of its work not being fully appreciated or given a sufficient prominence within the governments of its members. The ITU Convention and Regulations which cope with the requirements of international telecommunications have emerged piecemeal over many years. The result is that the history of almost every provision is essential to its being understood. There is a line of development to be understood providing the context for virtually every choice of words and expression. This knowledge has not, however, always been secured through well-informed and qualified persons being given the 'ITU responsibility' within governments, and there is a danger that provisions of the Convention and Regulations adopted for one reason come to be used for other purposes, and that the effects of well intentioned later principles can come adversely to affect earlier provisions. Ill-informed conference participants can be a menace.

Two elements can be seen to undergird the thrust of the Union in radio matters. First there is intention, born of the necessity that the use of radio frequencies be managed. Second, there is the adoption of a particular strategy or regime as the way towards carrying the intention into effect. The first can be traced right back to the earliest days of international action on radio, while the second is more modern, dating to decisions taken in 1947, when, like other international organizations, the ITU was picking itself off the ground and becoming part of the United Nations system of specialized agencies. An understanding of its history is necessary to grasp the powers, responsibilities and limitations of the ITU in all its work, not only relative to space communications.[3]

HISTORY

Although the ITU as such did not come into existence until 1932, on the radio side of its responsibilities its parentage can be traced to the International Radio-telegraph Union of 1906,[4] and on the telegraph side to the International Telegraph Union of 1865. Archaeologically we could even go back to 1848, the 'Year of Revolutions', when the telegraph was first used internationally. The ITU therefore offers the experience of almost a century and a half in the regulation of international communications. It is now the body through which international agreement is reached on radio frequency regulation, and through which problems can be solved, if necessary by compromise. It also has a considerable function in setting and in advising on technical parameters and specifications for communications services.

The very first organization dealing with international communications, the International Telegraph Union, was created at a conference held at Paris in 1865 by a Convention,[5] which followed the lines of a more restricted convention signed in Berne in 1858.[6] The Union's purpose was to provide uniformity in the international telegraph system in the areas of tariffs, codes, routing and similar operational matters. Regulations were adopted to govern such questions. In 1868 the Vienna Telegraph Conference added a permanent International Bureau to handle all the purely routine administrative work of the Union.[7] The Rome Telegraph Conference of 1871–72 agreed that the telegraph companies should be allowed to advise on the telegraph regulations, the first recognition of the role of private companies and operating agencies in international telecommunications regulation.[8]

A Conference in St Petersburg in 1875 revised the Telegraph Convention, and laid down that the Telegraph Regulations could in the future be revised by administrative conferences without the necessity of holding full-scale conferences of plenipotentiaries,[9] a practice continued by the modern ITU. In fact, from 1875 to the fusion of the International Telegraph Union with its radio equivalent to form the International Telecommunication Union in 1932 there were a whole series of administrative, and no diplomatic, conferences. The telephone was added to the interests of the Union at the 1885 Berlin Telegraph Conference,[10] but the telephone did not gain full import-

ance until the London Telegraph Conference of 1903.[11] Coincidentally, in that year also, radio first came under international scrutiny.

The international regulation of radio frequencies begins with a
Preliminary Radio Conference held in Berlin in 1903. As narrated in
Chapter 1, the first patent for wireless telegraphy stands in the name
of Guiglielmo Marconi, who founded a company in Britain in 1897 to
exploit his patents. In a short time the company had a near monopoly of
intervessel and ship-to-shore communications. In an attempt to complete that monopoly the Marconi Company refused to communicate
with any radio station other than those using the Marconi system.
This strategy had financial effects on the development of radio communications. It also had effects connected with the safety of life at
sea. France complained that a series of radio stations on her north
coast had been rendered partially ineffective by the ban, and the USA
cited a case where a Marconi-equipped ship had refused to answer
radioed queries as to the position of a derelict in the shipping lanes.

The final straw was more petty. A Marconi station refused to relay
a message from Prince Henry of Germany on board his yacht to
President (Teddy) Roosevelt. The German government decided that
something had to be done to sort out such difficulties, and called a
Preliminary Conference at Berlin in 1903 to identify the principal
problems relating to the international regulation of the new communications facility.

Nine countries attended the Preliminary Conference,[12] and there
was a wide-ranging debate covering topics which were later brought
together in a Final Protocol.[13] The conference agreed to propose to
their sponsoring governments various matters which should form
the basis of an international convention to regulate international
communication by radio, and these elements are still present in the current regulations. Thus, the originating cause of the Berlin conference
produced the principle that all radio stations engaged in communications (as opposed to broadcasting) should accept messages from all
other stations irrespective of the equipment used. It was also agreed
to propose that tariffs should be internationally determined. Most
interesting for our purposes, there was as early as this very first
conference, consideration of questions of interference. M. Bonomo,
the Italian delegate, drew the conference's attention to the need for all
countries to join in an international plan to ensure that there would
be no overlapping of frequency use, either because of the use of
similar wavelengths, or through the use of excessive power by one
radio station and thus *de facto* monopolizing the frequency it was
using.[14]

M. Bonomo's suggestion was taken up, and the *avant propos* of the
German government was modified so that the Final Protocol includes
Article V, which reads:

Le service d'exploitation des stations de télégraphie sans fil doit être organisé, autant que possible, de manière à ne pas troublés le service d'autres stations.[15]

The results of the Preliminary Conference were reported back to the participants' governments, and three years later the Berlin Radio Conference of 1906 was convened. Article V of the Final Protocol of 1903 was adopted virtually without modification to become Article 8 of the Convention signed at the Conference.[16] The Berlin Convention was wide-ranging. Even naval and military installations were covered by the provisions of art. 8(15) as to signal strength, and were also required to conform to other provisions of the Convention if they acted as relay stations for public correspondence.

Interference was a major question at both Berlin conferences. Apart from the simple question of signal strength, an attempt was also made to allocate frequencies to particular services – the device which is another main weapon in the international regulation of frequency use. In the Service Regulations annexed to the Berlin Convention, Regulation II reads:

Two wave lengths, one of 300 metres and the other of 600 metres are allowed for general public correspondence. . . . each Government may authorise the use at any coast station of other wave-lengths for the purpose of providing a long-distance service, or a service other than that of general public correspondence, established in accordance with the terms of the Convention, on condition that these wave-lengths do not exceed 600 metres or do exceed 1600 metres.

Public correspondence services were therefore to be at 300 and 600 metres. Long-distance communication and special services were to be available only at coast stations and had to be transmitted at under 600 or over 1,600 metres. Regulation III allocated the frequency of 300 metres to all ships carrying equipment capable of transmitting at that frequency, and Regulation XIX required all ship stations to adjust their receiving apparatus to the highest degree of sensitivity, and to make sure before transmitting that the coast station with which it wished to communicate was not already engaged in communication. If it were, the ship had to wait until there was a break in that communication. If, despite these precautions, interference occurred, the caller causing the interference had to cease transmission as soon as a coast station so requested. The exchange of superfluous signals and words was prohibited, and equipment trials and operator practice was permitted only insofar as interference with other stations was not caused (Berlin Radio Regulations art. V). There was also a general provision requiring stations to exchange signal traffic with the minimum expenditure of energy required for effective communication.[17]

In these ways everything that could then be thought of was done to minimize interference. Frequencies were allocated for services, power was specified, procedures were established. These techniques have continued throughout the development of what became the ITU, and remain major strategies in the securing of the best possible use of the radio frequency spectrum.

Naturally, the Berlin Convention borrowed procedural devices from the international regime which regulated telegraphic communication. Thus, by art. 2, the Convention and the Regulations could be modified by periodic conferences of plenipotentiaries, or by simple administrative conference in the case of the Regulations alone. In fact all conferences of the Radio-telegraph Union were conferences of plenipotentiaries, but the concept of the administrative conference, which has proved so useful within the ITU structure was therefore early embedded in the thinking on radio matters, and, as we have seen, was already present in the Telegraph Union. The new system for radio also borrowed the use of the International Bureau in concept and in fact. The International Telegraph Union had a permanent International Bureau dealing with its affairs since 1868, and it was to this body that the Radio Union entrusted the work of its own new International Bureau, set up in terms of art. 8 of the Convention and functioning in terms of arts. XXXVII and XXXVIII of the Radio Regulations.[18] To this Bureau, the forerunner of the present ITU Secretariat, was entrusted the duty of collecting, arranging and publishing information of every kind relating to radiotelegraphy; of circulating, in proper form, proposals for the modification of the Berlin Convention and the regulations; of notifying alterations adopted, and generally of carrying out any work bearing on matters of international radiotelegraphy of interest to the parties to the Convention. Art. 6 of the Convention required the members of the Union to keep each other informed of the names of the coast stations and their call signs over which they had jurisdiction, together with all the information required for the interchange of messages. By art. IV of the Radio Regulations the Bureau was also to be informed of the name, nationality, geographic position, call signal, normal range, system of radiotelegraphy, the nature of the receiving apparatus, the wavelengths used with the normal wavelengths underlined, the nature of the service provided, hours of service of each station, and charges. Such details would allow the orderly use of the radio spectrum, and facilitate traffic.

However, it may be noted that the International Bureau had no power to regulate the use of radio frequencies. Every station had to be licensed by the government within whose jurisdiction it was located, but if that government chose to license a station to utilize a frequency already in use, the Bureau could do nothing. The only recourse was to the arbitral procedures provided for under art. 18 of the Convention, which were not compulsory. Such a location of final authority in each sovereign state is still with us, and so remains for the reason which justified it 80 years ago. No state will entirely give up, to an international body, power to decide on such important matters as the frequency usage by radio stations within its jurisdiction. International agreement as to frequency allocation will in general be respected, but in the last analysis a state will please itself.

At the end of the 1906 Conference it was decided that a review of the Convention and the Radio Regulations should be conducted at a Conference to be held in London in 1912. In the intervening years the development of radio continued apace. However, the main use of the new facility remained in shipping, and the potentialities of the new medium in that arena were overshadowed and underscored for the 1912 Conference by the loss of the *Titanic* two months before the Conference opened. As the British Postmaster-General said in his opening address,[19] the disaster showed the need for the wider use of radio at sea. It also partially explains why, in an abrupt change of position, the British government was willing to countenance the new Convention providing for obligatory intercommunication between ships at sea regardless of the radio system used – a principle which was accepted by all other participants at the Conference. But, that apart, there is little advance in the 1912 Convention and its Radio Regulations.[20] One small development was the addition of three services to the regulations; radio beacons, weather reports and time signals.[21]

The first conference on radio communications after the First World War was held in Washington, DC, in 1927.[22] It faced a mammoth task. The 15 years since the London conference had seen the technical advances triggered by the War, the use of the short-wave frequencies, the development of both state and commercial broadcasting,[23] and a consequential vast increase in the number of radio stations. An indicator of the size of task can be seen in the fact that over 80 countries were represented and that some 64 organizations and companies sent observers. The Convention signed at the conclusion of the conference was much wider in scope than its predecessors.[24] Their concentration on public maritime correspondence is replaced with regulations applicable to all radio stations open to the international service of public correspondence and also to a large number of services not previously covered. The reason for the extension was pragmatic; it matters not one whit whether an interfering station is of one service type or another, interference is still caused. The majority of stations had, therefore, to come under international agreement. Art. X reads in part:

The term 'radio communications' applies to the transmission by radio of writings signs, signals, pictures and sounds of all kinds by Hertzian waves.

Further on in the same Article 'international service' was defined in a form recognizable as applicable to a strictly international service, but there was added:

An internal or national radio communication service which is likely to cause interference with other services outside the limits of the country in which it operates is considered an international service from the viewpoint of interference.

Unintentional or accidental interference was thus brought under the international arrangements.

Another advance was the establishment of the International Technical Consulting Committee on Radio Communications to advise on questions submitted to it, and to have a function on advising on problems of interference.[25] Further, the Radio Regulations adopted at the Washington Conference were much more thoroughly developed and comprehensive than those they replaced. Whereas in previous regulations it had been customary to state that gaps in the Radio Regulations were to be filled by the analogous provisions of the Telegraph Regulations, the new set aimed at a degree of completeness. As a result, the Washington Radio Regulations, consisting of 24 articles and eight appendices, are very detailed.

New procedures were adopted in art. 5 of the Regulations to cope with the problems of interference. In principle any frequency might be assigned to a station by an administration[26] on the sole condition that no interference was caused to the services of another country. But where interference might occur (and this would be in the majority of cases in many parts of the world) the principle of allocation of frequencies to services and not to countries is followed. Art. 5 goes on to contain an elaborate table which, beginning with frequencies in the band 10–100 kcs, and continuing up to 60,000 kcs, shows the allocation to the various services. Two broad areas – 23,000 kcs to 28,000 kcs and 30,000 kcs to 56,000 kcs – are unreserved, as are frequencies above than 60,000 kcs, but in terms of what was actually in use in the 1920s, most of the radio spectrum is allocated to various services. The whole approach to the problem of frequency allocation was therefore placed on a new footing, and this has been the core of international developments ever since. The shift to a comprehensive allocation of frequencies to services within the radio spectrum is in one sense a far cry from the rules of 1906, and yet it is also a logical development of the notion of the reservation of the 300 and 600 metre bands for public correspondence found in Regulation II of the 1906 Regulations.

The other major move at the Washington Conference was the taking of steps to amalgamate the Radio-telegraph Union with the Telegraph Union. This was easily agreed. The suggestion had first emerged at the Telegraph Union Conference at Paris in 1925, and it was because of a resolution of that conference that the matter arose at Washington. It made sense. The two Unions had cooperated for years, even to the extent of the Radio Union always having used the Telegraph Union's International Bureau to carry out the duties of its own 'Bureau', and it having, from 1927, used the Telegraph Regulations to fill in gaps in its own rules as we have seen. Accordingly, the Washington Conference passed a resolution calling on the participating governments to consider the combining of the two Conventions dealing with international communications, and set the next meeting of the Radio-telegraph Union for Madrid in 1932, when by sheer coincidence the International Telegraph Union would also be meeting there.

As a result of these moves the Thirteenth International Telegraph Conference and the Fourth International Radio-telegraph Conference met simultaneously in Madrid in 1932. Although they were juridically separate, it nonetheless proved possible to hold joint meetings, and finally a single Convention, permanent in nature, on the principles common to the telegraph, telephone and radio was signed.[27] In this way the International Telecommunication Union was created.[28]

The Madrid Convention itself built on its predecessors, although it made only minimal changes in the existing radio frequency service allocations. It was recognized that countries could more easily settle many of their problems in groupings smaller than the large-scale administrative conference, and art. 13 of the new Convention made allowance for such conferences. Nonetheless there was provision in the new Regulations carrying forward the allocation of frequencies to services and making a further increase in the services covered or recognized.[29] Thus European meteorological messages and a specific waveband for the distribution of information on crime and criminals were new 'services' for which frequencies were allocated and, at the behest of the USA, television services were defined as 'Visual Broadcasting of visual images, either fixed or moving, primarily intended to be received by the general public'.[30] No frequencies were, however, allocated for such services, as countries other than the USA were unwilling to see frequencies set aside for services in which they did not yet have any interest. Indeed, specific frequency allocations for television and facsimile services were not made until after the Second World War, because of fears that the USA would, by reason of its technical superiority, pre-empt all allocated frequencies.[31]

A major step taken at the Madrid Conference was the requirement to notify the International Bureau of all frequency assignments to stations capable of causing international interference. The International Bureau was to maintain a published list of assignments so that later assignments could avoid 'occupied' frequencies,[32] and a fresh notification was required for any increase in power of an existing station.[33] The Washington Regulations had made elementary provision for frequency tolerances, and this was refined and formally incorporated into the new regulations as Appendix I (Table of Frequency Tolerances and of Instabilities) and Appendix II (Table of Frequency Band Widths Occupied by the Emissions). This was both an important step forward at the time and is one crucial for satellites, the concepts of these appendices being now fundamental to the efficient use of the radio spectrum in satellite communications.

After the 1932 Conferences, and in accordance with art. 13 of the new Convention and arts. 7.5.3.4 and 5 of the Regulations, regional conferences dealt with European Broadcasting and Inter-American Broadcasting problems.[34] These need not, however, here detain us. Nor need the Administrative Radio Conference held at Cairo in 1938,[35] save to note that it set aside several channels in the band 6,500 kcs to 23,380 kcs in anticipation of the communications needs of intercontinental air routes, and that it added operating questions to the mandate of the International Radio Consultative Committee.[36]

As was the case in the First World War, great improvements were made in radio technology during the Second World War. Although other communication uses of radio did not diminish, broadcasting became important in the transmission of news and propaganda.[37] In addition the War had produced new services such as long-range navigational aids for both ships and aircraft, aeronautical and high-frequency broadcasting services and, at even higher frequencies, television, frequency-modulated broadcasting and radar were becoming available and clearly were to increase. The available radio was filled by stations, entirely apart from the deliberate jamming which occurred as part of the Cold War. It was quite clear under the pre-war ITU agreements that the mere notification of a frequency assignment to the International Bureau and its publication did not confer any exclusive rights to that frequency, but the ITU system did depend upon a willingness of governments to respect assignments in accordance with the Table of Frequency Allocations. Something had to be done to re-invigorate that willingness, and to revise the Table, which had become obsolete in the services it contemplated as well as the allocations it made. The ITU therefore had to be reactivated: it was and is a necessary organization. That was clear to both the emergent power blocs.

Following a Preliminary Conference of the Allied Nations held in Moscow in 1946, the USA arranged for the holding of three overlapping conferences, a Plenipotentiary Conference, an Administrative

Radio Conference and a High-Frequency Broadcasting Conference, all at Atlantic City in 1947. Seventy-four countries participated and the ITU was reconstituted.[38] The revision of the Radio Regulations proved to be a demanding task, although helped somewhat by technical advances which had made a further portion of the radio spectrum available. Thus while the Cairo Conference had allocated frequencies in the range 10 kc/s to 200 mc/s, the new Regulations dealt with 10 kc/s to 10,500 mc/s.[39]

In constitutional terms the Atlantic City Conferences saw major advances towards the present constitutional structure of the ITU. These were the constituting of the International Consultative Committee on the Telegraph, on the Telephone and on Radio (the CCIR) as permanent organs of the Union, the establishment of the International Frequency Registration Board (the IFRB), and the creation of the Administrative Council and the Secretariat all three also as organs of the Union (art. 4.3).

The Consultative Committees in their previous form were bodies convened from time to time as necessary. Art. 4.3(d)–(f) of the new Convention provided for them to be permanent organs of the Union, with the seat of the Union at Geneva as their seats also (art. 2). Permanent secretariats were assigned for each of the Committees and these were placed under the administrative jurisdiction of the Secretary-General of the Union. The respective duties of the Consultative Committees were laid down in art. 8 of the Convention and in Chapters 7–17 of the General Regulations annexed to the Convention (Annex 4 to the Convention),[40] and the intervals between Plenary meetings of each Committee were standardized at two years ordinarily.[41]

The creation of the International Frequency Registration Board (the IFRB) arose from the necessity of a new procedure for the orderly recording and processing of frequency assignments within the agreed allocations.[42] The IFRB had some similarities in concept with the US Federal Communications Commission (the FCC) which had been operating under the US Communications Act of 1934.[43] To this end art. 4.3(c) of the Convention added the IFRB as one of the permanent organs of the Union. By art. 6.2 the Board was to be staffed by independent members, nationals of different countries serving as 'custodians of an international public trust' not as representatives of their home states. The members of the Board were to be elected by each ordinary administrative radio conference, and the size of the Board was mutable. The USA had hoped that the members of the Board would be chosen on merit alone. However, it was clear that, following practices which were already emergent in the United Nations itself, regional representation would be adopted. In fact, what was settled on was an eleven-member Board with seats allocated amongst four regions as: Region A (the Americas) – three seats; Region B (West Europe and Africa) – three seats; Region C (East Europe and North Asia) – two seats; Region D (Rest of the World) – three seats.[44]

The Board's duties, to be further specified in the Radio Regulations, were indicated in art. 6 as being:

(a) to effect an orderly recording of frequency assignments made by the different countries so as to establish, in accordance with the procedure provided for in the Radio Regulations, the date, purpose and technical characteristics of each of the assignments with a view to ensuring the international recognition thereof;

(b) to furnish advice to Members and Associate Members with a view to the operation of the practicable number of radio channels in those portions of the radio spectrum where harmful interference may occur.

In point of fact, since 1928 and the Washington revision of the Radio Regulations, the old International Bureau of the Union had been attempting to maintain a Master Frequency List from the frequency assignments notified to it. But it had encountered difficulties. From the first, not all frequencies were duly notified to it, and that problem had massively increased during wartime. In addition, in compiling its List the Bureau had had no discretionary powers. In the case of conflicting notifications, the affected countries had to thrash the matter out between themselves. The new IFRB was not to be so hampered in its activities. It was made a full-time watchdog over the use of frequencies, staffed by experts and with power to make recommendations on its own initiative. In the last analysis, countries could still insist on using a frequency, but that did clearly involve standing against the international custodians of the radio spectrum. The Atlantic City Radio Regulations (art. 11.10) gave powers to the new IFRB to examine all assignment notifications and refer cases of conflict or failure to conform to the Table of Allocations and other ITU provisions. Power was also given to cancel frequency registrations not actively taken up within two years. A constant revision of the spectrum assignments and registrations was therefore intended, thus allowing many frequencies to be used which might otherwise be blocked by a careless (or deliberate) failure to 'de-register'. The old Master Frequency List was brought down to date by the new IFRB and a Provisional Frequency Board which preceded it.[45] By agreement in 1951, once the task was thought to be as completely accomplished as was possible, a new International Frequency List was drawn up fusing the old and the List which the new organ had prepared.[46]

The creation of the Administrative Council and the Secretariat, (and indeed some of the reasoning in the making permanent of the

Consultative Committees) was brought about through the decision that the Union should become a specialized agency of the United Nations. As it had been organized prior to the Second World War, the Union had in effect consisted only of conferences, meeting irregularly. Between the conferences, only the International Bureau had continuity of existence, and it was only a recording agency and without governmental or legislative responsibilities within the Union. In its review of the world organizations in 1946, the Economic and Social Council of the new UN noted that the ITU had no decision-making body operational between conferences and hence no governing body. This was perceived as a defect which would have to be remedied if the ITU were to be acceptable within the UN 'family'.[47]

The Atlantic City Conference confirmed a decision of the Moscow Preliminary Conference that the ITU should seek to become a specialized agency of the new UN. The Agreement between the two organizations retains the separate character of the ITU, while allowing it to act as the Specialized Agency of the United Nations for Telecommunications Matters.[48] The technical independence of the Union thereby recognized is a feature of the involvement of the UN system with matters of space communications. The United Nations, lacking technical skills and capacities, has had largely to content itself with exhortation; the partial realization of some of its suggestions has been achieved by other entities, such as the ITU, INTELSAT and other telecommunications organizations. At the same time, the UN connection is seen in the conforming of the ITU to the pattern of the Specialized Agency with some sort of standing governmental authority at its core and an organized Secretariat servicing the various organs of the Union.

Both to provide governmental continuity as an end in itself, and to meet the UN requirements, an Administrative Council was established as one of the six permanent organs of the new Union by art. 4.3(a) of the Atlantic City Convention. The Administrative Council was to meet at least annually at Geneva (art. 5.5) to oversee the work of the Union and to carry out such tasks as might be assigned to it (art. 5.10–11). As created in 1947, it consisted of 18 members elected by the Plenipotentiary Conference of the Union (art. 5.1). These were to be elected 'with due regard to the need for equitable representation of all parts of the world' (art. 5.1). To meet this requirement, the Conference decided that the membership should be drawn from the four regions into which the world was divided for the purpose of choosing the members of the IFRB. Five members of the Council were to come from each of Regions A (the Americas), B (Western Europe and Africa), D (Southern Asia and the Pacific), while three were to come from Region C (Eastern Europe and North Asia).[49] The Council elected under these arrangements soon proved that its creation was the most important constitutional innovation in the ITU made by the Atlantic City Conference, although for our purposes the IFRB has more impact. As a not unwieldy body the Administrative Council was able to concentrate much authority in its hands in the next few years.[50] Whether it has retained that authority with its growth in membership since the 1960s is, however, questionable.[51]

The General Secretariat (under a Secretary-General, a terminological concession to the UN) was the sixth permanent organ of the Union (art. 4.3(b)), and was constituted under art. 9 of the Convention. The Secretary-General was appointed by the Administrative Council to whom he was, and is, responsible. Under him was developed what has become the present Secretariat, with due regard to the normal UN geographic distribution of staff members. This body, servicing the permanent organs and the conferences of the Union, as well as discharging the varied duties which have been added to the ITU, is now considerably larger than the Bureau which preceded it.[52]

At the end of the Atlantic City revisions, therefore, the ITU had been transformed in its constitution. It had six permanent organs, the three standing Committees (respectively on radio, the telephone and the telegraph), the Administrative Council, the IFRB (which was permanently in session), and the Secretariat (also permanently in existence). Of these the Administrative Council was a complete innovation, grafted on as an annually meeting governmental organ overseeing the work of the Union. The main structures of the ITU as we have it today had been established.

Between the Atlantic City Conference in 1947 and the emergence of the space age with the launching of Sputnik 10 years later there was one institutional development which may be noted here. The Plenipotentiary Conference of the Union held at Buenos Aires in 1952 revised the ITU Convention,[53] and authorized the merging of the Consultative Committee on the Telegraph with that on the Telephone to form one organ. This was done in 1956, the new body, the International Telegraph and Telephone Consultative Committee (the CCITT), taking over the whole functions of its predecessors.

At Geneva in 1959 a further Plenipotentiary and Administrative Radio Conference revised the Convention[54] and the Regulations.[55] For the first time space needs were taken into account, but major long-range action was postponed to an Extraordinary Administrative Radio Conference to be held in late 1963. Suffice it to say that space was now firmly on the ITU agenda, and we will follow these developments separately below.[56] Here we will continue briefly with the institutional history of the organization.

In 1965 the Convention was again revised at Montreux.[57] The major step was there taken of reducing the membership of the IFRB to five.[58] A further revision of the Convention was made at a Plenipotentiary Conference held at Malaga–Torremolinos in 1973.[59] Finally, the current Convention was the product of the Nairobi Conference of August–September 1982.[60]

PRESENT CONSTITUTION

The ITU is the largest international organization in the world and is likely to remain so because, unlike many others, it is necessary.

Constituent documents

As I write (1988) the current basic document is the International Telecommunication Convention adopted at Nairobi in 1982.[61] The Nairobi Convention came into force on 1 January 1984 for those members who had ratified their signature of the Convention, or acceded to it prior to that date (art. 52). For other countries it comes into force on the date of deposit of ratification or accession as the case may be.[62] The Convention is complemented by the Administrative Regulations – that is, the Telegraph Regulations, the Telephone Regulations, and most importantly for this book, the Radio Regulations (art. 83) (see below).

The Union has had it in view for a number of years to split its constituent document into two separate instruments. One would be the Constitution of the Union and would contain the provisions which are of a fundamental nature. The other, 'the Convention', would contain provisions which might require periodic revision.[63] The Malaga–Torremolinos Conference of 1973, acting on Resolution 35 of the Montreux Conference of 1965, divided the provisions of the 1973 Convention on the basis of their fundamental nature or otherwise and this configuration was adhered to by the Nairobi Conference. The First Part of the Nairobi Convention contains Basic Provisions (arts. 1–52) on membership, the organs of the Union and general provisions relating to telecommunications. The Second Part (arts. 53–82) of the Nairobi Convention contains General Regulations, including provisions on such varied matters as the functioning of the Union, conference procedure, official languages, and finance. In the event of any inconsistency between the two Parts, it is those in the First Part, the Basic Provisions, which prevail (art. 41).[64] However, the fuller division of the ITU constituent document awaits the work of a group of experts set up and acting in terms of Nairobi Resolution COM8/5 by the Nairobi Conference, which resolved that the provision of the ITU Convention should be separated into two instruments, 'a Convention containing the provisions which are of a fundamental nature, and a Convention comprising the other provisions which by definition might require revision at periodic intervals'. A draft of the scheme recommended by the group of experts is to be submitted to the Administrative Council 'sufficiently in advance' to allow the Council to study the matter and circulate the draft with its recommendation (if any) to all members of the Union at least one year before the next Plenipotentiary Conference (which is to take place in Nice, France in 1989). That Conference will then have this constitutional question on its agenda. Work is proceeding on the task. Interestingly, Resolution COM8/5 contemplates that the amendment processes for the two instruments might differ and that that 'containing the provisions . . . of a fundamental nature' – the 'Constitution' – should require a special majority. That raises interesting questions for the future. However they are answered, it remains likely that there will be no major reconstruction of the ITU, irrespective of the forms in which the agreements constituting it are framed.

Membership

Membership of the ITU is open only to states,[65] and three routes to membership are involved. First, membership is open to members of the preceding Convention which sign and ratify, or accede to the new instrument (art. 1(a)). These are listed in Annex 1 to the 1982 Convention (art. 1(a)). States on that list which signed the Convention were required to ratify the Convention within two years of its entry into force on 1 January 1984 (arts. 45.1 and 52). For two years after the entry into force of the Convention, in the period between signature and ratification, a signatory state could continue to be a full member of the Union (art. 45.2(1)). Thereafter, while still a member, a signatory state which had not ratified the Convention lost its voting rights (art. 45.2(2)). A state listed in Annex 1 which did not sign the new Convention while it was open for signature could proceed only by accession to the new instrument. After that instrument came into force, the non-signing member ceased to have rights in the Union. Unless it specifies otherwise, an instrument of accession takes immediate effect upon its deposit with the Secretary-General of the Union (art. 46.2).[66]

Second, membership of the union is open to all members of the United Nations which accede to the ITU Convention as provided for under art. 46 (art. 1(b)). Again, accession has immediate effect (art. 46.2, above).

Third – originally introduced to provide for such cases as Switzerland although now the provision can have other application – art. 1(c) makes membership available to countries which are neither listed in Annex 1, nor are members of the United Nations, but which apply for membership and accede to the Convention, two-thirds of the membership of the ITU having approved the application.

These rules seem clear. However, it may be noted that the questions of membership of, and participation in, the Union has in the past produced argument, debate and dissension on a variety of grounds. One of these was the status of 'colonial' countries which some considered should be eligible for separate membership and others would have restricted in their voting participation.[67] Associate membership (which did not carry a vote or the right to be elected to any organ of the Union) was permitted for some of these non-self-governing countries, which included territories 'not fully responsible for the conduct' of their own international relations, and territories administered under the UN Trusteeship system.[68]

By the time of the 1973 Malaga–Torremolinos Convention most of the former associate members had gained independence from their former colonial powers and had become full members of the Union in their own right. The Malaga–Torremolinos Convention therefore provided for a membership of the Union which was restricted to fully sovereign states only, and the few remaining non-self-governing territories lost the option of associate membership. This has remained the position.[69]

Another ground of debate as to Union membership has been more political. At one end of the spectrum there was the careful consideration that was given to membership of the Union by West Germany and Japan at the Atlantic City Conference.[70] These countries were major users of the telecommunications services and had somehow to be retained within the international system. A solution was found in the unanimous adoption of Protocol II to the Atlantic City Convention, allowing them to become members once more on their accession to the Convention 'at such time as the responsible authorities consider such accession appropriate'. In the case of West Germany and Japan, the War had been lost and in a sense matters were closed. In the meantime the occupying powers would ensure compliance with the international procedures.

A different attitude was taken towards Spain, again at the Atlantic City Conference. General Franco's dictatorship was an object of aversion within the UN system, and this fed through to the decision that Spain and its territories could accede to the Convention and resume membership only when the United Nations itself had accepted Spain. In the meantime, the Union complied with the General Assembly recommendation of 12 December 1946, that Spain be debarred from membership in international organizations, even though Spain had been a member of both the Telegraph and Radio Unions from an early date, and indeed had hosted the immediately previous Plenipotentiary Conference of the Union at Madrid in 1932.[71]

Such allowing of political distaste for a regime to determine its acceptability as a member of the ITU or a participant in its activities has not been an entirely helpful precedent. The position is slightly different in the case of cable communications, but in the case of radio frequency usage, it would seem more sensible pragmatically to recognize that any radio interference is detrimental to the 'efficient telecommunication services' of which the Preamble of the ITU Convention speaks and seek to bring all states into the regulatory mechanisms of the ITU. As the Preamble also recognizes, a country has the 'sovereign right' to regulate its telecommunications. The purpose of the Convention is to produce a system under which all countries can maximize their efficient use of telecommunications and a significant natural resource be best used for the benefit of all. It is therefore better to have all sovereign administrations brought within the ITU system, than excluded from it. But that is not the majority view.

With regret, it must be noted that there is an increasing tendency, particularly on the part of the 'developing countries', to import other political questions into the ITU. Thus Cuba, by Doc. No. 290–E of the World Administrative Radio Conference for Space Telecommunications (WARC–ST, 1971) attacked the representation at that conference of Taiwan (for China), the 'Saigon regime' (for South Viet Nam), and the 'Seoul' authorities (for 'the Korean' people). It also protested that East Germany, North Korea and North Viet Nam had not been invited to the Conference. A more balanced protest of the exclusion of mainland China, East Germany, North Korea and North Viet Nam was made by Poland (Doc. No. 379–E). The result was a flurry of statements by various delegations,[72] which varied in their tone. Whether the work of the Conference was aided is doubtful. The China question was resolved by the decision of the Twenty-seventh Session of the Administrative Council to invite the People's Republic of China to take its seat in the ITU; this followed the United Nations decision of 1971.[73]

On occasion, political considerations have led to the Union determining by majority vote to restrict the participation of members, or even to delete signatures to the convention.[74] South Africa was denied the right to participate in ITU Plenipotentiary Conferences by Res. No. 45 of the Montreux Plenipotentiary Conference, and has not attended ITU conferences since. The exclusion of South Africa was extended to all meetings and conferences of the Union by Res. No. 31 of the Malaga–Torremolinos Conference which also confirmed Res. No. 619 of the Administrative Council to the effect that 'The Government of the Republic of South Africa no longer has the right to represent Namibia within the Union'. Res. No. PLEN/10 of the Nairobi

Conference 1982, resolved to continue the exclusion of the South Africa government from all ITU meetings and conferences. Following Res. No. 30 of the Malaga–Torremolinos Conference in 1973, Portugal was excluded from ITU meetings and denied 'the right to represent the African territories at present under its domination' because of its 'colonial racialist policy', which was condemned, and atrocities in the struggles in those African countries at that time. However, that Resolution was revoked by the Administrative Council following upon a postal consultation and vote by ITU members:[75] Portugal was thus readmitted in 1975 following the various changes in that country and in its relationships with its former African colonies.

Other governments have also come under disapproval at various ITU meetings. The possible expulsion of Israel from the Union because of its invasion of Lebanon bedevilled the first four of the six weeks of the Nairobi Plenipotentiary Conference in 1982, the official documents and minutes only partly revealing the extent to which the Conference was distorted by this question which was logically peripheral to it.[76] Nonetheless much time was consumed in argument and discussion, which finally produced Res. No. PLEN/1 of the Conference, condemning 'without appeal, the continuing violation by Israel of international law' and 'the massacres of Palestinian and Lebanese civilians'. Reference was also made to the destruction of telecommunications, but that element does seem not to have been the main justification of the Resolution.

The bare facts of the decision on Israel have just been laid out, but certain points must be made in relation to it as indicating possible future trouble, if not disaster, for the Union. First, it appears that there was a concerted attempt on the part of a well organized body of opinion to use the forums of the ITU to achieve a result for which the procedures of the United Nations itself had been insufficient – an abuse of the Union itself. Second, although the decisions on the three resolutions which dealt with the matter were taken by secret ballot, the eventual resolution condemning, but not expelling, Israel was adopted by a majority of only four. The figures were worse than that sounds, in that 132 delegations were present and voting, 61 being in favour, 57 against, with 9 abstentions and 5 invalid ballots.[77] Clearly the abstentions or invalid ballots could have turned the day. Third, that voting pattern occurred although the US delegation had made it clear that they had had instructions from the US State Department to announce the immediate withdrawal of the USA from the Union if the Israel vote were lost.[78]

In short, in 1982 a major disruption of the ITU for purely political reasons was only narrowly avoided, and that only by the expenditure of much argument and lobbying not just in Nairobi, but round the world.[79] This is not a welcome development within an organization aimed at technical matters, and whose function is close to essential for the health of world communications, broad- and narrow-cast.

So much for membership of the Union as such. It is restricted to states: but it is perhaps right here to reiterate a point made in the discussion of INTELSAT. It does seem inefficient that international organizations such as INTELSAT, INMARSAT and INTERSPUTNIK, which have a global responsibility, and even the territorially more limited organizations such as EUTELSAT and ARABSAT cannot be members of the ITU. As a result, when it really comes down to hard bargaining within the ITU mechanisms on such matters as orbital slots or frequency allocations, these organizations are dependent states arguing their case, and decisions which are crucial to them are made without their participation in argument or vote. On occasion it may be that the obvious state spokesman for an international organization will, for one reason or another, not make out as good a case as the organization deserves. Where there is an argument between the state providing the spokesman and the organization, a form of natural justice might indicate that the spokesman give way to one from another state. Bluntly, could the USA, beginning to engage in the 'separate systems' controversy with INTELSAT, properly and fully argue the organization's case at the First Session of the Geostationary-Satellite Orbit Conference in 1985? The USA is the headquarter's state of INTELSAT and remains its largest single member. It is the obvious spokesman for INTELSAT on such matters, and yet . . .? And why should the organization itself not appear within the decision-making forum to argue its own case? The First Session of the 1985–88 Geostationary-Satellite Orbit Conference has recommended that special consideration be given by its Second Session to the needs of 'multiadministration systems' but even that is not a wholly satisfactory position for such agencies as INTELSAT, INMARSAT, EUTELSAT and ARABSAT. Membership of the Union would be better.

I would suggest that the precedent of the associate membership of former years could provide a model for a status for international organizations (international legal persons created by states) which have operational capability in telecommunications. They could be required to contribute to the finances of the ITU (and are better placed than many ITU members – see below) in return for speaking and voting rights. Some might consider, of course, that to confer voting rights would be to go too far, and remain content that such organizations can participate in an advisory capacity in the business of the Consultative Committees, as they can do in terms of art. 68 3(1). I do not think this is sufficient. It is reasonable that such organizations have a vote and therefore a voice equal with the states which make up more than half of the ITU membership: their interest in the matters which the ITU decides demands that they play a part in the discussion and decision-making at least equal to those of states which, bluntly, are of limited competence in telecommunications. On the other hand, it might be that any rights of membership in the ITU given to international organizations should be somewhat restricted. One restriction would necessarily be in prohibiting organizations from being elected to the Administrative Council, or from nominating to it. Another would be to bar organizations from the nomination and election of ITU officials. To have one international organization involved in the appointment of the office-bearers of another would be a major step.[80,81]

Other participants

Be that as it may, the Union does not wholly consist of its members, organs, and employees. Provision is made for other entities and agencies to participate, in part at least, in the work of the Union. Thus under art. 32 of the Convention, international and regional telecommunications organizations which coordinate their work with that of the Union may be allowed to participate in the work of the Consultative Committees in an advisory capacity. The 'regional' organizations referred to are a recognition by the Union that the needs of certain parts of the world may be best served by a more local telecommunications organization, although in practice the ITU has tended to try to encompass such within its own structures. The 'international organizations' include such as INTELSAT, which has signed an agreement with the ITU by which each organization recognizes the other's competence. The procedure is for the international or regional organization to address a request for participation to the Secretary-General of the Union, who informs all members. The request is granted if a majority of replies received within one month are favourable (art. 68.3(2)). The organization may renounce its participation by notification to the Secretary-General, and the renunciation is effective after one year (art. 68.5).

In addition, the Union also permits the important categories of the recognized private operating agency and scientific and industrial organizations to participate in the work of the Consultative Committees. The latter are restricted to organizations engaged in the study of telecommunications problems or in the design or manufacture of equipment intended for telecommunications services (art. 68.4(1)), and they have provided invaluable input to the Committees' work. To participate, a private operating agency must request to take part in the work of a Consultative Committee and a scientific or industrial organization must request admission to meetings of the CCI study groups which particularly interest it, through the Secretary-General. The request is made through the state which has jurisdiction over the agency or organization, and the approval of its domestic administration is essential for success. Thereafter the Secretary-General informs all members and the Director of the appropriate CCI (art. 68.2, 3 and 4). Renunciation of participation is competent, taking effect one year from notice given by the agency or organization to the Secretary-General (art. 68.5). Representatives of private operating agencies and of scientific and industrial organizations can become chairmen of Working Groups of the CCIs and can also represent and act on behalf of a state member of the Union, provided that the member has so notified the CCI (art. 68.2(2)).

FINANCE

One major problem, which the ITU shares with the rest of the UN Specialized Agencies, is that of finance. In general, and at the urging of the UN itself,[82] most of the Agencies use versions of the UN's own financing system under which contributions from members are assessed on the basis of gross national product with a modifier for per capita income. The advantages of such a system, it is said, are that it places the main financial burden on those best able to shoulder it, while at the same time favouring the less developed nations.[83] However, as Codding has said, the ITU has stuck with 'a very primitive method of financing that remains basically what it was when it was introduced in 1868' at the Vienna Telegraphic Conference.[84] Perhaps revealingly, the only other UN Agency which has kept a similar form of financing is the Universal Postal Union, the only international organization older than the ITU.[85] It would seem that the relatively confined and concrete remit of each of the two organizations is best served by a system where contributions are left on a voluntary basis, where perceived benefit can also be related to expenditure, where the contribution relates to an agreed overall budget, and where no single contributor is in a position seriously to affect the work of the organization by threatening the withholding of contributions or withdrawal from it.[86] In addition, the ITU system allows non-governmental agencies to participate in some of the Union's work, making some financial contribution to the relevant budget.[87]

The ITU's budget is divided up into units. Six months prior to the coming into force of each new ITU Convention, each member is required to select the class of units which it will undertake to subscribe yearly throughout the period governed by the new Convention. The member cannot unilaterally select a lower class of contribution during the currency of that Convention, but it may change to a higher contribution class (art. 79 1(4)), and under exceptional circumstances ('such as natural disasters necessitating international aid programmes') the Administrative Council can authorize a reduction in contribution class when a member has established its inability to maintain the chosen class of contribution (art. 15.5). Members which fail to notify the Secretary-General timeously of their choice retain the class of contribution under the previous Convention (art. 79 (3)). This system has been under attack within the ITU, and the Malaga–Torremolinos Conference of 1973 remitted the question of the contributory system to the Administrative Council by its Resolution No. 8. There was much discussion, but the Council, unable to come to general agreement on the matter, submitted to the Nairobi Conference a Report, 'Contributory Shares for Defraying Union Expenses'[88] (Nairobi Conf. Doc. No. 9), which reviewed the arguments for moving to a UN fixed-scale system of assessment but made no recommendation on the matter. The Nairobi Conference finally decided to adhere to the existing Union practice of the voluntary choice of class of contributions.

That decision may seem fair enough, but the Nairobi Conference compounded what will be a major problem for the future. At Atlantic City in 1947 eight unit classes were set (1, 3, 5, 10, 15, 20, 25, 30 units) (Atlantic City Convention, art. 14.4). Article 13.4 of the Buenos Aires Convention of 1952 extended that to 14 classes (1/2, 1, 2, 3, 4, 5, 8, 10, 13, 15, 18, 20, and 25 units) and that remained the pattern through the Geneva Convention of 1959 (art. 15.5) and the Montreux Convention of 1965 (art. 16.4). The Malaga Convention added a new unit class of 1 1/2 (art. 15.2). Nairobi extended the number of classes further, from 15 to 19 (art. 15.2), with a further option of a member electing any number or units above the highest unit class (art. 15.4). The current classes of units of contributions are 1/8, 1/4, 1/2, 1, 1 1/2, 2, 3, 4, 5, 8, 10, 13, 15, 18, 20, 25, 30, 35 and 40 (art. 15.2). The newcomers are the 35 and 40 unit classes, and 1/4 and 1/8. The ratio of minimum to maximum contributory unit class is now therefore 1:320 – considerably increased from the previous range of 1:60, although it must be admitted that even that is nothing like the UN range which in 1985 was 1:2500. The argument is that the new small unit classes makes room for the poorer and smaller countries, allowing them ITU membership and voting privileges without placing an undue strain on their national resources. In this way, these countries are encouraged to come within the Union's structures, and this has the benefit of ensuring that their radio activities will conform to the general scheme created and maintained through the ITU channels. That is a weighty argument.

However, the result is that more than half the ITU members account for less than one-tenth of the ordinary budget, while less than one-tenth of the members produce more than one-half of the budget. This cannot be healthy, particularly with the ITU's new emphasis of providing technical assistance, converting the Union, in part at least, into a medium for aid: this invites the poorer countries to vote as they perceive their interest rather than in the interest of the best management of telecommunications.

It was clear that the introduction of new lower classes of units would have an effect on the ITU's financial balance. To that end, admission to the new 1/8 unit class was restricted. Regrettably, admission to the new 1/4 unit class was not. The 1/8 unit class could be selected only to those countries listed as least developed countries by the UN, and to others which apply for admission to the 1/8 unit class, and are accepted as eligible by the Administrative Council (Nairobi Convention, art. 15.2). Resolution COM 4/7 of the Conference indicated criteria on which the Council should act, including small population as well as low per capita gross national product (GNP).[89] In brief, the system requires that a country seeking admission to the 1/8 unit class shall have less than 1m in population. A base figure is then calculated for certain population sizes and compared with the actual per capita gross national product of that country. In the population band 0.75 million to 1 million the GNP per capita figure of $285 which is used by the UN in its determination of least developed countries is multiplied by 1 to produce a base figure of $285. In the population band 0.5 million to 0.75 million the UN figure is multiplied by 2 giving a base of $570; in the band 0.25 million to 0.5 million, it is multiplied by 3 ($855) and under 0.25 million the UN figure is multiplied by 4 ($1140). The base figure of the appropriate population band is then compared with the applicant's actual per capita GNP and, if that base figure is not exceeded, the country is eligible for admission to the 1/8 unit class. Of course, the sources of the figures used are very important. Population and per capita GNP figures must be taken from a UN publication, from a World Bank publication or from *The Statesman's Yearbook*. Where there are discrepancies, the lowest published figure is used. In May 1983 the Administrative Council made its first determination of those countries additional to the UN list which were to be entitled to select the 1/8 unit class. It accepted the cases of Belize, Grenada, Nauru, Tonga and St Vincent, as well as Kiribati and the Solomon Islands should these countries join the Union. Guyana was deferred for further consideration.[90]

Even before the Nairobi Conference it was clear that any extension of the classes of units of contribution might have serious effects. New minimal contribution classes might have a flow of countries into them: new high contribution classes were likely to be unattractive both intrinsically, and because the existing higher contributors would in any event have a proportionately increased burden because of reduction in actual contribution by the new low contributors. Add to that the ITU's new purpose of providing technical assistance to developing countries (art. 4.1(a)) making it likely that any increased contribution would go towards an objective which most of the developed countries feel should be dealt with on an interstate or commercial basis, and it is hardly surprising that no member has chosen either of the two new high unit classes, or the 'open' class above 40 units. There has been a Gadarene rush into the low classes.

Annex 5 of the Report of the Administrative Council to the Nairobi Conference[91] reveals that only France, the UK, the USA and the USSR were in the maximum unit contribution class under the Malaga–Torremolinos Convention. Indeed, the Soviet figure was greater, since in addition to that for the RSFSR, the Ukrainian SSR contributed 3 units and the Mongolian PR 1/2 unit, a global Russian total of 33 1/2 units. Of the other high class of unit choices, the Federal Republic of Germany contributed 25 units, the People's Republic of China and Japan 20 units, Australia and Canada 18 units, India 13, and Italy, The Netherlands, Sweden and Switzerland 10 units. At the other end of the scale, 83 of the 157 ITU members as at 31 May 1982 had chosen 1/2 unit class – the minimum available under the Malaga–Torremolinos Convention. These, a numerical majority of the Union's membership, contributed only 41.5 of the 429.5 units into which the budget was then divisible. The four countries in the maximum class took care of 123.5 units. The 14 members choosing 10 or more units were together responsible for 274 of the 429.5 total units.

During the currency of the Nairobi Convention, these imbalances have become much worse.[92] Now 34 members are in the 1/8 unit class, including those accepted by the Administrative Council under the procedures detailed above. Another 34 are in the 1/4 unit class, and a further 30 have selected the 1/2 contribution class. A total of 98 members therefore contribute less than 1 unit each. A further 28 members contribute 1 unit each; 1 contributes 1 1/2 units; 7 contribute 2 units; 5 contribute 3 units; and 4 contribute 5 units. At the top end of the scale, 6 contribute 30 units; 2 contribute 18 units; and 7 contribute 10 units. As mentioned above, no member has chosen the two highest classes, or the open class above them. Of the 168 members of the Union, 15 members contribute 286 units, and 126 contribute 55.75 units, of which 98 contribute 27.75 units. Apart from the flow into the 1/8 and 1/4 unit classes, the main changes from Malaga–Torremolinos to Nairobi are that China has reduced from 20 units to 10, Japan has gone from 20 units to 30, and West Germany has gone from 25 units to 30. Saudi Arabia has increased its take-up from 1 unit to 10 in accordance with an undertaking it gave to assist with the cost of the introduction of Arabic as a working language of the Union. As the Ukrainian SSR has dropped to 1 unit and the Mongolian PR to 1/4, the 'Russian' total is now 31 1/4 (formerly 33 1/2).

The total number of units into which the budget is now divisible is 392 1/4 units. Less than 10 per cent of the membership contributes 72 per cent of the ordinary budget; 75 per cent of the membership contributes 14 per cent of the budget, and 58 per cent of the membership contributes 7 per cent of the budget. This is not healthy on the face of the figures alone, nor when it is noticed that the 30 unit class is equivalent to 7 per cent of the budget – that is, each of the contributors at that class contributes an amount equal to that rendered by 58 per cent of the other members, and two contribute the same as 75 per cent. It remains, however, that no single contributor contributes even one-tenth of the budget, although the top six contribute a total of 46 per cent. The principle that the withdrawal of a single country should not irrevocably damage the Union's work through the loss of its contribution is still intact. However, the withdrawal of any four, say, of the major contributors (28 per cent of the budget) would be a financial blow from which the Union could not recover in its present form and with the complete range of its present functions, entirely apart from the practical damage which would result from their telecommunications services being outside the Union structures and machineries.

Finally, it should be said that finance is, as in the case of other international organizations, a matter which causes concern for the ITU's future. It is noticeable that the major contributors to finance succeeded in registering their concern in the Nairobi Conference, and in ways which drew the attention of other members to the problem. Additional Protocol I to the Nairobi Convention set financial limits for the expenditures on the operation of the permanent organs of the Union through 1989, and stipulates that thereafter funding shall be on a level funding basis.[93] Conferences, meetings of the CCIR and CCITT, and seminars are similarly given budget ceilings through 1989.[94] Thereafter, individual annual budgets for the CCIs and conference costings must be set by the Administrative Council, and approved in advance by a majority of members, in the light of a full statement of the case for any budget or costing which exceeds the previous ceiling by more than 1 per cent.[95] In addition, within the Convention itself, a new financial note is found in art. 7.2, which deals with the convening of administrative conferences, the mechanisms through which much of the technical responsibilities of the Union are discharged (for example, through the adoption of Radio Regulations). The article repeats the language of its predecessors in the line of Telecommunication Conventions and but now has an additional sentence:

When adopting resolutions and decisions, administrative conferences should take into account the foreseeable financial implications and shall try to avoid adopting resolutions and decisions which might give rise to expenditure in excess of credits laid down by the Plenipotentiary Conference.

The verbs 'should take' and 'shall try' are relatively uncommon in international agreements, and certainly are unusual in a financial context. It may be that their mood reveals the difficulty of their adoption, but their presence, along with Additional Protocol I, must be a warning to those Union members who would too lightly increase the financial burden taken up by others within the organization. There are limits: the usefulness of the organization is too great to be jeopardized by rampant free-loading, or decisions taken with insufficient regard to their financial implications. Even so, one views with some apprehension the voting power of the smaller contributing countries, which may not be accompanied with an accurate appreciation of what is possible. There was an attempt at Nairobi to switch to the UN system of financing, but that proposal, if successful in the future, would be unlikely to be effective. There is a possibility that if such a proposal went through, say, at the next Plenipotentiary Conference in France in 1989, that and other irrational decisions would lead to the end of the Union.[96] The UN financial system itself is creaking and the major powers are working in concert, irrespective of ideology, to control the financial problem within the UN. Something similar happened in Nairobi with the adoption on the last day of Additional Protocol I, which imposed budgetary ceilings 10 per cent less across the board on 1982 figures instead of a proposed 20 per cent increase.[97]

TECHNICAL ASSISTANCE

What has just been written should not be taken as criticism unimportant in the light of events and easily dismissed because the Union has continued to function. It is true that, even for the major contributors, the money which a state pays out on its membership of the ITU is relatively minor, in terms of international expenses as a whole. But each state member of the Union has but one vote in the ITU's deliberations, and one in the Administrative Council if it is elected to such a position. And the purposes of the Union have been changed by the Nairobi Conference in a way which demonstrated the voting power now available to the smaller and less developed countries. Throughout its existence the ITU has been an organization providing a clearing-house through which international telecommunications might be regulated to the benefit of all. It has acted as a broker and an information exchange, a place where arguments could be heard on telecommunications matters and decisions taken. It is a forum in which the use of radio spectrum has been managed to permit many stations to use a finite natural resource.

Nairobi extended the purposes of the ITU to the provision of technical assistance to countries requiring it. Technical cooperation and assistance has been added to the Preamble of the Convention, to the purposes of the Union (art. 4.1 and 2), to the budget (art. 15.1(c)) and appears within the articles dealing with the ITU's different organs. For example, added to the functions of the Administrative Council is the fresh requirement that it shall 'determine each year the policy of technical assistance in accordance with the objectives of the Union' (art. 8.4(1a)).

Prior to the Nairobi Convention, technical assistance and cooperation flowed through two main channels. One, often arranged through the agency of the Union, was on a direct country-to-country basis, or between a developing country and a manufacturing or advisory body within another member. This was the preferred option of many aid-giving countries. The other channel was through a programme operated by the Union, but financed voluntarily as a separate matter by certain members (although little was contributed) and through the United Nations Development Programme (UNDP). By contrast, Nairobi has brought such matters clearly within the competence of the Union as a responsibility of the Agency and as an ordinary matter of budgetary expenditure.

This is a significant development. It changes the Union from being a wholly service organization, within which the interests of members can be adjusted to achieve the best international telecommunications possible, into one which has also a duty to promote telecommunications within the developing countries as a matter both of policy and finance. These purposes are not necessarily incompatible with the prior functions of the Union, but the omens are not good. The change itself was wrought by a use of voting power to make the ITU into a source of finance, investment and technical skills which will supplement those available from other sources, such as UNDP. It will make the Union a place to which poorer countries can go (with their voting power) as of right and obtain what they think they need less expensively (perhaps) than through the traditional channels.

The question is whether this change in ITU purposes will work. It was noted above that the developed countries have not significantly increased their financial commitment to the Union through taking on higher unit classes of contribution. On the contrary, budget control and value for money were important matters at Nairobi and resulted in the budgetary ceilings laid down in Additional Protocol I to the Convention. Whether the aspirations of the developing countries can be met within such budgetary restraint and the financial reluctance of members is questionable. Furthermore, if the developed countries do perceive the ITU's new technical assistance duty as being merely a method by which their expertise and money is siphoned off, they will no longer cooperate with the Union as well as they have one hitherto. A minimal association with the Union would serve to safeguard the interests of the developed countries, and the availability of finance and personnel for the work of the Union and in particular for 'technical assistance' might diminish abruptly.[98]

In short, I have doubts whether the financial and technical assistance changes made by the Nairobi Convention will work as well as their proponents hoped. The ITU's mechanism does not allow the developing countries to take without the active cooperation of the developed countries which give. There is a distinct danger that the end result may be worse than the pre-Nairobi position. It may require a fresh Convention, and the notification of the new selection of contributory classes to make this plain, but I have fears.[99] It may be that, using their voting strength, the solution which the developing countries will seek to such problems in France in 1989 will be to put ITU financing on to the UN system, but that is subject to many difficulties, not excluding the rupture of the Union.[100]

STRUCTURE AND ORGANS

The Nairobi Conference made no substantial difference to the Union's overall structure. The organizational pattern developed over the previous Conventions remains with the Plenipotentiary Conference as the ITU's supreme organ, the Administrative Council as its ongoing governing body, and the four permanent organs – the General Secretariat, the International Frequency Registration Board (IFRB), the International Radio Consultative Committee (CCIR) and the International Telegraph and Telephone Consultative Committee (CCITT) (art. 5). However, changes were made in the composition and in the methods of election to certain of these bodies, and a review begun in another case.

The Administrative Council was increased in size from 36 to 41 seats (art. 8). The Council remains divided for election purposes into five regions, as was done first by the Geneva Plenipotentiary Conference of 1959. The new allocation is: Region A (the Americas) 8 seats (an increase of 1); Region B (Western Europe) 7 seats; Region C (Eastern Europe and North Africa) 4 seats; Region D (Rest of Africa), 11 seats (an increase of 2); Region E (Asia and Australia), 11 seats (an increase of 2). The arguments for the increase in the size of the Council were that it would allow a greater representation of the developing countries, and that it permitted a better compliance with the requirement of art. 8.1 as to geographic distribution within the Council. Less obviously, the increased element from the developing countries will, perhaps, allow them to ensure that the commitment to technical assistance to developing countries receives priority within the work of the Union. Nonetheless, there is a problem in increasing the size of the Council. Even with 36 members, the Council was becoming unwieldly: an additional five members will not help. Further, if these new members have major commitment to only certain of the Union's purposes, the lack of leadership by the Council discernible during the last decade can only be increased. In order to lead, the Council must speak with authority and carry with it the opinion of those countries which are leaders in international telecommunications. If it does not, the work of the Union, which remains fundamentally that of organizing cooperation within the international practice of telecommunications, will be substantially impeded. If the Council is perceived as ineffective, that will also inexorably lead to a lessening of the major contributors' financial commitment, a problem already touched upon above. And, if there is dichotomy between responsibility for the ITU's work vested in the Council, and the financial burden placed on members with a weakened presence in the Council, the Union could suffer.[101]

One cannot but agree with the comments by George Codding written prior to the Nairobi Conference.[102] It would have been preferable had the Administrative Council been cut down in size, and possibly also in responsibilities. As Codding states, and as is evident from the Report of the Administrative Council itself to the Nairobi Conference,[103] from the time it was created in 1947 the Administrative Council has become unduly involved in the actual work of the Union, rather than functioning as a supervisory body as it was intended. Were the Council smaller, it would perforce have to be considerably more flexible in its procedures, and therefore perhaps also more efficient. With reduced functions, the Council could also resume that supervisory role to which it is more fitted, both by its nature and, also one feels, by its actual composition. At Nairobi, an opportunity was lost.[104]

The IFRB was the other main organ which was thought likely to come under scrutiny at Nairobi. At Montreux in 1965 a number of countries proposed the abolition of the Board and the transfer of its functions to the General Secretariat of the Union. The reasoning behind these proposals was in effect that the function for which the Board had been created at Atlantic City in 1947 had been met as far as was possible. All that remained was the relatively mundane function of frequency registration for which a specialized organ was not needed. However, by 1965 the Union had a number of developing nations among its members, and these, perceiving that the IFRB might be useful to them in argument as to frequency use, as well as an important source of expertise and advice, managed to have the Board retained.[105] However, on grounds of cost, it was reduced in number from eleven to five (ironically the original US proposal in 1947).[106] Thereafter, the IFRB appears to demonstrate the validity of the case made above for the reduction in numbers of the Administrative Council. The 'reduced' body has been very active and useful in a variety of fields, ranging from frequency management seminars for telecommunications administrators from many countries, to the drafting of detailed allotment plans for pending administrative conferences. It has carried out its prior functions of maintaining the Master International Frequency Register and of notifying administrations of developments, as well as taking on board the newer problems relating to the opening up of space telecommunications, and even problems relating to the geostationary orbit. In short, for the last 20 years, the IFRB in its leaner form has arguably been a more efficient device despite the increased responsibilities which have been placed on it, particularly by the development of space telecommunications.

Even so, there was a question whether the Board's very existence might come under scrutiny at Nairobi. This did not, however, happen. On the one hand, the Board was given some extra funding, particularly for additional computer access (an increasingly important tool in its work) on a 5-year implementation basis – a recognition of its important role in international frequency management.[107] On the other hand, by Res. No. COM7/1 the Administrative Council was instructed to establish a panel of experts to review 'thoroughly and in the light of changing circumstances' the long-term future of the IFRB, with a view to the Council forwarding the panel's report, together with its own recommendations on the matter, to administrations by 1 July 1986.

Despite what has been said above about the development of the IFRB, there is much to be said for such a review. The Board has been in existence for 40 years, and during this period its role within the international regulation of telecommunications has altered and been modified to an extent reactively and in a piecemeal fashion to swiftly emerging requirements – space is a good example. While that development has generally been good, it was time for a reappraisal. Resolution COM7/1 requires a careful consideration of whether an alternative mechanism might better serve the foreseen needs of the Union, and a balanced account of the advantages and disadvantages of any alternative proposed. Such a review is not necessarily a threat to the IFRB, and it will be good to have a full examination of the Board as it has metamorphosed into its present form, and with respect to its present and potential future responsibilities.

Certain other developments at Nairobi are less welcome, both for the Union as a whole and for their potential impact on its work in relation to matters of space telecommunication.

One of these is the extension of the principle of separate nationalities among the elected officials. In the usual UN fashion it has been required since 1947 that, in the appointment of staff, due regard has to be given to 'equitable geographic distribution amongst the regions of the world' (art. 13.2). That generality has been refined and extended down the years. The Buenos Aires Convention, 1952 (art. 8.1(1)) required the Secretary-General and his deputies to have separate nationalities and, although requiring that that also applied to the Director-General of the CCI and the Deputy Director of the CCIR, it did not tie that to the other elected positions (art. 7.5). Later, the Secretary-General of the Union and his deputy together with the Directors of each Consultative Committee were required to be of different nationalities.[108] Art. 13.2 of the Nairobi Convention added the five members of the IFRB to the 'separate nationalities requirement'. Previously it was only 'desirable' that they should also be of different nationalities. Now nine separate nationalities are required in the nine elective ITU positions, despite the other principles enunciated in the Convention that the members of the IFRB do not represent their home states or the region which elects them (art. 10.3), that the elected officials are neither to accept or seek instructions from any authority other than the Union (art. 13.1(1)), and that members of the Union respect the exclusively international character of the duties of the elected officials of the Union (art. 13.1(2)). The impression given by this changed requirement is that either the members do not now trust the provisions concerning the independence of the elected officials, or that national spread is considered more important than excellence.

Similar factors seem to enter into the other notable development. Regrettably, by art. 6.2(i) and Additional Protocol VI of the Nairobi Convention, in future the Plenipotentiary Conference of the Union will elect the Directors of both International Consultative Committees. Hitherto the election of these officials has been a matter for the Plenary Assembly of the relevant Committee. The merits of candidates were therefore assessed by electors expert within the field of the CCI. That advantage is lost in the changed procedure, and there is instead a possibility (some would say a likelihood) that someone less suitable might gain election particularly, it has been said, because of the wish on the part of the less developed countries to see such posts being filled by persons from 'their constituencies'. That art. 13.3 imposes a requirement that the 'highest standards of efficiency, competence and integrity' shall be the paramount consideration in the recruitment of staff is not a sufficient safeguard, even if the requirement does apply to elected officials. In short, the Directors of each CCI, and those persons elected to the IFRB, should be appointed because they are of the highest calibre available. Questions of nationality should be of lesser importance. In an ideal world they would be irrelevant.

One other alteration to the Constitution by the Nairobi Conference deserves comment. Since 1949, a Coordination Committee has existed within the Union, although not as one of its permanent organs.[109] It was originally created by Resolution No. 48 of the Fourth Session of the Administrative Council as an attempt to control the problem of the Union's federal structure, with each of the organs having in effect its own secretariat and considerable autonomy. Personnel problems appear to have produced the suggestion for such a Committee, but it also was clear that the varied tasks of the organization did require such a body. Theoretically the Administrative Council should have taken an overview of the requirements of running the Union but, because it met too infrequently to be useful, some other device was also required. Accordingly, the Coordination Committee was created and was to be composed of the Secretary-General, the Chairman of the IFRB, the Directors of the CCIs and the Deputy Director of the CCIR. Until 1959 the Committee had only a coordinating and a purely advisory character, but, at the Geneva Conference of that year, it seems to have been felt necessary to clarify the Committee's responsibilities and to make it a main channel of advice on three matters – budget, the finance report and the annual report. The Committee could also make recommendations to the Administrative Council (Geneva Convention, art. 10.2(1)). By 1965 the problem re-emerged, arising originally from the question of which ITU organ should be responsible for selecting the computer system for the Union, but discussion broadened to other questions of the organs' interrelationships and the role of the Coordination Committee. At Montreux, therefore, the Coordination Committee was given its own article within the Convention (art. 11). The impression is clearly given that there was tension between the federal structures and the centralizing influence of the Secretary-General (or the central Secretariat). The result was odd. The Coordination Committee was to take decision unanimously, but the Secretary-General could act in 'urgent' cases (art. 11.2). The Malaga–Torremolinos Conference further tried to extend the Coordination Committee's authority and restrict that of the Secretary-General by still allowing him to take 'urgent' decisions but requiring that the reasons for such be reported to the members of the Administrative Council, along with the views of the other Committee members. In addition, the Committee's meetings were now required 'normally' to be held monthly (Malaga–Torremolinos Convention, art. 59). Nairobi has continued this arrangement. The Coordination

Committee 'shall meet' at least once a month, and also meets 'when necessary at the request of two of its members' (Nairobi Convention, art. 59.3). Perhaps the bones of a novel lie behind these developments.

SETTLEMENT OF DISPUTES

In common with other international organizations, the ITU Convention provides for the settlement of disputes arising within the organization and about the interpretation of its constituent documents. Only the first of these methods below has been invoked and, given the nature of the ITU, the others are unlikely to be employed. They therefore need only to be briefly summarized. By art. 50 of the Nairobi Convention, members may settle disputes on the Convention or the Administrative Regulations (that is the Telegraph, the Telephone and the Radio Regulations (art. 83)) through diplomatic channels, through other agreed procedures or (as it seems a last resort) by arbitration under the Convention.

Arbitration may be through a procedure set out in art. 82 or in terms of the optional Additional Protocol to the Convention. Under art. 82 procedures start by a party notifying the party with which it is in dispute that it is referring the matter to arbitration (art. 82.1). (The parties must agree to arbitration before it can proceed.) Within one month the disputing parties refer the matter to arbitration by individuals, administrations or governments and, if that cannot be agreed, the reference is to governments (art. 82.2). Individuals acting as arbiters must not be nationals of disputants nor be domiciled within them (art. 82.3). Administrations or governments must be parties to the disputed agreement, but not parties to the dispute itself (art. 82.4). In a two-party dispute, each appoints an arbitrator. Where there are more than two parties, each of the two 'sides' of the dispute appoints an arbitrator (art. 82.5 and 6). These two arbitrators appoint a third or, failing agreement, each appoints another, the selection of the third being by lots drawn by the ITU Secretary-General (art. 82.7). Alternatively, parties may agree on a single arbitrator (art. 82.8). The arbitrator(s) determine the procedure to be followed (art. 82.9). The decision (which may be by majority) is final and binds the parties (art. 82.10).

The Optional Additional Protocol to the Nairobi Convention on the Compulsory Settlement of Disputes is also available to deal with disputes between those members of the ITU which sign and ratify the Protocol. Under it, the procedures of art. 82 are invoked to compel the settlement of a dispute. To avoid frustration of the procedure by a failure of a party to appoint an arbitrator, the Secretary-General has power to appoint an arbitrator for a party after the expiry of three months from the start of the proceeding. In other respects, the procedure is the same. The Protocol is separate from the Convention and legally amounts to a separate treaty. Few states have ratified it.

RADIO FREQUENCY MANAGEMENT

We have already seen in brief outline how the ITU was formed through the International Radio Union being brought together with the International Telegraph and Telephone Union in 1932 in the Madrid Telecommunication Convention, and how it subsequently developed its present constitutional structure. It is now time to deal particularly with the mechanisms through which the modern Union carries out its remit in regard to the use of the radio frequency spectrum. The language used is that the ITU is responsible for frequency or spectrum management.

Three major mechanisms are involved. The first operates through the establishment of technical and operational standards. The second (which some would consider a subdivision of the first) deals with the allocation, or setting aside, of specific parts of the radio frequency spectrum for particular uses (and on occasion more detailedly for an area or particular radio stations within these uses). Third, through its procedures, the Union affords a degree of protection to particular stations, giving them a freedom from harmful interference. This mechanism acts through monitoring and making public to other users the assignment by states of particular frequencies to specific stations under their jurisdiction and is largely carried out through the agency of the IFRB.

However, there is an inaccurate inference in the above description – that the mechanisms to be described provide a clear solution for each particular instance. This is not the case. As David Leive points out, since the Atlantic City Agreements of 1947 the trend within the ITU has been 'towards more elaborate *procedures* but less clear legal *principles*'.[110] The mechanisms operate to facilitate the use of the radio spectrum, but, while they do speak of 'recognition' and 'protection', the extent to which they provide for what a municipal lawyer would consider a 'right' or a 'duty' is, in many instances, doubtful. It is better to consider the mechanisms as less concerned with 'rights and duties', and more as providing a set of ground rules within which those with competence within radio communications can arrive at compromises and arrangements. The ITU can then be presented as a place within which engineers and technicians (in the best sense of the word) try to get international and national systems of telecommunications up and running with a minimum of problems of system interference and system incompatibility. In such an endeavour it is best that politicians and lawyers unskilled in such matters are kept at a distance: their concerns, concepts and presuppositions are often not conducive to the settlement of such matters, whether amicably or not.

Technical and operational standards

Technical and operational standards are set through a variety of ITU mechanisms. In relation to geostationary satellites, station-keeping is a specified characteristic established through administrative conferences.[111] Standards relating to actual use of radio frequencies are developed and set through the CCIR, the International Consultative Committee on Radio. What is important is the emission characteristic permitted or required for all varieties of radio frequency emitted by any electrical equipment. It is not only radio transmitting stations which are, therefore, covered, but also such items as microwave ovens, electric motors, other electrical equipment used in industry, science and medicine, and the like. However, the regulation is most detailed of intentional radio emissions and specifications may exist for a variety of characteristics, setting levels and boundaries which may not be transgressed. Typical emission characteristics deal with field intensity, power level, modulation, bandwidth, permitted variation (frequency tolerances), and coding of signals. These are legally binding through their promulgation in CCIR publications and in the Radio Regulations and their appendices, and ultimately owe their legal force to the Convention itself.[112]

The ITU is also the forum through which international agreement is obtained as to such matters as call signs, emergency codes and the like, and through the CCITT the actual conduct of international telecommunication, network performance, interconnectivity, service operation and maintenance, as well as standards on the fixing of tariffs for services, are established.[113]

Radio Regulations: allocations

The second mechanism through which the use of the radio spectrum is managed is through the allocation of particular frequencies to particular functions. This is carried out through the adoption of international agreements worked out at international conferences arranged by the Union. These are the various Administrative Radio Conferences, and the resultant agreements have, in law, the technical status of treaties between the ITU members. The cumulative body of these agreements form the content of the Radio Regulations, which,

along with the Telegraph and the Telephone Regulations 'com-
plement' the ITU Convention.[114] In 1979 the Union held a major
conference – the World Administrative Radio Conference of 1979
(WARC–79) – which codified the work of its predecessors, the Regulations there adopted constituting the majority of the rules presently
in force.[115]

Like other Administrative Conferences, those dealing with radio
matters are governed by art. 54 of the ITU Convention. Such a conference may be a world administrative radio conference (WARC) or a
regional (RARC), or other limited administrative radio conferences
(ARC). An extraordinary administrative radio conference is coded an
'EARC'. Administrative radio conferences revise, in whole or in part,
the regulations dealing with radio, and an EARC may also give instruction by way of directive or additional tasks to the IFRB (art.
54.1(3)). A more limited administrative conference can consider only
specific telecommunications within the region or service that it is
dealing with, and its decisions must always conform with the general
Radio Regulations (cf. art. 32 of the Convention).

A world conference is convened on the decision of a Plenipotentiary Conference of the Union (art. 54.2(1)(a)), on the recommendation
of a previous world conference if the recommendation is approved
by the Administrative Council (art. 54.2(1)(b)), at the request of at
least one-quarter of the ITU members by individual request to the
Secretary-General (art. 54.2(1)(c)), or on the proposal of the Administrative Council (art. 54.2(1)(d)). A regional conference is convened
by similar measures, with the variation that a prior regional conference can also recommend the holding of the conference and that,
in the case of a conference by requisition, it is only the members of the
region concerned which can act (art. 54.3). The agenda and date of
each type of conference is set when it is arranged but may be modified by the Administrative Council, or with the approval of the
Administrative Council on the request of at least one-quarter of those
entitled to be involved in it – provided in each case that the majority
of those to be involved in the conference accept the proposed change
(art. 54.4). In the case of both types of conference it is, naturally,
competent for a preparatory session to form part of the proceeding,
and this, the use of a committee of experts, and the services of the
CCIR and the IFRB, have been relied on increasingly in recent
decades.

As indicated above, the Radio Regulations which are currently in
force are those adopted by the World Administrative Radio Conference, Geneva, 1979. This Conference was a codifying and consol-
idating exercise made necessary by the extensive modifications made
to the previous Radio Regulations (adopted also in Geneva, in 1959)
by further conferences in the intervening years. These further amend-

ments had not been wholly harmonized because of pressures on time on the individual conferences and by restriction on their terms of reference. Accordingly Res. No. 28 of the 1973 Malaga–Torremolinos Conference decided that a full-scale WARC was necessary to cope with what was fast becoming a patchwork of international agreement, with consequent unsatisfactory results. A group of experts set up by the Administrative Council in 1975[116] to study the questions which would arise recommended *inter alia* that the Radio Regulations be rearranged, and this opinion was endorsed in principle by the World Broadcasting-Satellite Administrative Radio Conference, Geneva, 1977 (which had had the matter specifically added to its agenda) in its Res. No. SAT-10.[117]

By Res. No. 801 (1977) the Thirty-Second Session of the Administrative Council arranged for the holding of the 1979 World Administrative Conference over ten weeks in Geneva, the length of the Conference being indicative of the size of its task. Although, as can be seen from the reservations made by many countries to the Final Acts, political questions of doubtful relevance to the Conference were not absent from the sessions, the result was the major revision of the Radio Regulations which presently govern not only satellite telecommunications – although further change has occurred and will continue to do so, especially through the programme of fifteen major conferences envisaged by the 1982 Nairobi Plenipotentiary Conference.

The Radio Regulations of 1979 are divided into Chapters which deal with specific areas; these are further divided into Articles which are subdivided into paragraphs. Reference to articles and paragraphs of the Radio Regulations is often made in a direct code: for example RR9–2, means Radio Regulations, art. 9 para 2. In addition, to aid in the convenient use of the Regulations the Secretary-General was entrusted with the task of adding a marginal consecutive numbering to the paragraphs, and such references are in the simple numerical form.[118] Despite its convenience, however, I will not generally use that system hereafter as it is better to refer to articles, so as to keep in mind the juridical element of the material. The Regulations are supplemented by Appendices, and each conference usually also adopts both Resolutions to deal with transitory questions or interim situations, and Recommendations which propose further action by either another session of the conference, or by another conference or body.

Since the Regulations form a treaty between the Parties to them, and among the members of the ITU, it is essential that the language in which they are couched shall be clear. To that end, certain words have a meaning defined in art. 1 of the Regulations themselves (RR1). Those specifically dealing with frequency management are set out in Section II of that article:

Allocation (of a frequency band): Entry in the Table of Frequency Allocations of a given frequency band for the purpose of its use by one or more terrestrial or space radiocommunication services or the radio astronomy service under specified conditions. This term shall also be applied to the frequency band concerned. (RR1–2.1)

Allotment (of a radio frequency or radio frequency channel): Entry of a designated frequency channel in an agreed plan, adopted by a competent conference, for use by one or more administrations for a terrestrial or space radiocommunication service in one or more identified countries or geographical areas and under specified conditions. (RR1–2.2)

Assignment (of a radio frequency or radio frequency channel): Authorization given by an administration for a radio station to use a radio frequency or radio frequency channel under specified conditions. (RR1–2.3)

,Complementary to these definitions are the following from the list of general definitions, selected as being terms which we require to use hereafter:

Administration: Any governmental department or service responsible for discharging the obligations undertaken in the Convention of the International Telecommunication Union and the Regulation. (RR1–1.1: also Nairobi Convention, art. 51 and Annex 2.)

Harmful Interference: Interference which endangers the functioning of a radionavigation service or of other safety services or seriously degrades, obstructs or repeatedly interrupts a radiocommunication service operating in accordance with [the Radio] Regulations. (RR1–7.4)

Interference: The effect of unwanted energy due to one or a combination of emissions, radiations or inductions upon reception in a radiocommunication system, manifested by any performance degradation, misinterpretation, or loss of information which could be extracted in the absence of such unwanted energy. (RR1–7.1)

Radiocommunication service: A service as defined in this Section [Section 3] involving the transmission, emission and/or reception of radio waves for specific telecommunication purposes. (RR1–3.1)

However, despite the pages of precise definition of terms in RR–1, it must again be noted that other terms in the Regulations are less defined, or are imprecise in their imposition of obligation or requirement of compliance. Administrations are rarely required at the instance of an organ of the Union to act in particular ways in order to effect compliance with the Convention, the Regulations, or the Master Frequency List by the modification of what they have permitted to stations under their jurisdictions. Put another way, the ITU has no effective policing mechanism: it is largely dependent upon the recognition by administrations that failure to behave in a responsible manner will result in 'harmful interference' for reasons of physics rather than of law.

As we have seen, the concept of setting aside a particular frequency for a particular use can be traced back to the Berlin Radio Convention setting the 300 and 600 metre bands as being appropriate for public correspondence purposes. In addition, by art. 4 of the Convention and XXXVIII of the Regulations, Parties were to inform the International Bureau of the International Telegraph Union of the technical characteristics of radio stations open to public correspondence under their jurisdiction. The Bureau was to publish a list of such stations. This raised a problem which is still with us, whether notification to the ITU gives any 'priority' in use of a particular frequency by a particular transmitter – 'vesting' a right to protection from interference by later developments, as it were. Correlatively there is the question whether the radio spectrum should be 'engineered', in the sense that every usable frequency should be dealt with and allocated in advance of requirement so as to distribute the spectrum asset fairly among all states (whether or not they have a current use for the allocation to them), or whether the mechanism should proceed on a basis of 'first come, first served'. This is a continuing problem, particularly in regions where spectrum use is congested (for example, in Europe) and, as we will see, it has emerged once more in a modified form in relation to the use of the geostationary orbit. It is also a problem which countries face in their internal decisions, especially where the right of broadcasting is permitted to commercial organizations.

The history of the ITU reveals a shifting approach to such questions.[119] Down to the Washington Conference of 1927, there was a helpful development of frequency allocation to services, and in the years after the First World War it might have been prophesied that shortly a division of the spectrum *a priori*, in advance of actual requirement, might be agreed, with wavelengths being distributed among the nations of the world, and protection guaranteed to them in their use. But the USA, which had supported such moves, changed its mind. This was not specifically occasioned by radio matters, but related to the League of Nations and the swift reversal of the Wilson policies. There was a distrust of international regulation and a desire to ensure that internal developments were not hindered by international considerations. Other countries also made it clear that they were unwilling to compromise and arrive at international agreement. Thus, the Table of Frequency Allocations adopted in 1927, although apparently comprehensive within the technical abilities of the times, made provision for regional variations and deviations. But no steps were taken to assign rights to particular frequencies to states, which would have been the next logical step, for they could not agree as to who should have what.

However, even if the concept of assignment of frequencies to countries proved to be impossible to agree, it remained the case that effective communication required an absence of harmful interference. As we have seen above, the Washington Convention therefore took up the matter, expanding definitions of international service to all stations capable of causing interference internationally, and requiring all stations, as far as possible, to operate without interfering with others.[120] The International Bureau was also to be the means through which states would notify each other of stations in use.[121] Further, under the Washington Radio Regulations,[122] administrations were required to assign frequencies to new stations under their jurisdictions so as not to interfere with existing stations whose frequencies had already been notified to the Bureau. If interference was likely, negotiation and even resort to arbitration was competent, and to facilitate matters further, the intended use of a frequency below 37.5 kc/s had to be notified to the International Bureau at least four months prior to the entry into service of the station concerned. In short, priority of use was recognized as affording a degree of protection from interference by later established stations. The principle of 'first come, first served', or the regulation of the spectrum *a posteriori* was dominant, and continues to be important.

Even in 1927, however, the *a posteriori* method of regulation was supplemented by a limited use of the *a priori* planning to cope with the problem of the congested region or area. In particular, Europe found that frequency congestion within its area was such that it had to adopt a system of agreeing a comprehensive plan for various of the allocation bands in order to avoid harmful interference and damaging disputes by the prior coordination of frequencies used by particular stations, and other regions found the *a priori* approach necessary under similar circumstances.[123]

With that experience, with the reconstitution of the ITU at Atlantic City in 1947, and with the creation of many new states in the succeeding years, pressure has grown for the adoption of *a priori* solutions to problems of frequency management. As the larger members of the Union in pre-1947 days have ceased to be responsible for the telecommunications requirements of areas of the globe, and as the new governments have taken up the responsibility, as frequency usage has increased and as technical developments have been achieved, so the rational planning of the use of radio frequencies to maximize the potential available has attracted more and more attention. The result has been the planning of certain frequencies both as to their use and as to which station shall use particular bands and at what power. The Radio Regulations of 1979 gathered together many of these plans, although certain regional plans remain outside their scope, and, of course, there have been later developments.

Art. 8 of the 1979 Radio Regulations contains the Table of Frequency Allocations. This Table of Allocations is fundamental to the

Union's work in the international regulation of the use of the radio spectrum. Over the years, certain principles seem to have emerged (their balance varying over the years) as to the way in which successive Administrative Radio Conferences have developed the Table.

1   The amount of spectrum and the bands within which frequencies are allocated generally correspond to what is considered to be the well-founded requirements of radiocommunications services at the time the decision is made, together with some account taken of foreseeable needs.

2   When, for good operational and technical reasons, some new service or requirement has to be fitted into a frequency banding which is already in use, the existing services operating in that band are adjusted either to give the new service equal status, or downgraded in status where their claim on the band has lesser long-term justification. In some cases, that can mean the deletion of a service from the frequency band concerned.

3   It is sometimes possible by agreement on a worldwide basis to make different allocations to a service in different parts of the world by reason of the technical characteristics and nature of a services and the propagation characteristics of radio in the bands concerned. The ordinary radio broadcasting bands show many examples of this being done.

4   It is, however, generally the rule to make allocations on a worldwide basis, the characteristics of the stations and services making sharing that much more easy.

5   Safety services and others which cannot share with other services (for example, broadcasting) are usually allocated on an exclusive basis.

6   In certain of the higher frequencies, the variable propagation characteristics of radio require a series of frequencies in order to maintain a radio circuit on a 24-hour basis.

Such are the generalized 'principles'; it must, however, be said that the Radio Regulations provide exceptions to them all.

Art. 8 of the 1979 Radio Regulations, then, deals with the allocation of frequencies. Its Introduction (RR8–1) art. 8 specifically refers back to the definitions of 'Allocation' of a service, 'Allotment' to an area or a country, and 'Assignment' to a station, established in art. 1 of the Regulations, and sets out equivalents for these terms in English, French and Spanish, which are the working languages of the ITU.[124]

Section I of art. 8 (RR8–2) divides the world into three regions for the purposes of frequency allocation.[125] Region 1 comprises Europe, Africa and the USSR including the Mongolian People's Republic, its west boundary running through the Atlantic down between Iceland and Greenland, crossing to roughly 40° W and then returning to 20° W just below the equator. Region 2 is America and Greenland, its west boundary running at 120° W from 60° S to 10° N, then west to 170° W when it runs up to off the Russian coast at 50° N and then up to between Alaska and Russia, and thence north. Region 3 is delineated as the area remaining, and comprises Iran, and east of a line from the Straits of Hormuz at 60° S and thence east to 120° W. Its northern limits are the boundary of Region 1, and it includes China and Pakistan.[126] In addition, the African Broadcasting Area, the European Broadcasting Area, the European Maritime Area and the Tropical Zone are also defined in RR8–3–6 for the purpose of the Regulations, some allocations being made in terms of these areas rather than as a Region as a whole.

Section II of art. 8 (RR–8 to 11) deals with categories of services and allocations, and describes the conventions which are applied in the Table of Regulations to indicate the status and extent of a particular allocation. Allocations are listed in an order and may be either worldwide, or restricted to a Region or an area as indicated in the Table. Additional comments may be contained in the Table, and footnotes are used to deal with particular matters affecting a broad statement of an allocation. In addition, in some instances a state, by Reservation to the Radio Regulations, has affirmed its intention of acting in a particular way, the contents of the Table notwithstanding.[127]

There are three main categories of radio services: primary services, permitted services and secondary services (RR–8(1)). Permitted and primary services have equal rights except that, in the preparation of frequency plans, the primary service has prior choice of frequencies (RR–8(3)). Where a band is indicated in the Table as allocated to a service 'on a primary basis' or 'on a permitted basis' to an area smaller that a Region, or to a country, that service has primary or permitted status only within the area or country (RR8–6). Secondary services are indicated either in the Table of Frequency Allocations itself, or in a footnote to the Table, where the allocation is effective within an area less than a Region, or within a specified country (RR8–1 and –5). A 'secondary service' is secondary in that it gives way to a primary or permitted service. In particular, two aspects are set out in art. 8(4) (RR8–4). Stations of a secondary service are prohibited from causing harmful interference to primary or permitted services whether these areas are already existing or whether they come into operation later (RR8–4(a)). The secondary service station cannot claim protection from harmful interference against a primary or permitted service which is already operational, or which later comes into operation (RR8–4(b)). However, as against other secondary services of any type, an existing secondary service station can claim protection from harmful interference from another secondary service station to which a later frequency assignment is made (RR8–4(c)).

Two other types of allocations exist, the 'additional allocation' and the 'alternative allocation'. An additional allocation is one made by a footnote to the Table of Frequency Allocations (RR8–9). It extends an allocation made within the Table to an indicated area or a country. The effect is that the particular allocation to a service operates within the indicated territory for the given service in addition to its normal service operation elsewhere and for that purpose in that area it has the same character as the allocation which is extended (RR8–9(1)). Stations operating under the additional allocation ordinarily have equality of right to operate along with the stations covered by the main allocation (RR8–9(2)). Special restrictions may, however, be imposed on an additional allocation in addition to the restriction implicit in its nature as an 'additional allocation' (RR8–9(3)).

An 'alternative allocation' is also made through a footnote to the Table of Frequency Allocations (RR8–10). For the area or country indicated, the alternative allocation replaces the allocation made in the body of the Table of Allocations (RR8–10(1)). An alternative allocation may be without further restriction, in which case stations operating in conformity with it have the same rights to operate as stations operating within the band as it is otherwise allocated within the Table (RR8–10(2)). Restrictions additional to the territorial restriction may, however, be set by the footnote which makes the alternative allocation (RR8–10(3)).

What legal effect has the agreement which is represented by these kinds of provision in the Radio Regulations?[128] The ITU is a body which takes decisions binding upon its members in a way that, for example, a Resolution of the General Assembly of the United Nations is not. First there is the purely legal side of things. The binding nature of the ITU Convention and the Regulations was first laid down in art. 9 of the Madrid Convention of 1932. Substantially its terms are identical to those of art. 44 of the Nairobi Convention 1982, binding the members 'to abide by the provisions of the Convention and the Administrative Regulations' in their own international telecommunications activities (art. 44(1)) and to impose these requirements on telecommunications entities under their jurisdiction (art. 44(2)). In both parts of art. 44 it is specifically stated that the obligation extends both to international telecommunications services and to services 'which are capable of causing harmful interference to radio services of other countries', with an exception in the case of state-operated national defence services.[129] The Regulations apply to military networks only if they are used for non-military purposes – for example, a Forces Broadcasting network. Apart from that case, states are free to make whatever assignments they choose to military stations irrespective of the Table of Allocations in the Radio Regulations. As a matter of convenience it is likely that frequencies assigned for military purposes will not conflict with the generality of allocations in the Regulations, but should this happen, an affected state can only seek to negotiate; it cannot base an argument on the Table of Allocations. With that exception, when an allocation to a particular service is therefore included in the Radio Frequency Table of Allocations, the Radio Regulations are ratified by a state and the Regulations come into force, the allocation has the status of an international treaty and all Signatories are bound to effective compliance. Inclusion in the Table by way of a footnote also gives treaty status to the footnote allocation, but its precise effect depends upon the wording of the footnote.[130] An 'allocation' made by other means does not have treaty force. In relation to Allocations, a Resolution of an Administrative Radio Conference indicates only the desirable nature of the suggestion it contains (the exact strength depending on phraseology) and imposes a degree of responsibility upon member states to observe or comply with the suggestion. Similarly, in relation to allocations, a Conference Recommendation expresses merely agreement that several local stations or administrations should work towards the concept therein expressed.[131] In addition, as with normal treaty processes, it is possible at the time of the signing of the Final Protocol of an Administrative Radio Conference for member states to make reservations to the adopted Regulations and the allocations they contain, with the result that the allocation objected to is not applicable in law to that state.

One curiosity exists. I have noted that, in law, the Convention and Regulations are binding upon the states member of the ITU. It would be the normal position in international law for subsequent versions of the Regulations, and indeed of the Convention itself, to become binding upon a state when that state ratifies the instrument concerned (and it comes into force). That remains the position in strict law. However, the ITU cannot work on a basis which might leave different members working under different 'versions' of either the Convention or the Regulations, and I am informed that, for example, in relation to the Radio Regulations, that the IFRB applies 'new' provisions to all countries when these provisions have come into force, whether or not every state member has ratified or acceded to the appropriate instrument. Certainly, for example, Chapter XIII and art. 69 of the Radio Regulations, Geneva 1979, make definite statements about the date and time of coming into force of the 1979 Regulations and specifically provide for the abrogation of their predecessors. Nonetheless, in law it might be arguable that 'prior' versions of the Regulations subsist between states which have not bound themselves to the 'new'.

However, in such matters it is necessary that all countries move in step, and the IFRB and the other organs of the ITU assume that the members do so. The alternative would be chaos.

But to return to the matter of the treaty status of allocations in the Table of Allocations (s. IV of art. 8 of the 1979 Radio Regulations), although I have stated above that their status is that of treaty, the further question arises as to the possible penalty for non-compliance with a treaty obligation. The penalties can only be categorized as 'weak',[132] but before we go into that in any detail it is necessary first to consider the mechanics by which the allocation to a service is, through the assignment of a frequency to a station, translated into something which ought to have international protection.

It is for a state, in exercise of its sovereign power, to assign the use of a frequency to a particular broadcasting station.[133] In so doing, the state should have regard to the Table of Allocations and the existing assignments which have been reported to the IFRB, the organ of the ITU which has that function. It is now time to consider how the IFRB goes about its business.[134]

As previously indicated when dealing with the history of the ITU, the IFRB is integral to the whole mechanism by which the use of the radio frequency spectrum is made as efficient as possible on a world-wide basis. The constitution and duties of the IFRB are laid down in art. 10 of the Nairobi Convention, while its functions and working methods are stipulated mainly in art. 10 of the Radio Regulations (Geneva, 1979) although other articles have relevance. Over the years, the balance of the IFRB's activities has altered, it early concentrating on the Master International Frequency Register and in scrutinizing and purging the Table of Frequency Allocations. An EARC held in Geneva in 1951 decided that the previous Master Frequency List should be replaced by an interim Master Radio Frequency Record for the purpose of clearing the ground and providing a new start.[135] This Record was replaced by the Master International Frequency Register by the 1959 Administrative Radio Conference which performed a major revision both of the allocations of frequencies to services, and of the procedures whereby an assignment was to be entered on the Register following upon IFRB scrutiny and actions. (The new Master International Frequency Register became effective on 1 May 1961.) In the 1960s more responsibility was exercised by the IFRB in relation to coordinating frequency assignments prior to their formal assignment by administrations, and also to ensuring that administrations were not protected in their assignments beyond the lawful time-limits stated in the procedures. These functions continue, but from the late 1960s the IFRB has had an additional and important role in the pre-parations for world and regional conferences and in assisting these conferences to take decisions informed by the best opinion. In addition, through undertaking studies at its own initiative, the IFRB has been able significantly to improve the extent to which the ITU has achieved its purposes in the area of radiocommunication.

In carrying out its particular functions the IFR Board, consisting of five members as indicated above, is aided by the IFRB Specialized Secretariat, first constituted under the Atlantic City Convention (para. 301) and now under art. 57.3(3) of the Nairobi Convention. This Secretariat began with five departments, but administrative evolution has reduced these to two – the Regulatory Department and the Engineering Department. There are three divisions within the Regulatory Department: the Frequency Registration, Publications and Administration Division; the Regulations Applications Division; and the Coordination and Agreements Division. The Engineering Department has two divisions: the Standards and Procedures Division, and the Systems Design Division. In addition, within the Engineering Division are units dealing respectively with Space Services, Fixed Services, Mobile Services, Broadcasting Services and Conferences.[136] Naturally the two Departments are interdependent in their execution of the responsibilities of the Specialized Secretariat.

In broad terms, when a state member administration of the ITU proposes to assign the use of a radio frequency at a specified power to a particular station, it is not entering blindly into an activity where it can do as it wishes. The Radio Regulations (and occasionally other World and Regional Plans adopted by the ITU and modifying the basic Regulations), provide a framework within which the proposed assignment must fit if it is to receive protection from harmful interference from other stations and, *mutatis mutandis*, is not itself to cause interference. The 'plans' which the Regulations and their modifications contain are intended to maximize the efficient use of the radio frequency spectrum so that every state shall, so far as possible, be able to make the best desired use of that natural resource. Such a 'plan' consists at its simplest of the Table of Allocations, which lay down permitted uses of particular frequencies. Other 'plans' deal with frequency assignments to particular stations (as in the case of agreed assignments and transmitter power to sound broadcasting stations in areas of the world where frequency space is congested), while yet others deal with particular services (for example, maritime mobile services) or with specific areas within the jurisdiction of ITU members (for example, Europe).

SPACE FREQUENCIES

So far we have been speaking of radio frequencies in general. It is now time to deal more particularly with the problems of the use and allocation of radio frequencies for space.

As stated at the beginning of the treatment of the ITU, radio is fundamental to the use of space. Without radio, most tracking and all telemetry and telecommand, as well as telecommunications links, would be impossible. A radio link degraded by interference is useless. Radio undergirds all, and it is therefore essential that satisfying arrangements are made at both the international and national levels so that the best use is made of the natural resource which is the radio spectrum. That principle was implicit in the pre-war Conventions and was stated as art. 42 of the 1947 Atlantic City Convention.[137] It appeared first in an expanded form with specific mention of space requirements as art. 33 of the Malaga–Torremolinos Convention 1973 and is now found as art. 33 of the Nairobi Convention 1982.[138] The change from 1965 to 1973 marks the emergence of the requirements of services for space telecommunications and other uses as a major element in telecommunications planning. As usual, however, there were some preliminary stages.

Early action

The need for allocations to space services became clear with the launching of the Sputnik series. Michael Aaronson reported that interference to UK ground-based microwave circuits occurred when Sputnik was overhead.[139] Andrew G. Haley stated that Sputnik interfered with the US official time signals sent round the centres of the Bureau of Standards, also by microwave, and pointed out that the then Radio Regulations had specifically assigned the particular frequency band used by Sputnik as the worldwide standard frequency for such time signals.[140] In addition, it soon became clear that there were additional problems since the early satellite series were not either fitted with a timer or given equipment which could be used to shut down transmissions from the satellite after their experiments had become useless due to solar degradation. Both the USSR and the USA had made that error,[141] and I have been informed that it took years before certain of the Vanguard series ceased transmission in what were otherwise very useful frequencies. It was their utility which had caused the selection of these frequencies in the first place: it was short-sighted in effect to pre-empt them for rubbish after the experiments on the satellites had become degraded and, as a result, the information transmitted was useless.

What was required was, therefore, the extension of the frequency allocation process to ensure that the various interests could be met.

The 1959 Geneva Conferences

As it happened the next regular meetings of the organs of the ITU were held in Geneva in 1959, in two major conferences, a Plenipotentiary Conference charged with the revision of the Convention itself and a separate Administrative Radio Conference dealing with the Radio Regulations.[142] Space communications were on the agenda of the Administrative Radio Conference and some steps were taken to alter the procedures for notification and registration of frequency assignments which had a bearing on space needs. More important might have been the steps taken at the Conference to deal with the allocation of spectrum space to meet space requirements. However, the allocations settled on by the WARC were made for the purposes of space research only, and were more a ratification for interim purposes of the existing uses of spectrum space by the satellite-launching states rather than an attempt fully to envisage future developments and provide for the needs of all states, or even all uses. Those important questions were postponed.[143] The Conference did, however, adopt definitions of the new radio services necessary for the regulation of radio in outer space, and did allocate 13 bands, but on a shared channel basis only to them. These new services were defined in paras. 70 and 71 of the revised Radio Regulations (Geneva, 1959) respectively as:

Space Service: A radiocommunication service between space stations and

Earth–Space Service: a radiocommunication service between earth stations and space stations

with appropriate definitions of 'earth station' and 'space station'.[144]

However, it must be repeated that the allocation for these new services were on a shared basis and were for research purposes only.[145] This could not carry the emerging demand for long. The radio needs of geophysical, navigational and meteorological satellites, to say nothing of telecommunications satellites, could not be classified satisfactorily as for research except in their very early stages, but assigment of frequencies for these services (as opposed to research) was not contemplated within the language of the allocations. It followed that assignments for such requirements lay outside the scope of the Regulations and therefore could not be registered with the IFRB. The ITU mechanisms therefore could not protect assignments to these different types of satellites uses from harmful interference, nor could they be used to allow a state properly to plan its future use of the radio spectrum in space activities not engaged in purely in order to forward research into space.

The 1959 Conference was therefore of limited use in space matters: but it was at least a start, and one group with a 'passive interest' was given more favourable treatment. These were the radio astronomers, whose case for protection was different from those of other radio users in that they could not negotiate to change their frequency usage. Radio astronomy has to make use of certain frequencies which are set by nature, and are not open to amendment by compromise.[146] Only the hydrogen line was in fact allocated on an exclusive basis for the radio astronomy service in 1959, but footnote action was taken on nine other frequencies.[147] Although these nine were not precisely what the astronomers had asked for, they were considered to be sufficiently close to their preferred frequencies to be acceptable. However, it did soon become obvious that this well-intentioned effort by the 1959 WARC was insufficient and further steps were taken as soon as possible at the next appropriate WARC, that of 1963.

The World (Extraordinary) Administrative Radio Conference, Geneva 1963

Recommendation No. 35 of the 1959 Administrative Radio Conference postponed long-range action on the question of allocating spectrum bands for space activities to an Extraordinary Administrative Radio Conference which, it was proposed, should be held in late 1963, a date sufficiently in the future to permit the results of space research programmes, the extent to which frequencies might usefully be shared by different types of activities, and other technical information to become available. Various bodies as well as state telecommunications departments and recognized operating agencies were active in the intervening period, which symptomatically corresponded also with a surge of activity on matters of space law by such organizations as the United Nations (both the General Assembly and the Committee on the Peaceful Uses of Outer Space), the International Astronautical Federation and the Institute of International Law.

The Extraordinary Administrative Radio Conference (EARC) was duly held in Geneva from 7 October to 8 November 1963, the work of the Conference being mainly discharged through seven committees, of which the most important for our purposes were Committee No. 4 – Technical; Committee No. 5 – Frequency Allocation; and Committee No. 6 – Regulations. Each of these Committees was presented with a large number of proposals, the most complete set of which were put forward by the Delegation of the United States.[148]

The eventual Final Acts of the EARC consisted of amendments to the Radio Regulations and appendices (effective from 1 January 1965), together with various declarations, reservations and statements by the various delegations.[149] The amendments to the Regulations sufficed to meet the requirements of most space telecommunications for a number of years and were, under the circumstances, quite imaginative. One is reminded of the enterprising allocation of specific frequencies for international aviation requirements made by the Cairo Radio Conference of 1938.

The 1963 EARC allocated, on a shared or exclusive basis, frequencies totalling 6076.462 Mc/s for the various kinds of space services. The totals for radio astronomy were not as complete as might have been hoped, with only the band 1400–1427 Mc/s (the hydrogen line band) being allocated with no footnote exceptions. But many other bands were more or less freed for the use of radio astronomers, and it was a question for future settlement how far their work might be impeded by stray interference from the few stations which were permitted still to broadcast at or near those frequencies. In the realm of 'active' radio uses, tracking, telemetry, telecommand, radio navigation and radio-location, meteorological satellites and space research were all fairly well served for the time by the allocations made in 1963. Thus, while the 1959 ARC had made only about 1 per cent of the then usable spectrum space available for space needs, the 1963 EARC made available some 15 per cent.

However, there was great difficulty in getting frequencies allocated for telecommunications satellites. We find that, at the very first conference specifically directed towards space matters, problems emerge occasioned by suspicion that the 'developed nations' will seize the advantage of any allocations, leaving other nations to make do with the less useful frequencies. The Report of the Chairman of the US Delegation[150] reveals that four members of that delegation were detailed to tell members of the other delegations of the US objectives in establishing a global commercial communications satellite network. Apparently many other delegations expressed fears that the USA and other leaders in satellite telecommunications technology would pre-empt all the frequencies that the EARC might allocate for telecommunications services, leaving none for those countries which lagged behind in this area of attainment.[151] And apart from that unwillingness on the part of some to make any allocations at all, there was no unanimity among delegations anxious to make allocations as to which exact frequencies should be set aside for space telecommunications services.

The Western nations did tend to sing in harmony if not exactly the same tune, but the views of the USSR differed from those of the West, both as to the size of the allocation (1125 Mc/s less than the US wanted) and as to the positioning of the frequencies in the radio spectrum. In addition, while the USA sought to have at least 100 Mc/s or allocations set aside on an exclusive basis, the USSR considered that all frequencies allocated to space telecommunications should be allocated on a shared basis.[152]

Eventually after much exploratory talk and negotiation compromise was achieved in most fields, only the question of the 100 Mc/s having to be decided to be allocated on an exclusive basis by a secret ballot of the Conference. The fears of other countries were in measure allayed, and the new arrangements were incorporated in revisions to the Table of Allocations of the Geneva Radio Regulations 1959, sometimes in global and sometimes in regional terms.[153] However it must be stressed that the fears were allayed, not dispelled. From the 1963 EARC can be dated the demand that, in space matters at least, an engineered spectrum should be aimed at, with states being allocated rights to spectrum space whether or not they would or could use them. This demand was to grow directly proportional to the emergence of new (but small and/or poor) states within the international community. That cavil apart, however, the work of the EARC was notable and was welcomed, for example by the General Assembly of the United Nations, which within a month of the conclusion of the EARC stated that it welcomed

> . . . the decisions of the Extraordinary Administrative Conference held in October and November 1963 under the auspices of the International Telecommunication Union, on the allocation of frequency bands for space communications and procedures for their use as a step in the development of space radio communications.[154]

On the procedural side, other modifications were made by the 1963 EARC to the 1959 Radio Regulations. Thus under art. 20 and Appendix 9, the Secretary-General of the ITU was required periodically to publish a 'List of Stations in the Space Service and the Radio Astronomy Service' akin to those published for other services. But the most important change in procedure came with revisions to art. 9 and Appendix 1 and the addition of a new art. 9A and Appendix 1A to the Regulations.[155] The purpose of the changes, particularly in the new art. 9A and Appendix 1A, was to make special provision for the registration of frequency assignments for telecommunications satellites. In particular art. 9A and Appendix 1A dealt with the notification and registration of frequencies in the frequency bands between 1 and 10 Gc/s, which are the frequencies which were then (and to a degree still are) shared between earth and space services. Studies by the CCIR in preparation for the 1963 EARC had shown that interference might be caused between services during these frequencies which were within a 'defined coordination distance'.[156] Assignments of frequencies within the shared earth–space bands to stations lying within such a defined coordination distance were required to be coordinated with other states lying within that distance. Under certain conditions, such coordination might be carried out through the IFRB. The frequencies to be used by a space link were required to be notified to the IFRB 24 to six months before use.[157] In the case of ground services (which *ex natura* were more likely to be permanent in their use of the frequencies) notification to the IFRB had to be made two years before use.[158] Notifications were examined by the Board, and, if in conformity with the Convention and Regulations and had been coordinated as required, were registered. If coordination had failed, the assignments were examined to see whether harmful interference would in fact result. If there was no danger of interference, the assignment was registered as notified. If there was a likelihood of interference, the IFRB was instructed to try to effect a compromise.[159]

The periods of time within which an assignment in the earth–space bands had to be notified according to the 1963 arrangements and reflected the requirements imposed by the new technologies and media. At the time, ordinary assignments to radio stations were notified 90 days before to 60 days after use. In contrast, in the earth–space bands the requirement for terrestrial microwave links was one of notification at least two years before use, while space service assignments had to be notified two years to six months before use. These requirements took account of the difficulty of planning a space service and the awkwardness of changing assignments. A satellite cannot be retuned as required. In addition, the planning process for a space service simply took longer because the engineers and technologists were working in new areas in which theory might not square with practice. With these requirements in effect requiring states to give adequate notice of their intentions, planning could be carried on in the knowledge of what the terrestrial parameters of interference were likely to be. They also took account of the fact that (in 1963 at least) it was contemplated that the terrestrial arrangements would be of greater permanence than the satellite, which might last three or four years and then be replaced by another satellite to which the frequency assignment could be different.

Overall, therefore, there was much change accomplished in the 1963 EARC. The major omission was a compulsory procedure for the coordination of satellite systems *inter se*, as opposed to the earth–space elements now covered by the new art. 9A. A procedure similar to that of art. 9A was indicated by Resolution 1A adopted by the Conference, involving advance publication of proposed developments, and comment and negotiation by interested parties, but it only carried the status of a Resolution and was also insufficiently specific in the data which was involved. More was to be accomplished at the next space conference.

The SPACE WARC, 1971

In its Recommendation No. Spa 9, the 1963 Space EARC recommended that the Administrative Council should annually review both the progress in space radiocommunication made by member administrations, and regular reports on space communications to be made by the several ITU organs. The Administrative Council was also asked, when it considered it appropriate in the light of these reviews, to recommend that a further Administrative Conference be convened at a date it should determine 'to work our further agreements for the international regulation of radio frequency bands allocated for space radiocommunication by the 1963 Conference'. In accordance with these instructions, Resolution No. 632 of the Twenty-Third Session of the Administrative Council called for the convening of such an administrative radio conference in late 1970 or early 1971, and asked administrations to send proposals for the agenda and work of the Conference to the Secretary-General. In arriving at its decision, the Administrative Council had in view a variety of factors including the results and working of four years of the 1963 EARC's decisions as well as the effect of other intervening ITU Conferences whose work impinged on the development of satellite facilities and services and which had further modified the 1959 Radio Regulations – the EARC held in Geneva 1966 which prepared a revised allotment of frequencies for the Aeronautical Mobile (R) Service, and the WARC held in Geneva in 1967 which dealt with revisions to allotments for the Maritime Mobile Service.[160] In addition there were technical and organizational considerations. The years since 1963 had seen significant improvements in radiocommunication technology and space technology, as well as the institutional and organizational developments of the 1960s. In the area of telecommunications, INTELSAT was well on the way to establishing itself in its interim form, and there were already moves within the USA directed towards the development of national satellite communications systems. In other areas, remote sensing and weather satellites and programmes, navigational services, land-mobile and maritime mobile systems were on the drawing boards and even broadcasting satellite services were already being contemplated.

The World Administrative Radio Conference for Space Telecommunications met in Geneva from 7 June to 17 July 1971, and in its Final Acts adopted a Partial Revision of the Radio Regulations together with various Resolutions and Recommendations.[161] The revisions to the Radio Regulations came into force on 1 January 1973. Since many of its decisions are now either superseded by technical and other later developments, and others remain within the present Radio Regulations, it is unnecessary to do more than summarize certain of the developments made by the 1971 Conference.[162] Four elements may be noted:

1 the recognition by the Conference of the equality of right of all countries to the use of frequencies allocated for space services and the geostationary orbit for the purpose;
2 the size of the spectrum range brought under regulation;
3 changes made to interference and coordination concepts;
4 alterations to procedures for the processing of assignments by the IFRB.

First, by Res. No. Spa2–1 the 1971 Space-WARC resolved that the registration with the ITU of frequency assignments for space radiocommunication services and their use should not provide any permanent priority for a country or group of countries, nor should such registration create an obstacle for later-comers to satellite ability. Indeed, countries registering assignments with the ITU for their space needs should take all practicable measures 'to realize the possibility of the use of new space systems by other countries or groups of countries so desiring'. And the ITU should itself take into account the views thus expressed. The bases on which these matters were resolved were that 'all countries have equal rights in the use of both the radio frequencies allocated to various space radiocommunication services and the geostationary satellite orbit for these services', that both these frequencies and the geostationary orbit are limited natural resources and ought to be most effectively and economically used, and that different countries were likely to require and be able to start their usage of the new technologies at different times.

Resolution No. Spa2–1, therefore, contains a further round in the battle between the developed and the developing countries, together with echoes of the arguments for and against *a priori* planning and allocation as opposed to the 'first come first served' principle. Developing countries were afraid that the best frequencies and satellite positions might be snapped up by those countries already able to make use of them, and that later system and requirements might suffer accordingly. Most developed countries wanted to establish their systems as efficiently and economically as possible, and wanted to go ahead without being restricted by what another country might want to do perhaps a decade or more in the future. The argument is still with us.

Second, the most noticeable development made by the 1971 Space-WARC was the massive increase in the spectrum range which was made subject to regulation. The new Table of Allocations extended the bands dealt with from 40 GHz to 275 GHz, almost a sevenfold increase. Within that figure, the portion of radio frequency spectrum usable for space radiocommunication was massively extended, the new allocations being some 35 times more than that available under the allocations made by the 1963 Space-EARC (Spa1). Ten new services were also defined which, in part, made use of the new allocations. These were the aeronautical mobile-satellite service, the maritime mobile-satellite service, the mobile-satellite service, the aeronautical radio navigation-satellite service, the maritime navigation-satellite service, the broadcasting-satellite service, the standard frequency-satellite service, the time signal-satellite service, the earth exploration-satellite service and the amateur-satellite service. Additional spectrum space was also made available to the fixed-satellite service, the meteorological-satellite service, the radio navigation-satellite service, the space research service and the radio astronomy service. In the new services it is noticeable that many of them were merely known and existing terrestrial services which were being recognized as capable of being provided also through a new technique, the satellite link – time signal and standard frequency services being obvious examples.

New allocations were made, especially within the newly regulated spectrum bands. However, in general the allocations which the 1963 Space-EARC had made for the fixed satellite service and the space research service were maintained, and a number of the new allocations for aeronautical mobile, maritime mobile and radio navigation purposes also lay in the pattern of preliminary moves made in 1963. The broadcasting-satellite service and relative allocations, and the amateur-satellite service with its allocations, were new developments which have since burgeoned. Similarly the earth exploration-satellite arena, including both remote sensing and meteorological satellite services was able to develop significantly on the base of the 1971 allocations for such purposes.

The second element deserving of special note among the work of the 1971 Space-WARC is technical. The Conference revised the existing Regulations and established a series of criteria for frequency-sharing between space and terrestrial systems and between space systems. These, largely contained within art. 7 of the 1959 Radio Regulations related to the calculation of interference in terms of emission strength, polarization, isotropic radiation, signal noise temperature and power flux density. In addition, in art. 1 of the Radio Regulations, definitions were added to 'equivalent isotropically radiated power (e.i.r.p.)', 'equivalent satellite link noise temperature', 'coordination distance', 'coordination contour' and 'coordination area', these all being concepts integral to the requirements and procedures which were to be required under the new arrangements from 1 January 1973 – the date on which the 1971 decisions became operative. In particular, new appendices were added to the Radio Regulations. Annex 18 to the Final Acts added a new Appendix 28 to the 1959 Radio Regulations and laid down a complex procedure for the determination of the coordination area round an earth station in frequency bands between 1 and 40 GHz where there was sharing between space and terrestrial radiocommunication services based upon the calculation of coordination distances from the station in all directions of azimuth and their reduction to scale on an appropriate map in the form of a coordination contour.[163] Annex 19 to the Final Acts added a new Appendix 29 to the 1959 Radio Regulations establishing a method of calculation by which the degree of interference between geostationary satellite networks sharing the same frequency bands might be established. These two appendices therefore provided a mechanism through which the effect of a proposed new satellite system might be evaluated in terms of its coherence with existing satellite and terrestrial systems. It remained to create a method whereby the resultant information might be processed and made available for international scrutiny and action.

The third category of decisions and recommendations for which the 1971 Space-WARC is notable comprises the changes that were made in the procedures for notification, checking, dissemination, coordination and protection of new assignments to satellite sevices. This was a fundamental matter, for, (unlike a terrestrial service) once a satellite system is put in orbit there are (usually) few opportunities significantly to alter its frequency usage – it is both financially and technically necessary to get it right first time. Working with the newly revised definitions and technical criteria in revisions of Appendices 1, 1A,[164] and a newly adopted appendix 1B,[165] as well as the previously mentioned new Appendices 28 and 29, the Conference revised arts. 8 and 9[166] and adopted a new art. 9A.[167]

The purpose of these changes in the Regulations was to provide a procedure whereby, prior to the use of a frequency by a satellite network or networks, various stages would be ordinarily gone through which would achieve the end-result of the best use of the radio spectrum for space requirements. Loosely, these stages were (and are) the publication of relevant information in the IFRB Weekly Circular well in advance of the intended development; the coordination of the proposal with other geostationary satellite networks or systems; the further coordination of the new space radiocommunication service with terrestrial radiocommunication services; the notification of the frequency assignment to the IFRB; the examination of the frequency notice by the IFRB; and the eventual recording of the assignment in the Master Register. Harmful interference was therefore avoided so far as possible in the initial planning stage of a satellite system, through affording both the protection of reception aboard the satellite from interference from terrestrial sources and the protection of ground reception stations from interference from other satellite sources. In addition, other forms of interference were guarded against through the requirements of coordination within the coordination area.

The coordination procedures for both the space–space and space–terrestrial coordination requirements were contained within art. 9A.[168] Committee 6 of the 1971 Space-WARC, which dealt with the matter, established some six principles which it then sought to embody in the final text of the new article. These were first, that the data relating to the advance publication procedure must be publicized worldwide through the IFRB Weekly Circular, whether or not geostationary satellites were to be involved in the proposed new service or system (art. 9A 1(1)–(4)). Second, all administrations have the right both to comment on the proposal and to be involved in attempts to iron out difficulties with the proposal before the technical coordination procedures began (art. 9A 1(4)–(6)). Third, the data intimated must have a specified minimum information content in order to permit the calculation of the increase in noise temperature which would be occasioned by the new system (art. 9A 1(1) and App. 1B). Fourth, when the coordination procedure was to begin, it would be directed both towards those member administrations which would be affected (by interference) and also to those who had commented on the proposals, whether or not they were to be affected in a physical or technical sense (art. 9A 1(6)). The administration, beginning the coordination procedure(s) should notify the IFRB which would publicize this notification through its Weekly Circular (art. 9A 1(1), (3)). Last, any other administration which considered that a notified or coordinated system under its jurisdiction had not been taken into account in the steps taken was to have the right to insist that it be brought into the coordination procedure (art. 9A 2(4)), provided that the complaining administration was willing to provide sufficient information about its system to allow it properly to be given consideration within the coordination process for the new system (the effect of interrelated provisions of Art. 9A 3).

As analysed by the IFRB itself,[169] the new art. 9A made important innovations on the procedures set out by the 1963 Space-EARC. These, however, relate to the normal satellite telecommunications services and did not affect the procedures regarding the broadcast satellite service – a different kettle of fish which was left to be dealt with by separate regional Administrative Radio Conferences.[170]

The complex requirements of the new art. 9A as to publication of the data on a proposed system in advance rendered the new procedure much more public and allowed administrations throughout the world to comment, discuss, negotiate and in the last analysis, if necessary, change their own plans (Cf. art 9A 1(6)(a)–(c)). In addition, the assistance of the IFRB itself might be asked for (art. 9A 1(7)). Again, art. 9A foreshadowed the coordination between geostationary satellite systems to operate on a frequency by freqency basis among all member administrations having assignments to ground stations or to satellites, whether or not such assignments were already recorded in the Master Register and whether they were, or had been, subject to previous coordinations. In such cases, much technical data and its evaluation was involved, and the IFRB could be called in to help administrations which found it difficult to cope with such questions, either through difficulties with the coordinating administration or for other reasons (art. 9A 4). This was a useful innovation, especially since the coordination procedure would involve any administration which wished to be involved, since all commenting administrations would be included. It was (and is) therefore quite possible that an administration not itself particularly experienced in space radiocommunication might find itself discussing matters in which its experience and competence were inadequate. Again, while the 1963 Conference had adhered to a concept of 'harmful interference', the concept of harm is perhaps unduly subjective, and was, therefore, replaced by one of a 'permissible level of interference', which could be related to objective standards set by the CCIR or agreed between the parties.[171]

The timing of all such steps was also improved. The initial notification of intention to establish a satellite system was to be sent to the IFRB not earlier than five years before the scheduled service date (art. 9A 1(1)). Coordination procedures could then occupy the next few months or years. The assignment of a frequency thereafter to an earth or space station is required to be notified to the IFRB if the use of the frequency is capable of causing harmful interference to any service of another administration, if the frequency is to be used for international radiocommunications, or if international recognition of the use of the frequency is desired (art. 9A 6(1)). Similar notice is required for decisions on signal reception (art. 9A 6(2)). However, an actual frequency assignment following upon successful completion of any required coordination process must reach the IFRB not earlier than three years before the date on which it is to be brought into use (art. 9A 8(1)). The effect of this apparently curious formulation was intended to prevent administrations from pre-empting spectrum space by pressing ahead with the procedures in order to 'bank' a number of frequencies for potential future use (a problem found in terrestrial assignments in the past). The three-year cut-off meant that notified assignments were a little more likely to reflect actual development of space services, and therefore the efficient use of the radio spectrum was aided. The minimum period of notification to the IFRB was set at 90 days prior to the date at which the assignment was to be brought into use (art. 9A 8(1)).[172] It was not, however, possible for a recalcitrant participant in the coordination procedure to hold up submission of a frequency assignment to the IFRB by refusing to permit the satisfactory completion of the coordination procedure. Subject to safeguards to avoid the abuse of the possibility, an assignment could be notified to the IFRB 150 days after making the unsuccessful request for coordination to other administrations provided that the assistance of the IFRB had been sought in the matter (art. 9A 5).

A notice of assignment in the space radiocommunication field was then subjected to a special procedure within the IFRB itself. The Board examined the notice with respect to whether it contained the required basic data in terms of the new Appendix 1A (art. 9A 9), and publicized the notice in the weekly Circular (art. 9A 10). Thereafter the notice was examined for conformity with the Convention, the Table of Frequency Allocations, its coordinations and the probability of harmful interference to other assignments already on the Master Register (art. 9A 13). A favourable finding meant that the assignment was recorded on the Master Register with its date of receipt noted (art. 9A 16(2)). A finding unfavourable with regard to harmful interference was possible, but an administration could insist on its notice being re-examined and the assignment recorded in the Master Register if the assignment had been in use for 120 days and no complaints of interference had been received. However, if interference was caused to an assignment already entered on the Master Register the newcomer administration was required immediately to eliminate that interference (art. 9A 26(3)).

It must be noticed that, in the space section of the Master Register and unlike the other parts of the Register, the 1971 Space-WARC laid emphasis upon the date of receipt of the notice of assignment. That date, entered in Column 2d of the Register, becomes for space purposes virtually the equivalent for other services of an entry in Column 2a – the date on which the IFRB enters a non-space assignment on the Register. In space matters, therefore, an element of priority is given in order of the date of receipt of an assignment which is not accorded in the terrestrial assignments. This again reflects the need for a degree of certainty in planning a space system, which a terrestrial system does not require.

Overarching all the decisions of the 1971 Space-WARC was the increasing need to provide for the efficient use of the space spectrum allocations. This is a problem not confined to space matters, but in that arena it is of heightened importance. For that reason therefore, the new art. 9A provided for the modification, cancellation and review of entries in the Master Register. Where the use of a frequency had been suspended for 18 months, the notifying administration was required within that period to notify the IFRB of the suspension and the date on which the assignment would resume use (art. 9A 29(1)). The IFRB itself had power to investigate cases where it thought that such an interruption had occurred (Art. 9A 29(2)) and if no reply was received within six months the station was deemed to have been out of use for two years (art. 9A 29(3), see below). Permanent discontinuance of the use of a recorded assignment was to be intimated to the Board within 90 days by its notifying administration and the entry on the Register would be removed (art. 9A 30). Again, where an assignment had been recorded but not brought into use in accordance with its notified characteristics, it was to be deleted from the Register or might be suitably modified (art. 9A 31). Finally, where during a co-ordination, it was found that a registered entry had not been in use for two years it was to be disregarded (art. 9A 14). The IFRB was therefore to act as a policeman, although duties were laid on states to keep the Register up-to-date. Where an administration failed to supply the IFRB with requested information within 45 days an entry was made in the Remarks column of the Master Register – a *nota niger censoris* as it were (art. 9A 32). Naturally, the implementation of all these changes and development took time, and the Final Acts of the 1971 Space-WARC also contain interim measures in both the Resolutions and Recommendations there adopted. But it is on the basis of the 1971 decisions that the immense strides in satellite technology were taken and the major networks, notably the INTELSAT system, were established. The tracks of future development were, except for the broadcast services, established in 1971.

The World Broadcasting Satellite Administrative Radio Conference, 1977

Chronologically the next ITU conference with major importance for space telecommunications was that which dealt with questions of the Broadcasting-Satellite. Matters of orbital position and of frequency allocation were dealt with there, and at the related Regional Administrative Radio Conference of 1983 which dealt with the matter for the Americas (Region 2). In so far as they are important for telecommunications, these conferences are considered below (Broadcasting Satellites). I hope to consider questions of broadcast services and the control of content, copyright and similar matters in a separate work on Space Law.

The Radio Regulations, Geneva 1979

The Radio Regulations of 1979 are the major set of Radio Regulations in force at the time of writing. They have been subject to modification by various of the administrative radio conferences which have been held subsequently, but in essence they are the framework within which the allocation of frequencies is set out and the procedures laid down for the notification and registration of the assignment of a radio frequency by an administration whose state is a member of the ITU.

The effect of Space-WARC 1971 had been to revise the Geneva Radio Regulations of 1959 to such an extent that thereafter the Regulations were properly referred to as the Radio Regulations and the Additional Radio Regulations including the Partial Revision of the Radio Regulations.[173] In addition, apart from space matters since the 1959 Radio Regulations other Conferences had revised other portions of the Regulations. As a result, it was increasingly necessary that a further revision should occur in an attempt to draw together the various revisions and iron out incoherences, as well as extensively to rethink what had been decided. Accordingly Res. No. 28 of the Malaga–Torremolinos Plenipotentiary Conference of the Union determined that a World Administrative Radio Conference should be convened in 1979 to revise as necessary the Radio Regulations en bloc and instructed the Administrative Council to make preparations.

The World Administrative Radio Conference was therefore held in Geneva from 24 September to 6 December 1979, a 74-day marathon involving some 2,000 delegates and observers from 142 member states of the ITU and 30 international organizations working through 800 plenary or committee meetings, exclusive of smaller meetings.[174] Whereas the 1927 Washington Conferences had dealt with spectrum lying within the band 10 KHz to 30 MHz, the 1979 Conference dealt with that between 9 KHz and 400 GHz.[175] More than 12,800 proposals made by administrations related to the Table of Frequency Allocations and the associated Radio Regulations. That the Conference was able to accomplish all that it did in 11 weeks is remarkable.[176]

Space matters formed only a part of the work of the 1979 WARC. Modifications were made to the decisions made at the 1971 Space-WARC, but the main lines then adopted were adhered to, as were the principles it had adopted to the allocations to be made for space purposes. Thus Recommendation 2 of the 1979 Conference refers specifically to those principles and to the need that they identify for full consultation and cooperation in space to permit the optimal use of those spectrum bands which are most suitable for space purposes.[177] Recommendation 2, however, also recognized that problems might occur in the future and that a further Space-WARC would be required to cope with the matters. Particular problems were foreseen to be, or to derive from, the following:

1 Frequency bands appropriate for use in space are limited in number and size.
2 Orbital positions are limited and those most suitable for many uses (particularly telecommunications) are even more limited.
3 By right of their sovereignty, administrations must be able to establish systems they consider necessary.
4 The size and cost of networks require that their operation and development be as little hindered as possible.
5 Technology evolves rapidly, so efficient use of spectrum space requires state-of-the-art technology and unused assignments should be relinquished.
6 Difficulties in finding frequency and orbital space may be encountered despite consultation and coordination under the Regulations.

All these are identified by Recommendation 2 of the 1979 WARC as probably requiring a fresh look at the space arena in due course, and the Recommendation is not exhaustive. There will also be developments in other areas, and therefore Recommendation No. 13 of the 1979 Regulations puts it on the agenda of the Administrative Council to decide, as from 1990, whether it is necessary to convene a further WARC to undertake either a general or a partial revision of the Radio Regulations. This matter will therefore come up annually from 1990, but it seems likely that there will be early rather than late action on the matter.

In the meantime, the Table of Allocations contained in art. 8 and the procedures of art. 11 and Appendix 4 of the 1979 Radio Regulations carry the major administrative and regulatory strain in telecommunications matters. (Broadcasting satellites are separately regulated in art. 15 and Appendix 30 to the 1979 Regulations.) The Table of Allocations can be quickly dealt with. The 1979 Conference made further allocations mainly in the bands over 960 MHz and up to 40 GHz for various of the space services. These additions were thought to be adequate for some years and, I gather, have generally proved to be so. Below 960 MHz some additional provision was made for such as the amateur satellite service, and some bands were cleared for passive space use (such as radio astronomy and remote sensing satellites). Minor criticism can probably be made of these allocations, more particularly in some areas than in others, but the allocations seem broadly to balance states' stated requirements. More important, perhaps, is the machinery by which specific assignment is made and particular frequencies utilized.

As indicated above, the major doctrinal question in space matters within the ITU has been whether the spectrum should be engineered so as to vest claimant states with a portion of the spectrum for their uses whether or not they are in a position to make use of their allocation, or whether the best use is made of the spectrum by permitting those countries which can use it to do so subject to coordination with other users. The 1979 arrangements adopted this latter course.

Art. 11 and Appendix 4 deal with the procedures for notification, checking, dissemination, coordination and protection of new assignments to satellite services. Broadly speaking, the factors indicated above in Recommendation No. 2 of the Conference were which determined the modifications and changes made to the 1971 system. States ought to be able to establish such systems as they chose (and in the absence of international agreement would proceed to their mutual detriment). However, spectrum space is finite, ideal orbital positions limited, and a space system costly and difficult to alter once decisions have been transmuted to hardware. Technological advance is swift, allowing efficiencies of use to be brought into effect – although for economic reasons not always as swiftly as invention itself might allow. However, the structures of the international procedures are directed to the best use of both orbital position[178] and radio frequency spectrum.

As in the 1971 schema, the nub of the procedures is the advance publication of data regarding a proposed satellite network. With appropriate variations, the procedures are the same for the satellite network and for other necessary coordinations. We will here outline that applicable to the satellite system and refer readers to the appropriate sections of the Radio Regulations for the other cases.[179]

When a satellite system is intended to be established, publication and various procedures must occur before the administration begins to coordinate its system with other systems (art. 11.1(1)). To effect advance publication, data relating to the proposed system must be sent to the IFRB by the administration concerned, or by one acting on behalf of the group involved (for example, for INTELSAT).[180] The information, which is fully specified in Appendix 4 to the Regulations,[181] must be supplied not earlier than five nor later than two years before the date on which it is intended to bring the system into operation (art. 11.1(1)). This data is publicized worldwide in a special section of the IFRB Weekly Circular, whether or not geostationary satellites are to be involved in the proposed new service (art. 11.1(1), (2), (4)). A circular telegram is also sent to all administrations informing them that the Weekly Circular contains the notified data, and giving the proposed freqency bands and the orbital location of any geostationary satellite in the proposed system (art. 11.1(3)). All administrations are therefore notified of the proposed new system and are afforded sufficient data to permit them to consider their own positions. Any administration which considers that the new system may cause unacceptable interference to its own existing or planned system is required to send its comments to the notifying administration and to the IFRB within four months, and it is assumed that an administration which fails to do so has no basic objection to the system (art. 11.2). This 4-month period is an extension of the 90 days prescribed by the 1971 WARC (Annex 8, Regulation 9A 1(4)), and has been necessitated by the growing pressure on administrations.

Thereafter a notifying administration is required to engage in consultations directed to the resolution of any difficulties which have been brought to its attention and provide any additional data which a commenting administration requires to decide whether it has an actual objection to the proposed network (art. 11.3(1)). It must be stressed that these 'difficulties' relate to the whole or part of the proposed system. Questions of particular frequency use are to be dealt with under the procedures for the coordination of assignments which takes place after 'difficulties' have been dealt with. In dealing with difficulties the Regulations list three possibilities which are hierarchically arranged, each having to be exhausted before moving to the next potential solution. In the case of a network using a geostationary satellite, all possible means of accommodating all parties are to be explored, including the possible relocation of satellites, although (as is only just) first the proposed system (the newcomer) is required to be considered. Only then may proposals be made and another administration requested to relocate one or more of its satellites or to alter the emission characteristics, frequencies or other technical and operational characteristics in use. Then, if that attempt fails, all affected administrations are to seek mutually acceptable adjustments to their networks (art. 11.3(2)). In all these proceedings the IFRB can be asked for its expert assistance (art. 11.3(3)).

The notifying administration must inform the Board of the receipt or non-receipt of comment after the 4-month period from publication in the IFRB Weekly Circular, and of progress made in the resolution of any difficulites. Additional information on these matters is also to be submitted to the Board as appropriate. The Board intimates such information in its Weekly Circular and by circular telegram to all administrations as before (art. 11.4).

Where difficulties have to be resolved following upon the initial publication by the IFRB, the notifying administration must, if necessary, defer the commencement of coordination procedures or (in appropriate cases) the notification of frequency assignments for six months after the initial publication. Coordination may be begun with other administrations which have responded favourably to the new system or with which difficulties have been settled (art. 11.5).

However, administrations are expected to bear in mind that actual assignments are to be submitted to the IFRB not earlier than three years or less than three months before the assignment is brought into use and to arrange the timetable for coordinations accordingly (art. 13.3(1) with fn. 1496.1).

The coordination process for a frequency assignment is substantially the same for the coordination of assignments to a space station on a geostationary satellite or an earth station communicating with such a station in relation to stations of other geostationary satellite networks (art. 11 s. II, paras. 6–15), the coordination of assignments to an earth station in relation to terrestrial stations (art. 11 s. III, paras. 16–22), and the coordination of an assignment to a terrestrial station for transmission to an earth station (art. 11 s. IV, paras. 23–30). In order not to be repetitive, we will here outline only the first procedure – coordination for assignments to geostationary space stations or earth stations working with such stations.

Coordination is the technical process directed towards the compatible use of radio frequencies to minimize interference with other users' services, and proceeds on the basis that a frequency assignment which has been duly recorded in the Master Register is entitled to protection from interference caused by later assignments. Coordination of each frequency assignment for a system is required, a more particular operation than the general questions of 'difficulties' dealt with under the 'advance publication' of proposals just dealt with. In a geostationary satellite system, an administration intending to bring into use a frequency assignment must coordinate it with any other administration whose assignments might be adversely affected (art. 11.6(1)). With whom to coordinate ought to be clear, but any other administration which considered that a notified or coordinated system under its jurisdiction had not been taken into account in the steps taken has the right to insist that it be brought into the coordination procedure (art. 11.9). The complaining administration may be asked to provide sufficient information to allow it properly to be given consideration within the coordination process for the new system. Lack of data will probably lead to a continuing disagreement and future difficulty; accordingly by art. 11.13 the IFRB may be requested to effect coordination in a variety of cases, including: where an administration has refused to reply to communications; where there is a disagreement over the acceptability of a given level of interference; or where there is any other reason for the parties not reaching agreement on coordination (art. 11.13(1)). The Board, which has power to request data, assesses the interference, informs the parties and requests a decision. If no decision is forthcoming, it is to be deemed that the administration with which coordination has been sought has undertaken not to complain in respect of interference to its sevices and that its own space radiocommunication stations will not cause interference to the use of the new assignment (art. 11.13 and 14, condensed).

Coordination is required where:

1   the assignment is in the same frequency band as recorded assign-ments or as other assignments which have been duly coordinated under the Regulations;
2   recorded assignments have to be taken into account for coordi-nation purposes because they have already been intimated to the IFRB as undergoing coordination procedure before the instant assignment has been so notified;
4   recorded assignments have already been notified to the Board but are themselves exempt from coordination requirements (art. 11.(2)).

Assignments exempt from the requirement of coordination are:

1   those which will not raise the equivalent satellite link noise tem-perature above a stipulated threshold value;[182]
2   those where the assignment is a modification of a previously co-ordinated assignment but remains within a value already agreed within its coordination process;
3   assignments for earth stations within the service area of an exist-ing network, but do not add to interference caused by that network;
4   an assignment to a new receiving station where the notifying administration states it accepts interference resulting from exist-ing notified and recorded assignments;
5   assignments to earth stations which need not be coordinated with other earth stations using frequency assignments in the same dir-ection whether Earth to space or space to Earth (art. 11.6(3)).

Naturally, coordination requires that data is given to the admin-istration with which the coordination is sought, and the required content of this data is listed in Appendix 3 (art. 11.7(1)). Although the request may state all or any of the frequencies involved, each assign-ment is taken as a separate matter (art. 11.7(1)). The copy of the request for coodination, the data supplied with the request and a list of the administrations with which coordination has been sought (art. 11.7(2)). An administration which considers its assignment exempt from coordination requirements may similarly inform the IFRB of relevant data complying with Appendix 3, or straightaway invoke the procedures of art. 13 (which relates to the notification of an assignment and its recording in the Master International Frequency Register). If it uses art. 13 and notifies the assignment to the Board, the IFRB immediately informs all administrations by circular tele-gram (art. 11.7(2)).

The assignment of a frequency to an earth or space station is required to be coordinated if the use of the frequency is capable of causing harmful interference to any service of another administration, if the frequency is to be. used for international radiocommunications, or if international recognition of the use of the frequency is desired (art. 11.6(1)). Notification of an actual frequency assignment following upon successful completion of any required coordination process must reach the IFRB not earlier than three years nor less than three months before the date on which it is to be brought into use (art. 13.3(1)), a provision which prevents administrations from pre-empting spectrum space by pressing ahead with the procedures in order to 'bank' a number of frequencies for potential future use (a problem found in terrestrial assignments in the past). The 3-year cut-off ensures that notified assignments reflect actual development of space services, and therefore aids in the efficient and equitable use of the radio spectrum. A recalcitrant participant in the coordination procedure cannot delay submission of a frequency assignment to the IFRB by refusing to permit the satisfactory completion of the coordi-nation procedure. Unless the IFRB has been asked for assistance in the dispute, the administration setting up the system is required to defer notification of an assignment for which coordination is dis-puted by six months from the date on which the IFRB publishes notification of its requests for coordination in the Weekly Circular, unless that would lead it to breach the 3-month low limit for notice under art. 13.3(10 (art. 11.15).[183]

Thereafter the procedures of art. 13 apply, dealing with the notifica-tion of frequency assignments and their examination by the IFRB and their being entered in the Master International Freqency Register.

Any frequency assignment made by an administration which is to be used for transmission or reception by an earth or space station must be notified to the IFRB if

1   its use is capable of causing harmful interference to any service of another administration;
2   the frequency is to be used for international radio-communications;
3   the assigning administration wishes to obtain international recognition of the use of the frequency (art. 13.1(1)).

In addition, any frequency or frequency band which is to be used for reception by a particular radio astronomy station may be notified and that information will be placed on the Master Register, thereby gain-ing a degree of protection from interference (art. 13.1(2)).

Notification of an assignment must, as noted above, be sent to the IFRB not earlier than three years or later than three months from the date on which the frequency is to be brought into use (art. 13.3), except for space research assignments in bands exclusively allocated to that service or in bands where space research is the primary service. In the case of space research, notification should be given prior to use, but in any case not later than 30 days after the assignment is brought into use (art. 13.3(1)). In all cases, a notification which is late will, when recorded, be annotated to indicate that the time-limits have not been complied with (art. 13.3(2)), and that can diminish its right of protection as against another recorded assignment which has been timeously notified.

Notification of an assignment must contain at least the basic data as contained in Appendix 3 to the Regulations,[184] and a notice which is defective in these respects will be returned to the notifying administration (art. 13.4). A complete notice will be accepted, and the Board includes its data together with the date of receipt in the Weekly Circular within 40 days, or as soon as possible thereafter, giving reasons for any delay to the administration concerned (art. 13.5).

Notices are considered by the IFRB in the order of their receipt, and a finding regarding a notice may be postponed only if the Board lacks sufficient data to give a decision (art. 13.7). However, the Board may not act on a notice which has a technical bearing on an earlier notice until it has arrived at a finding on that earlier notice (art. 13.7). The Board examines each notice for its conformity with the Convention, Table of Allocations and other provisions of the Radio Regulations. Any required coordinations are checked. If coordinations have not been effected, the notice is examined with respect to the probability of harmful interference to assignments which have already been accorded a favourable finding and duly entered in the Master Register (art. 13.8). Where the problem lies with a space station which the Board has reason to believe is not in regular use, the Board consults the administration responsible for the assignment to that station and, if the assignment has not been in use for two years, it is disregarded for the exercise and that assignment must be newly notified and examined by the Board before it is brought back into use (art. 13.9).

Where the Board's examination of a notice of an assignment is favourable, the assignment is recorded in the Master Register, and the date on which the approved notice was first received by the Board is entered in Column 2d of the Register. It follows that it is important for an administration to act swiftly once it has decided on an assignment, and this is in some tension with the other provisions requiring that an assignment be not notified earlier than three years before it is to be brought into use. This tension is a matter of some concern, but it is difficult to see what other procedures might be more satisfactory.

Finally, of course, once an assignment has been entered on the Master Register with a favourable finding by the IFRB it is entitled to the protection which is afforded by the ITU agreements and procedures. Thus its conformity with the Table of Allocations has been confirmed, and its priority of claim to use of that frequency is established. Later assignments by other administrations will have to be coordinated with it. Provided that other administrations comply with their obligations under international law, its degree of protection is therefore high. Although there have been modifications in detail, the 1979 Radio Regulations remain the kernel of radio frequency use in general, and in particular are the cornerstone of radio within the sphere of space communications.

Developments regarding Frequencies to WARC–ORB 1985–88

Since 1979 the ITU has been carrying forward a rolling programme of examination and adjustment of procedures and of frequency allocations throughout the whole range of services and of the radio spectrum, as well as across the ITU areas. One of the difficulties of writing about the matter is that the area does not remain static long enough to assure accuracy at the date of publication, given the lead-time for the production of books. A major step will, however, be taken at the special World Administration Radio Conference dealing with the use of the geostationary orbit and the planning of space services using it, which held its first session in Geneva in 1985 and the second session of which takes place in autumn 1988 (WARC–ORB 1985–88). Before turning to the WARC–ORB Conference we may note that, in general matters of radio frequency usage, the 1979 procedures have not been departed from, and it will suffice quickly to mention the major administrative radio conferences which have dealt with space matters.

In 1983 came the second conference dealing specifically with the direct broadcast satellite. It made provision for Region 2. It and its 1977 predecessor established procedures not dissimilar to those required in the case of the satellite networks as detailed above. We will discuss the broadcast satellite service briefly below, and simply note that their main interest in the general development of ITU regulation of satellites lies in the adoption of a 'Plan' for broadcast satellites under what countries are allocated frequencies and orbital slots. They are a major example of what can be done through applied planning in the use of the natural resources of radio and the geostationary orbital position.

In 1987 the question of mobile services was considered on a world-wide basis by the World Administrative Radio Conferences for Mobile Services (WARC–Mob). One hundred and eight countries sent delegations to the conference which met for five weeks in September and October 1987, and other organizations such as INTELSAT sent observers. Naturally the Conference was of major importance both to INMARSAT and the International Maritime Organization (IMO) which were also accorded observer status. Terrestrial mobile services, mobile satellite services radio navigation and radiodetermination satellite services were all under review. A new Chapter IX (to be referred to as N IX) was added to the Radio Regulations to cope with new equipment developments, and includes a requirement that such equipment on board ship shall be maintained on-board by a qualified radio-electronics officer, which is perhaps a comment on the success of other satellite facilities and the ease with which they can be used by persons untrained in electronics. A further conference will review the 1 GHz–3 GHz band allocations in 1992. In addition, alterations were made to the allocations and to the status of allocations for radiodetermination space services, alterations which differ in each of the three major ITU areas.

OTHER DEVELOPMENTS PRIOR TO WARC–ORB 1985–88

So far, we have been dealing with the procedures whereby frequencies are allocated to particular satellite services through the ITU and its mechanisms, and whereby procedures are laid down directed towards allowing countries to secure their telecommunications interests. The matter is one of compromise and agreement with considerable weight being given to priority of action. However, in one special form of telecommunication, significant steps have been taken towards a complete world Plan, setting aside orbits and frequencies and stipulating for transmission characteristics which each country will be able to use when it so wishes. This is in contrast to the 'first come, first served' attitude of many states in other telecommunications activities, but it seems to be necessary in the field of the broadcast satellite.

BROADCASTING SATELLITES

Introduction

Strictly speaking, broadcasting satellites lie outside the scope of simple telecommunications, being directed towards the large-scale dissemination of information whether their signals are received by the end-user direct from the satellite or through some central point followed by cable to the customer. Nonetheless, the solutions adopted within the ITU framework for the international regulation of their use of orbits and of frequencies have impact on the arrangements made for space telecommunication and may in measure provide an indication of future developments even in telecommunications matters.

A broadcast satellite transmits a signal which is intended for direct reception by the public either through individual reception on a simple domestic installation with a small antenna or through community reception via a larger antenna and distribution thereafter by ground links to the public in a limited area (Radio Regulations, 1979, art. 1 3.18 with 5.14 and 5.15). The broadcast satellite services cause complex legal problems as to copyright and control of content. These are under active discussion in many forums, and I hope to discuss them later in another work. Here we are interested in the technical arrangements that control questions of radio frequencies and orbit.

As has been said there have been two World Broadcasting Satellite Administrative Radio Conferences, both held in Geneva – one in 1977 dealing largely with Regions 1 and 3 and the other in 1983 dealing with Region 2 (the Americas).

The World Broadcasting Satellite Administrative Radio Conference, 1977

The first specific Broadcasting-Satellite WARC, the World Administrative Radio Conference for the Planning of the Broadcasting-Satellite Service in the 11.7–12.2 GHz band (12.5 Ghz in Region 1) Geneva, 10 January to 13 February 1977, was held in accordance with Resolution No. 27 of the Malaga–Torremolinos Plenipotentiary Conference of the Union and Resolution No. Spa2–2 of the 1971 Space WARC.[185] It took up matters identified by the 1971 Space WARC as requiring further and major consideration. In its Resolution No. Spa2–2 'Relating to the Establishment of Agreements and Associated Plans for the Broadcasting-Satellite Service' the 1971 Conference had noted certain reasons why the broadcast-satellite services should be closely controlled, and the possibility of either a world or regional plans was identified as a solution. These reasons were, first, that the best use ought to be made of the geostationary orbit and of the frequency bands allocated to the broadcasting-satellite service. Second, such services were, and are likely, to result in great numbers of directional antennae which could be an obstacle to the later changing of the location of the transmitting satellites (particularly if, as was only prudent, these antennae were fixed and not easily repositioned).

The configuratrion of each system would therefore be relatively un-adaptable. Third, such services might cause harmful interference over a large area of the world. Last, other services having allocations in the same band would be likely to use these allocations before the broadcasting-satellite services were established, thus pre-empting potential assignments and possibly causing later dissatisfaction and dispute. Space WARC therefore resolved that the broadcasting-satellite service should be established and operated in accordance with ITU world and regional agreements and that plans should be adopted at conferences at which all administrations could participate (Res. No. Spa2–2). The Administrative Council was asked to act in the matter. By its Resolution No. 27 the Malaga–Torremolinos Plenipo-tentiary Conference of the Union sharpened that request, calling for the holding of a World Administrative Radio Conference for the Planning of the Broadcasting-Satellite Service in the 11.7–12.2 GHz band (12.5 GHz in Region 1) to be held not later than April 1977. Its reasoning was that there was an urgent need to bring into use certain frequencies within that band for terrestrial services for which alloca-tions had also been made. If the broadcasting satellite requirements were dealt with through their prior consideration and through alloca-tion to specific countries, then the terrestrial needs could then be safely met.

It was found, both during the preparatory work (particularly by the CCIR and in pre-conference seminars) and during the Conference itself, that the needs of Region 2 (the Americas) differed from those of the other ITU Regions. In general, countries in Region 2 did not wish to see the adoption of a fixed plan, but rather thought (under the influence of the USA) that the best approach was an evolutionary one, guided by certain principles as to equality of states, equality of services and equitable rights of access to the relevant geostationary orbital positions. Within such guidelines, technology could progress and allow the development of satellite clustering and cross-beam geometry – as indeed it has. A flexible approach could therefore be adopted, bearing in mind both technical and economic factors as they interact with questions of national sovereignty and interstate agree-ments. A better use might therefore be made of the relatively compressed orbital locations which most suited to broadcast satellites in the Americas. The countries of Region 2 went so far as to incorporate these principles into what became Annex 6 'Planning principles in region 2' and Annex 7 'Use of the spectrum/orbit resource' of the Final Acts of the conference. The Region then asked if they could sort out the detail of their own problems on their own at a later date. The Conference agreed that Region 2 should settle its 'domestic' affairs separately, made interim provision for Region 2 in art. 12 of its Final Acts and resolved that the necessary Conference for Region 2 should occur not later than 1982.[186] Region 2, however, was not immediately allowed to go home: its administrations had to be involved in various decisions taken in 1977 in order to prevent further interference problems between the Regions.

For Regions 1 and 3 somewhat different considerations were paramount, and they preferred the adoption of a plan affording to each country guaranteed access to a portion of the geostationary orbit and to frequencies for their broadcasting satellite service purposes. They agreed the general principles raised by Region 2 – the equality of countries, the equal rights of services within each region, and equitable rights of access to the geostationary orbital position – but general principles can lead to different results in practice, and so it was in 1977. The eventual provisions of both the Articles and Annexes of the Final Acts dealing with the Plan for Regions 1 and 3 are extremely technical, and it appears that major use was made of the computer and programmes of Telediffusion de France in order to arrive at the final results.[187] In short, it was agreed that there should be national coverage for each spot-beam based on a polygonal footprint, the vertices of which must lie within the national frontier; where there was a deliberate crossing of national frontiers, the agreement of countries so 'spilled-over' should be required. In the eventual Plan, which is contained in art. 11 of the Final Acts, most countries are assigned four or five channels, while some groups of countries obtained 'superbeams'.[188] Clustering of orbital positions was also arranged so that cross-interference in highly congested portions for the geostationary orbit (for example, that most useful for Europe) was avoided. Requirements and principles for calculation were set for such matters as power flux density, signal polarization, signal noise and temperature (Annexes 3 and 5). Naturally, and in line with the existing requirements of the Radio Regulations outlined above, procedures were also established for the advanced publication of information on planned fixed satellite services and for the coordination of planned services and for their examination by the IFRB and eventual inclusion on the Master Register (arts. 5, 6 and 7). Attention was also paid to the problems of interference between the Regions, including between Region 2 and Regions 1 and 3 (arts. 9 and 10).

There were therefore three important elements in the 1977 Conference: the allocation of frequencies within the Band 11.7–12.2 GHz and 12.5 GHz in Region 1 for the Broadcasting-Satellite Service; the procedures for coordination and registering duly assigned frequencies; and the matter of assignment of orbital positions on the geostationary torus to particular countries. This last matter will be considered separately as it forms a part of the problem of the geostationary orbit: suffice it to say here that a uniform 6 degree orbital spacing was decided on. The two former matters are, perhaps, somewhat specific, and have been subsumed into the revisions of the Radio Regulations made in the 1979 WARC. Nonetheless, it should be noted here that the Plan for Regions 1 and 3 adopted by the Conference represented both a compromise, and also a decision by the affected countries to adopt planning as a basis of action as opposed to evolution.

Most countries had asked for four or five channels, and some larger countries had asked for special consideration for parts of their territory. That a Plan was able to be drawn from the welter of claims and materials owed much to the work of a small number of men, and extensive use of computers. However, it may be that the Plan was adopted too early. Given the technical progress made by the time of the 1983 Conference for Region 2, one wonders whether the 1977 Conference was premature. As it is, the 1977 Plan for Regions 1 and 3 will persist until January 1994, although there were provisions for its amendment. The major missing element of the 1977 Plan was agreement on feeder links for satellites. For Regions 1 and 3 that will be dealt with through the ordinary procedures under (now) the 1979 Radio Regulations, together with the variation of procedures, required by Resolution 102 of the 1979 WARC regarding the special coordination of feeder links to satellites in the same orbital position.

The alterations and additions to the 1959 Radio Regulations made by the 1977 Conference duly came into force for Regions 1 and 3 and, where appropriate, Region 2, at 00.01 GMT on 1 January 1979. They were, with minor modification, incorporated into the 1979 Radio Regulations by art. 15 of those Regulations, Appendix 30 to the Regulations now containing the detail of the provisions outlined above. As indicated, the 1977 Plan for the broadcasting-satellite service in Regions 1 and 3 remains in force until January 1994.

The Regional Administrative Radio Conference for the Broadcasting-Satellite Service, 1983

The 1983 Regional Administrative Radio Conference for the Planning of the Broadcasting-Satellite Service in Region 2 was held from 13 June to 17 July in Geneva.[189] It was attended by 35 administrations although the conference itself identified some 56 countries or geographical areas which were to be taken into consideration. Care was taken to preserve the interests of non-attenders. One particular problem was the Caribbean where there are a large number of small countries many of which were unlikely themselves to be able to take up an allocation of an orbital position were one allocated to them. The eventual result was that five Caribbean beams were planned, intended to provide common programming coverage for the bulk of the Caribbean countries (including mainland countries such as Belize, Surinam and Guyana) if they so desire in the future. One common 'Andean' beam is also provided in the Plan for Bolivia, Colombia, Ecuador, Peru and Venezela for common programming for these countries, again if they so wish.

The provisions made for advance publication of a proposed broadcasting-satellite system, for coordination, for the various necessary calculations and for notification to the IFRB and for its concerns and responsibilities are, with minor variation, of the usual form. However, there was one untoward effect of the deferral of action in Region 1 from the 1977 Broadcasting-Satellite WARC. The amendments which the 1983 RARC proposed should be made to the Radio Regulations had themselves only the status of recommendations to the 1985 WARC–ORB1, the First Session of the World Administrative Radio Conference which is dealing with access to the geostationary orbit.[190] As indicated above in dealing with the ITU in general, a regional administrative conference has no power to amend the Radio Regulations themselves. Having been adopted by the Union at a World Administrative Radio Conference these can only amended by a similar World Conference. There was no problem, however, and the amendments proposed by the 1983 RARC were duly approved in 1985 and the Radio Regulations amended accordingly as part of the Final Acts of WARC–ORB1.[191] Like those of the 1977 WARC–BS Conference, the 1983 arrangements expire in 1994.

There were, and are, major differences between the 1977 Plan for Regions 1 and 3 and the 1983 Plan for Region 2 which deserve some comment. In part, these differences are simply a matter of improved technology in the six years between the two Conferences, and in part they are matters of experience and cooperation. The most notable difference is that the 1983 satellite Plan has an associated Plan coping with the question of feeder links for the broadcast-satellites using the 17.3–17.8 GHz band.[192] This is extremely important, particularly when a large number of satellites are clustered on the 'best' positions of the geostationary torus for the Region. Improved technology has also contributed to the drawing up of the feeder-link Plan. A further 'improvement' in the 1983 RARC is that the spacing of the satellite locations is non-uniform: by contrast the arrangements for Regions 1 and 3 have the satellite 'slots' equally spaced at 6 degrees separation. Again, this is largely a matter of improved technology allowing better station-keeping by satellites, satellite clustering and the use of cross-beam technology. The possibilities so opened up, coupled with the somewhat smaller portion of orbital torus which is ideal for the Americas, meant that what might have been an acute problem in allocating orbital slots was reduced to the status of a major problem.

Other technical improvements in the 1983 RARC arrangements are that the channel spacing has been reduced (14.85 MHz as opposed to 19.18 MHz), power flux density at the edge of the coverage area or footprint has been reduced, and improved transmission potential has allowed transmitter power to be reduced at the expense of only a small increase in the diameter of receiving antennae (c. 15cm, or 6ins).

THE GEOSTATIONARY ORBIT

Arthur Clarke's seminal article of 1945 pointed out that most of the populated regions of the earth could be linked by radio transmission using a relay system of satellites stationed on the geostationary orbit at a separation of 120°.[193] However, the land masses of the earth are not equally distributed, nor are the centres of population evenly distributed within the various land territories. The positioning of satellites on the geostationary orbit has therefore become a matter of concern. As in the case of radio freqency use, so in the case of geostationary torus there has been a divergence of opinion. Those capable of making use of the orbit through putting up satellites or procuring others to do so on their behalf naturally wish to proceed with a minimum of obstruction. That could produce a scramble to get satellites into the most orbital positions. Those likely to be left behind for technical or economic reasons would therefore prefer a more orderly allocation of geostationary orbital positions (commonly referred to as 'slots'). They also have an interest in service from satellites, but would wish that interest not too unduly affected by the more entrepreneurial activities of their larger brethren. In short, once more we have the apostles of 'free access' and 'first come first served' against the proponents of the 'planned' or 'engineered' geostationary orbit. It is the latter, led by Algeria, China, India and Kenya, who seem to be winning.

The question of the use of particular orbits presupposes a legal framework. Ordinary (or terrestrial) public international law gives jurisdiction to a state over its territory and over the airspace above its territory.[194] The Outer Space Treaty and other space treaties[195] clearly assume that a different jurisdictional regime applies in outer space. Although therefore there is no formal treaty or other authoritative statement on the matter, states appear to accept that there is a division somewhere between outer space and airspace. I do not intend here to discuss the technicalities of past discussions there on this and will simply assume that, at the level of the geostationary orbit, one is dealing with outer space and that the segment of international law – space law – applies.[196]

The earliest statement I have found of the importance of the geostationary orbit is that by Richard G. Gould who began an article in 1967: 'There is a natural resource, as precious to all the nations in the world as coal or oil or pure water.'[197] Others have challenged whether a point in space can be a 'natural resource' but that language and the attitude which it implies has passed into the law.[198] It does indicate the importance of the point at issue.

It is to be hoped that, later in 1988, major steps will be taken through the adoption, under ITU auspices, of a treaty regime which will regulate the use of the geostationary orbit for a good many years and which will contain principles which will continue to apply even when the particular detail of the 1988 agreement has been departed from (see Chapter 9). But this happy expectation has taken time to appear. It represents another instance of the ITU's ability to provide a forum for discussion and decision in matters important to telecommunications.

As Gould pointed out,[199] the decision by interim INTELSAT to proceed with the development of its global satellite system by basing it on synchronous satellites meant that the geostationary orbital torus was likely to be congested were nothing done to regulate its use. Technological development both in satellite station-keeping and in frequency and transmission efficiencies have permitted the closer stationing of satellites than was forseen 20 years ago, but even the best projection of future technical achievements forsees a limit to what can be done. Regulation is necessary. Regulation as to the allocation, assignment and use of the radio spectrum has already been considered. The 1959 World Administrative Radio Conference, meeting two years after Sputnik 1, made allocations for space research purposes and adopted definitions of 'space service', 'earth–space service', 'space station' and 'earth station'. It also recommended the holding of what became the first Extraordinary Administrative Radio Conference (EARC) in 1963 which was devoted to space matters, and thereafter the requirements of space have formed a major part of the discussions of successive radio conferences, of both regional and world status.

The question of orbits has been less discussed, in international forums, although it is a thread which runs through the early ITU discussions on space. In the 1980s, with many countries launching geostationary satellites for many purposes (not only for telecommunications) the question has become urgent.[200]

In its Resolutions Nos. 1721 (XVI) and 1802 (XVII) of 1961 and 1962 the UN General Asssembly, seeking to indicate the ways in which space exploration and communications satellites in particular should proceed, spoke of non-discriminatory access to space for all nations. However, at the 1963 Extraordinary Administrative Radio Conference although there were calls for a planned approach to the use of radio frequencies in and for space, it was thought more important not to delay technical development and that regulation should wait upon experience. That position underlay the decisions taken, but from 1963 can be dated the increasingly vociferous demand for a 'planned approach' to space matters, in which planning of the use of orbits is an element. At the 1963 EARC various countries complained that equitable participation in the benefits of space was not being achieved. The less developed countries were already anxious that, unless specific principles were adopted, by the time that they were able to participate in space both frequency and orbital congestion would have severely limited the benefits of space available to them. By Recommendation 10A of the 1963 Conference, international agreements based on justice and equity were called for to preserve the interests of countries less able to make significant strides in space technology.

The 1971 Space WARC took more concrete steps, 'taking into account' that the geostationary orbit is a natural resource to be efficiently used and 'having in mind' that different countries will start their use of the orbit (and frequencies) at different time, partly because of their 'readiness of technical facilities'. Therefore the 1971 Conferences resolved that registration or use of a frequency did not establish a permanent priority and that countries should take all practical measures to maximize the possibility of other new space systems.[201] It also adopted a recommendation for the holding of a further conference, one of its bases being that 'the possible positions for a satellite whose main purpose is to establish telecommunication links are limited in number and that certain positions are more favourable than others for certain links'.[202] These steps, although directed towards radio frequency matters, showed that the interest of the world community in orbital use was increasing. Less obvious, though of equal or even greater importance, was the new art. 9A placed in the Radio Regulations. We have discussed this above and simply note here that the requirements of advance publicity for new developments and the giving of a right to all countries to comment on a proposed development for the first time permitted countries formally to raise the orbital position of satellites in the proposed system. There is no data on what use was made of this possibility. But all that was required by art. 9A was consultation: there was no guarantee that representations would have any effect. The claims of other countries, and in particular those of the less developed states, were therefore entirely dependent on the goodwill of the developed nations and their willingness to modify their plans to accommodate what might seem a hypothetical situation. The demands for a planned approach, correlative to that for an 'engineered spectrum', therefore grew.

At the Malaga–Torremolinos Plenipotentiary Conference of the ITU in 1973 the Union first made express provision for the geostationary orbit in its constituent document, adding a special paragraph on space to the customary article on the 'Rational Use of the Radio Freqency Spectrum' and in it requiring that members 'bear in mind that . . . the geostationary satellite orbit' is a limited natural resource, to be 'used efficiently and economically so that 'others may have equitable access' to it (Malaga–Torremolinos Convention, art. 33.2). In addition art. 10 of the Convention added to the duties of the IFRB that of recording orbital positions on the geostationary torus assigned by countries (art. 10.3(b)), the giving of advice 'with a view to the equitable, effective and economical use of the geostationary orbit (art. 10.3(c), cf. art. 33.2, above) and other duties connected with the use of that orbit as might be assigned by competent conferences (art 10.3(d)). The geostationary orbit was therefore no longer on the agenda only of ITU conferences: it had entered into the work of the Union, and was recognized as presenting particular problems in meeting the requirements of equitable access and the needs of all nations.

At this point a rogue factor entered. In 1976 a meeting of 'equatorial states' – Brazil, Colombia, Congo, Ecuador, Indonesia, Kenya, Uganda and Zaire – adopted a Declaration at Bogota, Colombia, to the effect that, by reason of their terrestrial position, they each had sovereign rights over that part of the geostationary orbit above their terrestrial territory.[203] Although there is good reason to trace the genesis of this claim to a joke during a seminar in the Institute of Air and Space Law, McGill University, Montreal, it was a claim which was insisted upon. Thus Doc. No. 121 of the 1977 Broadcast-Satellite Conference contained a claim by Colombia, Congo, Ecuador, Gabon, Kenya and Uganda that the positioning of a geostationary satellite over any of these countries was subject to the prior consent of the country concerned, that the satellite would be subject to the appropriate national legislation, and that the equatorial countries would be obliged to enforce their national laws 'with all the consequences that this might entail, resorting to any technical or other means at their disposal'. A more extensive statement and explanation was made by Colombia, Ecuador, Kenya, Uganda and Zaire in Doc. No. 165 of the Conference. Australia, in Doc. no. 181, produced a strong defence to the effect that the recognition of national sovereignty over portions of the geostationary orbit was not a matter before the Conference, but involved other treaty instruments such as the Outer Space Treaty of 1967 and was thus currently subject to consideration in COPUOS. In Doc. No. 295, West Germany, Austria, Belgium, Canada, Denmark, the USA, France, Ireland, Italy, Japan, Luxemburg, Monaco, Norway, New Zealand, Holland, Portugal and the UK supported the Australian statement.

Chronologically, and one might have thought intellectually, the World Broadcasting Satellite Conference of 1977 was next in importance. However, it and the 1983 Regional Administrative Conference dealing with the broadcast-satellite sevice in Region 2, can be taken together. Their detail is discussed above. Their importance here lies in the decision to allocate particular positions or slots on the geostationary orbit to particular states on a basis of equality and in the making of allocations even to states not represented at the conferences.

At the 1979 World Administrative Radio Conference the decisions of the 1977 Conference and its Plan were incorporated within the Radio Regulations. However, in 1979 the question of the geostationary orbit was again a matter of discussion and negotiation, not to say dispute. The equatorial countries – adherents to the Declaration of Bogota – persisted in their claims, adding that by their use of their claimed sovereignty they would be able to protect the rights of developing countries to access to orbital positions which might otherwise be lost through more rapid action by the developed countries.[204] Resolution No. 2 of the Conference took up the point that priority of use did not give a permanent right to the use of an orbital position which had first been made at the 1971 WARC (Res. Spa2–1, above) and added a specific statement that all countries have equal rights in both space frequencies and the geostationary-satellite orbit.[205] Resolution No. 3 'Relating to the Use of the Geostationary-Satellite Orbit and to the Planning of Space Services Utilizing It', called for the convening of a special WARC to deal with the question not later than 1984. In particular also, the Resolution called for *a priori* planning in measure at least, and for procedures to help attain a 'guarantee in practice for all countries [of] equitable access to the geostationary-satellite orbit and the frequency bands allocated to space services'. It was also indicated in the Resolution that 'attention should be given to the relevant technical aspects concerning the special geographic situation of particular countries'. Accordingly, the preparatory work for WARC–ORB 1985–88 was begun.

Before we turn to that, however, mention must also be made of the Nairobi Convention of 1982. At that conference, questions of the geostationary orbit were again raised and the worries of the developing countries that they might be left out were once more aired and more than aired. Art. 33.2 of the Convention was extended. It repeated the words of Malaga–Torremolinos regarding the efficient and equitable use of the geostationary orbit and added that, as members bore in mind the duty of the efficient use of the orbit, they must also take 'into account the special needs of the developing countries and the geographical situation of particular countries' (art. 33.2). In addition, the equatorial countries reserved their position of sovereignty over the geostationary orbit, reaffirming the stance they had taken at the 1979 WARC.[206]

To summarize, then, in the period 1973–82 concern about the use of geostationary orbit grew. States without the capacity to put their own satellites in orbit were worried lest they be left out of the reckoning. States on the threshold of implementing their own programmes were concerned lest their plans be prejudiced as the advanced technological nations gobbled up the best orbital positions. The developing countries, now with the evidence of the planning approach that was enshrined in the 1977 and 1983 Broadcasting-Satellite Conferences, considered that *a priori* planning of the geostationary orbit was the best way to secure their interests. The equatorial countries had also the curious 'weapon' of the Bogota Declaration. Further, we must not overlook the general willingness of the less developed countries to use their numerical majority to vote into treaty, resolution or recommendation language which they felt would help them. This was a development in many international organizations at the time, and it should be no surprise that it was found in the ITU, notably in Nairobi, as earlier described. The inclusion of a reference to the 'special needs of the developing countries' in art. 33.2 of the Nairobi Convention is an example of the way in which those countries have sought to protect their interests and justify their demand for a special position in many international organizations. The previous formulation of art. 33.2 that referred to the countries having access to frequencies and to the geostationary satellite orbit equitably in conformity with the Radio Regulations and 'according to their needs and the technical facilities at their disposal' (Malaga–Torremolinos Convention, art. 33.2) was, at their insistence, eliminated. 'Equitable access' was therefore glossed to mean not only equitable in the even-handed sense, but to gain a tinge of what in other connections is called 'positive discrimination'. The less developed countries were therefore attempting to enshrine their 'rights' together with some sort of 'duty' on others to help them attain their 'rights' and to refrain from actions which might pre-empt them.[207] Complete, detailed *a priori* planning of the geostationary orbit and of the appropriate frequencies would be a way to implement such an aim.

Conversely, the developed countries were unwilling to be fettered. *A priory* planning was unwelcome as it restricted their actions. Not only might such planning be a major inconvenience but a serious problem was involved. As the differences between the 1977 and the 1983 Broadcasting-Satellite Conferences illustrate, a Plan adopted in the light of present technology can quickly become a hindrance to precisely the efficient use of orbital position and spectrum usage that the ITU seeks if technology progresses beyond what the planners had envisaged.

Nonetheless, the ITU was committed to a special WARC to deal with the questions of the geostationary orbit.

WARC–ORB 1985–88

The 'World Administrative Radio Conference on the Use of the Geostationary Satellite Orbit and the Planning of the Space Services Utilizing It' met for its First Session in 1985, and will meet for its Second Session in autumn 1988.[208] Much preliminary work went into the First Session,[209] and the intersessional preparations of the participants are proceeding.[210]

At the outset, the Conference was faced by the claim by the adherents of the Bogota Declaration of sovereignty over the geostationary orbit. As in 1979 and reaffirmed at Nairobi, the claim was made on their own behalf as a matter arising from their 'special geographic situation' and also on the grounds that these countries would be able to protect the rights of developing countries in their access to orbital positions. For example, Kenya recommended additional principles which would have favoured the developing countries, given an equatorial country preferential right to its own orbital segment, and required its prior authorization to the placing of a satellite in geostationary orbit over it.[211] However, the Conference determined that, as questions of sovereignty over the geostationary orbit were not on the agenda established for it by the Administrative Council, it could not be dealt with, and it was referred to the UN's Committee on the Peaceful Uses of Outer Space.[212] There these questions rest.[213]

As for the other work of the 1985 Conference, it made a Report to the Second Session of the Conference dealing with a variety of matters on planning methods and the areas to be decided in 1988.[214] However, an important element for us, the Report and the Addendum to the Report (Docs. 328, 324(Rev. 1) and Corrigendum 1, 356(Rev. 1) and 345(Rev. 1) – which inserted into the Report statements of agreement as to planning and of directives to the IFRB a Resolution and a Recommendation – were 'adopted' by a Plenary meeting attended by less than half the delegations;[215] and some countries made sweeping reservations of their positions.[216] But this appears to be a difficulty in theory only: work on the 1988 Session has proceeded as if the Addendum were properly adopted.

Five matters in WARC–ORB–85 can be isolated as being of general importance for telecommunications:

1   the compromise that was achieved between *a priori* and *a posteriori* approach;

2   the guarantee to 11 countries of some access to the geostationary
      orbit;
3   recognition of the need for improved procedures;
4   the introduction of the concept of the multilateral planning
      meeting;
5   the recognition of the special international role of the multiadmin-
      istration operational agency.

First, at WARC–ORB–85 neither the proponents of total planning
nor the proponents of a free-for-all approach gained full acceptance
of their arguments. There has been a compromise with some detailed
planning and a *laissez-faire* approach governed by the normal exigencies of coordination and consultation through the prior publicity and
notifications requirements of the Radio Regulations outlined above.
Some planning will take place, but the detailed planning of frequency
allotment will apply only to the fixed satellite service on the expansion bands which were allocated to that service in the 1979 Radio
Regulations. Other services remained unplanned (except of course
for the broadcasting satellite service already arranged), and the fixed
satellite service also remains unplanned apart from the expansion
bands.[217]

Second, while no specific orbital positions were assigned, it is intended that, in order to supply service within a particular geographic
area, particular states will be given a right to a slot within prescribed
arcs of the geostationary orbit and to frequencies correlatively allocated to these arcs.[218] Precise positions and frequencies are not to be
guaranteed, but access to some portion of the orbit and to the necessary frequencies will be reserved for each state. In this way, states are
guaranteed access to the geostationary orbit when they require it.
This access is more limited than some would have wanted, and the
'spare' slots within each arc will be available for others to use in the
normal way, but a balance is therefore held between the rights of
states to proceed as and when they will, and the rights of other states
to their right freely to use the resources of outer space.

Third, there was a general recognition of the need for improved
procedures in the assignment and coordination of national developments as well as within the IFRB part of the process, and, this,
interlinking with modified planning procedures is to be tackled.[219]
This, a technical matter, is obviously of crucial importance for the
future development of the geostationary orbit. Fortunately, and in a
way a compliment to the ITU's past work, what is needed is an
improvement in procedures. There are good models to work on already within the Radio Regulations.

Fourth, a new mechanism is to be considered, the multilateral planning meeting (MPM), which would convene periodically in order to

deal with particular areas or particular problems. These appear to be less than administrative radio conferences, but more than informal workshops: their nature and authority remains to be determined in 1988.[220] Most of the work of the MPM would be to deal with coordination matters and act through consensus and discussion, with no place for a determination of an issue by voting. However, if such mechanisms can be developed and work efficiently and fairly, they could be the way in which to keep the regulation of space telecommunications timeously in touch with the development of technology. The present system is cumbersome. As mentioned, the development between 1977 and 1983 in broadcast-satellites permitted major change between the two conferences dealing with that matter. Again, the suggestion for WARC–ORB–85/88 came in 1979. The multilateral planning meeting may be swifter.

Last, the importance of the multi-administration system was recognized at the First Session of WARC–ORB, although not without difficulty and not as many would have wished. This is crucial for the orderly development of telecommunications.

The ITU is an organization in which states combine and compromise, discuss and negotiate to maximize their own interests. We have seen that. But it is precisely because in its rule-making activities the ITU is composed of states that that might now work to the detriment international organizations such as INTELSAT and others which provide international (and domestic) services. They have no specific state spokesman (or as some said to me, reliable spokesmen)[221] but, however the future goes, clearly within the context of a planned, unplanned or mixed system of regulation such as the 1985 WARC–ORB has settled for, some mechanism, role or consideration has to be given to the multi-administration telecommunications systems. As matters stand in law in mid-1988, although these organizations are international legal persons, because they are not states they cannot participate directly within the ITU mechanisms. There was a risk (particularly in the light of the changing attitudes of certain major states), that they might be left on the sidelines while states agreed what was in their purely national interests. This was a matter of particular concern to INTELSAT[222] and also to countries dependent on INTELSAT for their domestic and international communications.[223] A proposal by Switzerland, backed by many, was therefore directed to ensure that multi-administration systems were fully incorporated in the future planning process.[224] Other countries – notably the UK, USSR, and USA (which had a special interest also in its 'separate systems') – were anxious that language was not used which might give an undue preference to the multi-administration system over other national systems, although the USA stated that it considered the multi-administration system as part of national provision.[225] The eventual result was para. 3.2.6 of the Report which states in part that the planning to be undertaken shall:

> a) take into account the requirements of administrations using multi-administration systems created by inter-governmental agreement and used collectively without affecting the rights of administrations with respect to national systems;
> b) take account of the specific characteristics of such multi-administration systems in order to enable them to continue to meet the requirements of Administrations for international services as well as in many cases, for national services.[226]

Whether this will work out satisfactorily is a matter which time will show. INTELSAT, INMARSAT, EUTELSAT and ARABSAT have submitted a paper to the intersessional planning process which outlines the needs of their own and similar systems,[227] and INTERSPUTNIK was likely to have associated itself with that paper. Certainly there is a good argument that the multi-administration systems be given a place within the planning process: that is almost axiomatic. What is disturbing is that the recognition of their role and of their importance has been diluted. Matters may go well in WARC–ORB–2, or they may not: the Report to the Second Session of the Conference by the First Session leaves the options open and no doubt there will be much discussion, argument, and debate. However, what may happen takes us into the realm of uncertainty – and that is for the next chapter.

NOTES

1   M. Aaronson, 'Space Law', *International Relations*, April 1958, pp. 416–27, notes that interference to UK ground stations in the microwave network occurred when Sputniks were overhead. In the USA, interference was caused in the time-standardizing link between Washington and Boulder, Colorado.

2   The oldest is the Universal Postal Union: see G.A. Codding, *The Universal Postal Union: Coordinator of the International Mails* (New York: New York University Press, 1964).

3   The best history of the ITU down to its date is, G.A. Codding, Jr., *The International Telecommunication Union: An Experiment in International Cooperation* (Leiden: E.J. Brill, 1952; rep. New York: Arno Press, 1972), cited hereafter as 'Codding,' op. cit. Professor Codding also wrote the ITU centenary book, *From Semaphore to Satellite* (Geneva: ITU, 1965), which functions as an official history of the Union. D. Leive, *International Telecommunications and International Law: The Regulation of the Radio Spectrum*, (Leyden: A.W. Sijthoff; Dobbs Ferry, N.Y.: Oceana, 1970) is best within its more restricted aim, although there have been later developments. See also for earlier times: J.D. Tomlinson, *The International Control of Radiocommunications* (Michigan: J.W. Edwards, 1945) (his dissertation of Geneva, 1938), and O. Mance, *International Telecommunications* (Oxford: Oxford University Press, 1943).

4   Technically such a Union was never established, but I follow Codding, op. cit.

(p. 6ff) in using this title to denote the various signatories of the adherents to the first Radio-telegraph Conventions.

5   International Telegraph Convention, Paris, 17 May 1865, 130 CTS 198; BFSP vol. LVI, 295. The Parties were Austria, Baden, Bavaria, Belgium, Denmark, France, Hamburg, Hanover, Italy, The Netherlands, Portugal, Prussia, Russia, Saxony, Spain, Sweden–Norway, Turkey and Wurtemburg.

6   Telegraph Convention between Belgium, France, The Netherlands, Sardinia and Switzerland, Berne, 1 September 1858; 119 CTS 461.
    In fact, there were many bilateral, trilateral and quadrilateral treaties during this period. Indeed, there is a special part of the indices to the Consolidated Treaty Series (CTS) is devoted to the postal and telegraph treaties in the series: Index Guide, vol. 2. Special Chronology, compiled by M.A. Meyer. The Paris Convention was regulatorily more thoroughgoing than these other prior arrangements.

7   International Telegraph Convention, Vienna, 21 July 1868; 1366 CTS 292; BFSP vol. LIX, 322. Newcomers to the Telegraph Convention by then were Great Britain (on behalf of India), Greece, the Holy See, Luxemburg, Persia, Serbia, Switzerland and the United Principalities of Moldavia and Wallachia.

8   International Telegraph Convention and Regulations, Rome, 14 January 1872; 143 CTS 415; BFSP vol. LXVI, 975; 1872 Parl. Papers vol. lxx (C. 547).

9   International Telegraph Convention and Regulations, St Petersburg, 22 July 1875 (Old Style 10 July); 148 CTS 416; BFSP vol. LXVI, 19; 1876 Parl. Papers vol. lxxxiv (C. 1418).

10 Regulations in Execution of the International Telegraph Convention of 22 July 1875, Berlin, 17 September 1885; 165 CTS 212; BFSP vol. LXXVI, 597.

11 Regulations Annexed to the Revised International Telegraph Convention, London, 10 July, 1903; 193 CTS 327; BFSP vol. XCVII, 736.

12 Austro–Hungary, France, Germany, Great Britain, Italy, Spain, Russia and the USA.

13 Preliminary Conference at Berlin on Wireless Telegraphy, Proces-Verbaux and Protocole Final. (1903) Cd. 1832. UK State Papers, vol. CXX, 1904. The Final Protocol is also printed, 194 CTS 46; BFSP vol. XCVII, 467.

14 Memoir, Preliminary Conference at Berlin on Wireless Telegraphy, Proces-Verbaux and Protocole Final (1903) Cd. 1832. UK State Papers, vol. CXX, 1904, at p. 54.

15 Preliminary Conference at Berlin on Wireless Telegraphy, Proces-Verbaux and Protocole Final. (1903) Cd. 1832. UK State Papers, vol. CXX, 1904; 194 CTS 46.

16 Radio-telegraphic Convention, Final Protocol and Regulations signed at Berlin, 3 November 1906, 1906 Parl. Papers HC 368; UK State Papers, vol. CXXXVIX, 1906; 203 CTS 101; BFSP vol. XCIX, 321.

17 Berlin Radio Regulations art. XXVIII. Under normal circumstances this would not exceed 1 kw. (Berlin Radio Regulations art. VI(2)(c)).

18 The rationale for using the International Telegraph Bureau was economy, but the use of the Telegraph Bureau required the consent of the Swiss Government, which oversaw it, and the approval of the International Telegraph Union (Radio Regulations art. XXXVII). As the membership of the two Unions was to a degree common, this approval was forthcoming.

19 Quoted by Codding, op. cit. at p. 98.

20 International Radio-telegraph Convention, Final Protocol and Service Regulations signed at London, 5 July 1912. (1915) UKTS No. 10, Cd. 6783; UK State Papers, vol. LXXXI, 1913; BFSP vol. CV, 219; 216 CTS 244.

21 London Radio Regulations, art. 45.

22 Y. Stewart, 'The International Radiotelegraph Conference of Washington', *American Journal of International Law*, vol. 22, 1928, p. 28.

23 Commercial broadcasting in the UK began on 14 November 1922; see *The Times*, 15 November 1922, for the curious conditions under which the broadcasters operated.

24 Radio-telegraph Convention and General Regulations signed at Washington, 25 November 1927; 84 LNTS 97; Manley O. Hudson, *International Legislation*, vol. III, 1931 (Washington, DC: Carnegie Endowment for International Peace, 1925–7), 2197; *American Journal of International Law*, vols. 22–4, 1928–30, Supp. 40 (Convention only).

25 Washington Convention, art. 17; Washington Radio-telegraph Regulations, art. 33. This body (now the CCIR of the ITU) is very influential.

26 The technical term used in the ITU for a government department or service responsible for licensing a radio station and ensuring that the provisions of the ITU Convention and Regulations are complied with. See ITU Radio Regulations, 1979, art. 1.1.1.

27 Telecommunication Convention, General Radio Regulations, Additional Radio Regulations, Additional Protocol (European), Telegraph Regulations and Telephone Regulations, Madrid, 9 December 1932; 151 LNTS 4; Hudson, *International Legislation* op. cit., vol. VI, 1932–34, p. 109.

28 See also J.M. Arto Madrazo, 'The Establishment in Madrid of the International Telecommunication Union (1932)', (1983), 50 *Telecommunication Journal*, vol. 50, 1983, pp. 6–8.

29 Madrid General Radio Regulations, art. 7.3; the Table of Allocations is art. 7.7.

30 Madrid General Radio Regulations, art. 1(28).

31 Cf. attitudes as to geostationary orbital positions in the later 1970s and 1980s.

32 The procedures for notification are contained in art. 7.5 of the 1932 Regulations.

33 Madrid Radio Regulations, art. 7.5.

34 European Broadcasting Convention, signed at Lucerne 19 June 1933, Hudson, *International Legislation*, op. cit., vol. VI 1932–34, p. 345; cf. European Broadcasting Convention (with Copenhagen Plan protocols etc), Copenhagen, 15 September 1948, (1950) UKTS No. 30; Cmd. 7946. H.B. Otterman, 'Inter-American Radio Conference, Havana, 1937', *American Journal of International Law*, vol. 32, 1938, p. 569.

35 H.S. LeRoy, 'Treaty Regulation of International Radio and Short Wave Broadcasting', *American Journal of International Law*, vol. 32, 1938, p. 719.

36 Cairo Radio Regulations, art. 33.1.

37 Codding, op. cit., pp. 180–4, 'War and Telecommunication'.

38 International Convention on Telecommunications, Atlantic City, 2 October 1947; (1950) UKTS No. 76, Cmd. 8124; 63 Stat. 1399, TIAS 1901.

39 See the synoptic table in J.H. Glazer, 'The Law-Making Treaties of the International Telecommunication Union through Time and in Space', *Michigan Law Review*, vol. 60, 1962, 269 at p. 315.

40 On the discussion of the new system, see Codding, op. cit., pp. 297–8.

41 Chapter 10 of the General Regulations annexed to the Atlantic City Telecommunications Convention (Annex 4 to the Convention).

42 For detail, see Codding, op. cit., pp. 241-52.

43 See the discussion of the Communications Satellite Corporation, Chapter 2 above. The notion of a controlling of supervisory body of experts was a US proposal, though developed by the Russians.

44 See Codding, op. cit., pp. 247-9.

45 See Codding, op. cit., pp. 252-64, esp. pp. 260-4.

46 'Extraordinary Administrative Radio Conference, Agreement for the Preparation and Adoption of the New International Frequency List for the Various Services in the Bands between 13 kc/s and 27,500 kc/s with a View to Bringing into Force the Atlantic City Table of Frequency Allocations' (Geneva: ITU, 1951).

47 The ITU and its predecessors had stood aloof from the League of Nations: G.A.
    Codding, 'The Relationship of the League and the United Nations with the
    Independent Agencies: A Comparison'. *Annales d'Etudes Internationales*,
    pp. 1165–87.
48 Printed as Annex 5 to the Atlantic City Convention, above, the Agreement has
    been annexed to all subsequent versions of the ITU Convention. The relation-
    ship of the two institutions is the subject of art. 39 of the Nairobi Convention,
    1982 (the current Convention, cited below) and the Agreement is Annex 3 to
    that Convention.
49 Codding, op. cit., pp. 290–1. Complaints that this system was inequitable or
    unsuitable were met by the increase in the membership of the Council by
    subsequent revisions of the Convention. It was not until the Malaga–Tor-
    remolinos revision of 1973 (cited below) that the language requiring equitable
    representation in the Council was changed from 'all parts' to 'all regions' of the
    world, a closer reflection of ITU practice.
50 Codding, op. cit., p. 259.
51 See generally, G.A. Codding and A.M. Rutkowski, *The International Telecom-
    munication Union in a Changing World* (Dedham, Maryland: Artech House, 1982)
    pp. 139–58, esp. pp. 156–8.
52 Codding, op. cit., at 219 notes that in 1945 the staff of the Bureau consisted of 30
    persons; 28 Swiss, one Frenchman and one American.
53 International Telecommunication Convention, Buenos Aires, 22 December
    1952; (1958) UKTS No. 36, Cmnd. 520. UK State Papers 1957–58, vol. XXIX.
54 International Telecommunication Convention, Geneva 21 December 1959;
    (1958) UKTS No. 74, Cmnd. 1484. UK State Papers 1960–61 vol. XXXII.
55 By now these were very bulky and were published by the ITU itself.
56 See section 'Space Frequencies', p. 357.
57 International Telecommunication Convention, with Final Protocol, Additional
    Protocols I to III and Optional Additional Protocol, Montreux, 12 November
    1965 (with Additional Protocol of 12 October 1965); (1967) UKTS No. 41, Cmnd.
    3383.
58 Montreux Convention, art. 13.2(1): the reasons for this development are dealt
    with below in the consideration of the present ITU constitution.
59 International Telecommunication Convention, Final Protocol, Additional Pro-
    tocols, Resolutions, Recommendations and Opinions, Malaga-Torremolinos, 25
    October 1973; (1975) UKTS No. 104, Cmnd. 6219; TIAS 8572.
60 International Telecommunication Convention, with Final Protocol, Additional
    Protocols I to VII and Optional Additional Protocol, Nairobi, 6 November 1982;
    (1985) UKTS No. 33, Cmnd. 9557. The Convention together with the Resolu-
    tions of the Conference are also published by the Secretariat of the Union,
    Geneva 1984.
61 International Telecommunication Convention, with Final Protocol, Additional
    Protocols I to VII and Optional Additional Protocol, Nairobi, 6 November 1982;
    (1985) UKTS No. 33, Cmnd. 9557. The Resolutions of the Nairobi Conference
    are available in the version of the Convention published by the Secretariat of
    the Union, Geneva, 1984. The ITU Convention will be revised by a further
    Plenipotentiary Conference in 1989.
62 The effective date for the UK is 15 November 1984, the date of deposit of the UK
    instrument of ratification.
63 On this see Codding and Rutkowski, op. cit., p. 204.
64 The Malaga–Torremolinos Convention had the same provision.
65 See below for a proposal as to international organizations.
66 The failure of countries, often through dilatoriness, to process ratifications
    through their appropriate mechanisms or to accede to a new ITU Convention

can have significant effects. The Malaga–Torremolinos Convention, which had similar provisions (art. 45), came into force on 1 January 1975 (art. 52). A Note by the Secretary-General to the Broadcasting Satellite Conference, Geneva 1977, dated 21 January 1977, shows that 50 states which had had membership of the ITU under the previous Convention had by that date not ratified or acceded to the 1973 Convention, although six had indicated action was pending: Documents of the Broadcasting Satellite Conference, Geneva 1977, Doc. 28 (Rev. 1)–E. That meant that nearly half of the members of the Union could not vote at that Conference. At the time of the First Session of the World Administrative Conference on the use of the Geostationary Orbit in August 1985, 14 members had forfeited their voting rights through having neither signed (and hence making themselves unable to ratify) nor acceded to the Nairobi Convention: Documents of the WARC–ORB/85 Conference, Doc. 45 (Rev. 1)–E.

67   See on the Atlantic City discussions generally, Codding, op. cit. pp. 208–14, 214–19 on Mongolia and the Baltic States, pp. 222–3 on Pakistan and India: pp. 275–83 deal with the Plenary debates on the colonial countries.

68   See art. 1.4 of both the Atlantic City and Buenos Aires Conventions; art. 1.3 of the Geneva and Montreux Conventions. Cf. Codding, op. cit., p. 280.

69   Nairobi Convention, art. 1.1.

70   See Codding, op. cit., pp. 280.

71   Atlantic City Convention, Protocol III; Codding, op. cit., pp. 219–22 (on Spain's exclusion from the 1947 Conference), pp. 279–80 (on Protocol III).

72   South Viet Nam (Doc. No. 393–E), Rumania (Doc. No. 401), the USA (Doc. No. 403–E), South Korea (Doc. No. 405–E), China (Taiwan) (Doc. No. 406–E), Chile (Doc. No. 409–E).

73   Administrative Council, 27th Sess., 1972, Res. No. 693.

74   As happened in the case of Southern Rhodesia by decision of the Administrative Council in 1966 following approval indicated by a majority of members (Administrative Council, 21st Sess., 1966, Res. No. 599): the ITU action followed Res. No. 2022(XX) of the UN General Assembly.

75   Administrative Council, June 1975, Res. No. 765. Wallenstein, op. cit., vol. 1, p. 50 notes that 98 members voted in favour of readmission – 68 per cent of the eligible membership.

76   Nairobi Conference, Docs. 120 (Rev. 2) + Corr. 1, 123 and 205; Minutes of the Plenipotentiary Conference of the ITU, Nairobi 1982, 261–90, Minutes of the 15th, 16th, 17th and 18th Plenary Meeting, (Nairobi Conference Docs. 456–E, 457–E, 458–E, 459–E).

77   Minutes of the Eighteenth Plenary Meeting, Doc. No. 459–E, para. 1.22.

78   See Testimony of M.R. Gardner, Head of the US Delegation to the Nairobi Conference: *Oversight of the Unispace and International Telecommunications Union Conference*, Hearings before the SubCommitte on International Operations of the Committee on Foreign Affairs, US House of Representatives, 98th Cong. 1st Sess., 22 February 1983, at pp. 3 and 5.
      The same source indicates that of the four votes margin, one was cast by the USA, and another also by the USA acting for Haiti which may make the margin three.

79   See Gardner Testimony, op. cit., note 78 above.

80   The involvement of the UN in the appointment of judges to the International Court of Justice is not comparable to the involvement of, say, an INTELSAT delegation on the election of the Director of the CCIR.

81   Under the Montreux Convention of 1965, the last which permitted associate membership, such members were subject to the indicated restrictions, but in addition they had no vote in ITU conferences or organs.

82   UN GAOR 2472 XXIII (1968); 2190 XXI (1966); 311 IV (1949).
83   It must be observed that the UN system is itself under stress; cf. the Report of the Committee on Contributions, Supp. No. 11 (A/37/11). In December 1986 a revision of the system was decided upon, with the consent of the major contributors including the UK, USA and USSR, who were concerned about the scale of contributions and the UN failure to exercise proper budgetary control.
84   G.A. Codding, 'International Constraints on the Use of Telecommunications: The Role of the International Telecommunication Union' in L. Lewin (ed.), *Telecommunications: An Interdisciplinary Survey* (Dedham, Mass.: Artech House, 1979), pp. 1–37 at p. 18.
85   G.A. Codding, *The Universal Postal Union* (New York: New York University Press, 1964).
86   Cf. recent experience with UNESCO and, to a lesser extent, with the UN itself as indicated above.
87   Thus in 1979, 48 private organizations contributed 45 1/2 units of the CCIRs and 49 units of the CCITT's expenses (G.A. Codding and A.M. Rutkowski, *The International Telecommunication Union in a Changing World* (Dedham, Mass.: Artech House, 1982) p. 188. As to 'units' see below.
88   Nairobi Conference, Doc. No. 9.
89   Res. COM 4/7 also instructed the Council at each session to review small countries not on the UN list of least developed nations which may encounter difficulties in contributing in the 1/4 unit class and which request review, so as to decide which of them may be considered as being entitled to contribute in the 1/8 unit class. This appears to be incompetent given the terms of art. 15.5's ban on a country dropping its contribution class. That indicated the Council could authorize a reduction only in exceptional circumstances, such as 'natural disasters necessitating international aid programmes'. Res. COM 4/7 is in much weaker terms.
90   Administrative Council Doc. No. 6038 was the working document in the matter.
91   Nairobi Conference, Doc. No. 65 at pp. 155-9.
92   Hereafter I use figures for emphasis.
93   Nairobi Convention, Additional Protocol I, 1.1 and 2.
94   Ibid., Additional Protocol I, 2.1.
95   Ibid., Additional Protocol I, 2.2 and 7.
96   Cf. Oversight of the Unispace and International Telecommunications Union Conference, Hearing before the Subcommittee on International Operations of the Committee on Foreign Affairs, US House of Representatives, 88th Cong. 1st Sess., 22 February 1983, at 3, 4 and 11, Testimony of M.R. Gardner op. cit. and comments by Committee members.
97   Gardner Testimony, op. cit., at p. 4: Minutes of the 32nd Plenary, Doc. No. 512–E at sect.1, Minutes of the 33rd Plenary, Doc. No. 513–E at 2.24–8.
98   Thus it was the suggestion of the US Head of Delegation to the Nairobi Conference that the US restrict its contribution to areas not including technical assistance. See Testimony of W.R. Gardner, op. cit., at p. 4. Cf. The US reservation to the Final Acts of the Nairobi Conference, Reservation No. 70; The USA is not alone here: see the UK Reservation, Res. No. 84, para. III; the Byelorussian SSr, Ukrainian SSR and USSR Reservation, Res. No. 79; that by Bulgaria, Hungary, Mongolia, Poland, German D.R. and Czechoslovakia, Res. No. 73; Federal Republic of Germany, Res. No. 57; Greece, Res. No. 62, and others which express either doubt about finances, or unwillingness to shoulder extra expenditure if others reduce their share.
99   See also the Reports of the Administrative Council to the Nairobi Conference,

'Implementation of Resolutions', etc. relating to the Technical Cooperation Activities of the Union (Conf. Doc. No. 4) and 'The Future of ITU Technical Cooperation Activities' (Conf. Doc. No. 47).

100 See the start and end of section on 'Finance', p. 332 above.

101 See Codding and Rutkowski, op. cit., pp. 151–8. Cf. for example, the West German proposals for the work of the Nairobi Conference (Conf. Doc. No. 16) para 2; the US proposals (Conf. Doc. No. 15) USA/15/6–12; the UK proposals (Conf. Doc. No. 18) G/18/3–7; and the Canadian proposals (Conf. Doc. No. 26) CAN/26/7–9, 20–6.

102 Codding and Rutkowski, op. cit., pp. 156–8.

103 Nairobi Conference, Doc. No. 5.

104 At Nairobi there was a US proposal for the creation of a five-member advisory committee to act in the intervals between the annual Advisory Council meetings. This proposal was lost on the ground that it was élitest. See Questions submitted to Diana Lady Dougan from Chairman Fascell and Responses thereto, Appendix 2 to: *Oversight of the Unispace and International Telecommunication Union Conference*, Hearing before the SubCommittee on International Operations of the Commitee on Foreign Relations of the Committee on Foreign Affairs, House of Representatives, 22 February 1983, 88th Cong., 1st Sess. at p. 95.

105 On the question of help to developing countries, the IFRB has specific instructions to take into account the needs of members, and the specific needs of developing countries in dealing with geostationary orbit questions (art. 10.4(c)).

106 See Codding and Rutkowski, op. cit., pp.121–2.

107 Nairobi Convention, Res. PLB/1: see also 'Extended use of the Computer by the IFRB' a joint document from the Secretary-General and the Assembly Council (Nairobi Conf. Doc. No. 33) and the Report of the same title to the Plenary by Working Group PL/B (Conf. Doc. No. 280).

108 Cf. art. 13.2 Malaga–Torremolinos Convention 1973; art. 12.1, Montreux Convention 1965; Art. 11.1 Geneva Convention, 1959.

109 See Codding and Rutkowski, op. cit., pp. 174–8.

110 D.M. Leive, *International Telecommunication and International Law: The Regulation of the Radio Spectrum* (Leiden: Sijthoff; New York: Oceana, 1970) p. 22.

111 As to administrative conferences, see below in relation to radio frequencies.

112 See below. Radio Regulations, 1979, arts. 4, 5, 27, and 28 as well as the Recommendations and Reports of the CCIR are particularly important.

113 This book does not consider the CCITT's work in any detail, important though it is within international telecommunication. Suffice it to say that the CCITT meets every four years and its recommendations run to some 30 books.

114 Nairobi Convention, art. 83. Down to the Nairobi Convention the words were that the Regulations 'complete' the Convention: cf. Malaga–Torremolinos Convention, art. 82. Down to Montreux the Convention used also to stipulate specifically that the Regulations had the same force and duration of the Convention itself (cf. Montreux Convention, art. 11.1), although elsewhere providing for the Regulations to remain valid, subject to revision until the entry into force of new regulations to replace them under the Convention (cf. Montreux Convention, art. 26).

115 Final Acts of the World Administrative Radio Conference, Geneva, 1979, published by the Union at Geneva, 1980. Amendments have occurred.

116 Administrative Council Res. No. 768, 1975.

117 Cf. Final Acts of the World Broadcasting-Satellite Administrative Radio Conference, Geneva, 1977, Geneva, nd, Part II, 77, instructing the Secretary-General to publish the Radio Regulations in the new form, Res. No. SAT–10, which urged members to use the new format appearing at pp. 139–40.

118   Cf. WARC–79 Final Act, FA–1. This practice was first adopted in the 1959 revision of the Regulations.
119   Codding and Rutkowski, op. cit., pp. 260–5.
120   Washington Convention, art. 8.
121   Ibid, art. 6.
122   Washington Radio Regulations, art. 5.
123   Europe has continued to order its affairs by a considerable measure of *a priori* planning, the Copenhagen Plan of 1948 on maritime mobile stations being a major example of such work.
124   Nairobi Convention, art. 16.1(2). The French text of the Convention prevails in the case of a dispute (art. 16.1(3)). The Nairobi Conference added Arabic to the other official languages of the Union (Chinese; English, French, Russian and Spanish) (art. 16.1(1)).
125   These correspond to those used for the purposes of election to the IFRB and the Administrative Council (see above).
126   For detail see RR8–2. Great circle lines are used to link nominated points, though the boundary between Regions 1 and 3 refers to state territories.
127   See the Final Protocol to the Final Acts of the World Administrative Radio Conference, Geneva, 1979; eg. (selectively) Iran (Statement No. 10), Pakistan (No. 13), Greece and Yugoslavia (No. 14), Japan (No. 16), Nigeria (No. 17), United States (No. 38), Ireland and the UK (No. 48, adding Ireland to a footnote), Thailand (No. 60).
128   Cf. Leive, op. cit., pp. 144–207.
129   States retain entire freedom with regard to national defence services – art. 38 of the Nairobi Convention has long roots.
130   It may be noted that the 1979 Radio Regulations as adopted contain no less than 487 footnotes, dealing with a variety of matters.
131   Cf. Edward Wenk, Jr., *Radio Frequency Control in Space Telecommunications*, a Report prepared by the Legislative Reference Service, Library of Congress, at the request of L.B. Johnson, Chairman, for the use of the Committee on Aeronautical and Space Sciences, US Senate, 8th Cong. 2nd Sess., 19 March 1960, pp. 81.
        Of course, the Resolutions and Recommendations of an Administrative Radio Conference are also used to guide the future work of the Administrative Council and of the CCIR and CCITT, but this use is not what we are considering at present. We are dealing with the legal force of an 'allocation' in the Radio Regulations.
132   Cf. Leive, op. cit., index s.v. 'Penalties: enforcement weak'.
133   I omit the problem of the 'pirate radio station'.
134   For much of the following detail see the Report of the International Frequency Registration Board to the Plenary Meeting of the World Administrative Radio Conference, Geneva 1979, Doc. Nos. 64–69.
135   Codding, op. cit., pp. 376–81.
136   There is also a project management team considering the best structure for the future having regard to the requirements of the IFRB for use of the ITU computer systems.
137   Subsequently arts. 43, 45 and 46 respectively of the Buenos Aires Convention 1952, the Geneva Convention 1959, and the Montreux Convention 1965.
138   Art. 33(2) in both cases deals with space. The article also states the requirements to make efficient and economic use of the geostationary orbit.
139   M. Aaronson 'Space Law', *International Relations*, 1958, pp. 416–7; reprinted in *Legal Problems of Space Exploration – A Symposium* prepared for the Committee on Aeronautical and Space Sciences, 87th Cong. 1st Sess., S. Doc. No. 26, 1961 at p. 221.

140   A.G. Haley, 'Space Laws: the Need for Agreement on Astronautical Radio Allocation', *Proceedings of the Canadian Astronautical Society*, vol. 1 no. 2, July 1959.

141   A.G. Haley, *Space Law and Government* (New York: Appleton, Century, Crofts 1963) at pp. 168–71.

142   On the 1959 Conferences, and looking ahead to the subsequent developments which they set in train, see: *Policy Planning for Space Telecommunications*, Staff Report prepared for the Committee on Aeronautical and Space Sciences, US Senate 86th Cong. 2nd Sess., 4 December 1960. See also generally, Edward Wenk, Jr. *Radio Frequency Control in Space Telecommunications*, a Report prepared by the Legislative Reference Service, Library of Congress, at the request of L.B. Johnson, Chairman, for the use of the Committee on Aeronautical and Space Sciences, US Senate, 86th Cong. 2nd Sess, 19 March 1960 (hereafter cited as 'Wenk,' op. cit.).

143   Wenk, op. cit., pp. 81–2.

144   Geneva Radio Regulations, 1959, Ch. I,:

> 'Space Station: a station in the earth-space service or the space service located on an object which is beyond, or intended to go beyond, the major portion of the earth's atmosphere, and which is not intended for flight between points on the earth's surface. (para. 72)

> Earth Station: a station in the earth-space service located either on the earth's surface or on an object which is limited to flight between points on the earth's surface. (para. 73)

145   Wenk, op. cit., Appendix G, pp. 149–223, prints the revisions to the Table of Allocations made by the 1959 ARC.

146   See Wenk, op. cit., pp. 61–79, 'Spectrum Demands for Radio Astronomy', especially pp. 1–4 which summarizes points made by various organizations and includes tables of the principal emission lines which radio astronomers then used. Cf. M.A. Stull and G. Alexander, 'Passive Use of the Radio Spectrum for Scientific Purposes and the Frequency Allocation Process', *Journal of Air Law and Commerce*, vol. 43, 1977, pp. 458–534.

147   Wenk, op. cit., pp. 79–81.

148   Report of the Chairman of the US Delegation to the Extraordinary Administrative Radio Conference of the International Telecommunication Union. Geneva October–November 1963. US Department of State, Telecommunications Division. TD Serial No. 949 IU Doc. No. 61, cited hereafter as 'US Report on EARC 1963'. Also printed 110 Congressional Record 166–74, 9 January 1964 and (1964) III ILM 224–32. Cf. 'In the Matter of An Inquiry into the Allocation of Frequency Bands for Space Communications', FCC 61–652, docket no. 13522 and attached draft of 'Preliminary Views of the United States of America on Frequency Allocations for Space Radiocommunications', adopted 17 May 1961 and released 19 May 1961, printed in *Communications Satellites Part I*, Hearings before the Committee on Interstate and Foreign Commerce, US House of Representatives, 87th Cong. 1st Sess, July 1961 (1962) at pp. 22–69, and 'Preliminary Views' as finally adopted, printed ibid, at pp. 233–277. Cf. also *Policy Planning for Space Telecommunications*, Staff Report prepared for the Committee on Aeronautical and Space Sciences, US Senate 86th Cong. 2d Sess., 4 December 1960.

149   Final Acts of the Extraordinary Administrative Radio Conference, Geneva 7 October–8 November 1963.

150   US Report on EARC 1963, 6.

151   Cf. fears expressed at the Madrid (1932) and Cairo (1938) Radio Conferences that the advanced nations would pre-empt frequency allocations for new services.

152 See US Report on EARC 1963, pp. 13–22, dealing with the work of Committee No. 5 – Frequency Allocation.

153 Wenk, op. cit., p. 81: 'For a variety of reasons related to past practice, geography, propagation characteristics and spectrum crowding, allocations are sometimes made by regional rather than global agreement.'

154 UNGA Res. 1963 (XVIII) Part IV, para. 2.

155 See the Final Acts of the Conference. The Partial Revision of the Radio Regulations is printed (1964) III ILM 92.

156 The technical procedure for the calculation of a coordination distance around an earth station was laid down in Recommendation 1A of the 1963 EARC, and is now found as Appendix 28 to the Radio Regulations, Geneva 1979 as modified. Recommendation No. 711 of the 1979 Regulations invited the CCIR to continue with studies on the matter.

157 Geneva Radio Regulations 1959, as amended, art. 9, s. 1.4(1) para. 639 AL.

158 Ibid., art. 9, s. 1.3(1) para. 491.

159 Ibid., art. 9, ss. IIB; art. 9A ss. II.

160 The Final Acts of the 1966 EARC and the 1967 WARC were published by the ITU, Geneva in the respective years.

161 'Final Acts of the World Administrative Radio Conference for Space Telecommunications', published by the ITU, Geneva 1971. This Conference is frequently referred to in technical literature as Spa2, thereby distinguishing it from Spa1, the 1963 EARC. Thus Res. No. Spa2–1 is the first resolution adopted by this WARC.

162 For more detail see the Final Acts of the Conference, and the Summary Record of the World Administrative Radio Conference for Space Telecommunications prepared by the International Frequency Registration Board, *Telecommunication Journal*, vol. 38, 1971, pp. 673–82.

163 Annex A to the new Appendix 28 provided a flow chart as a guide through the maze of the new procedures.

164 Appendix 1A was entirely replaced by Annex 14 to the 1971 WARC Final Acts.

165 Annex 15 to the 1971 WARC Final Acts.

166 Annex 7 to the 1971 Space–WARC Final Acts.

167 Annex 8 to the 1971 WARC Final Acts.

168 In what follows citation is to example rather than exhaustive: there are many routes through art. 9A dealing with different problems and configurations.

169 'Summary Record of the World Administrative Radio Conference for Space Telecommunications prepared by the International Frequency Registration Board', *Telecommunications Journal*, no. 38, 1971, pp. 673–82 at pp. 680–1.

170 Res. No. Spa2–2, relating to the establishment of agreements and associated Plans for the Broadcasting-Satellite Service. As described below, a Conference for all regions bar Region 2 was held in 1977, and a separate Conference for Region 2 in 1983.

171 See notes 639A0.1 Spa2 and 639AP.1 Spa2 to respectively arts. 9A 3(2) and (3).

172 Notifications for the space research service could be made up to not later than 30 days after the assignment is brought into use (art. 9A 8(1)).

173 Cf. ITU Convention, Malaga–Torremolinos 1973: 'The provisions of the Convention are completed by the following Administrative Regulations: – Telegraph Regulations, – Telephone Regulations, – Radio Regulations, – Additional Radio Regulations' (art. 82).

174 'World Administrative Radio Conference' *Telecommunication Journal*, vol. 47, 1980, pp. 4–10. See also (UK) Home Office Radio Regulatory Department publications 1. 'Preparation for the World Administrative Radio Conference' April 1978; 2. United Kingdom Proposals for the World Administrative Radio Conference 1979', January 1979; 3. Supplementary United Kingdom Proposals for

the World Administrative Radio Conference 1979', March 1979, and 4. 'Report on the World Administrative Radio Conference 1979', March 1980. Cf. 'Report of [Glen O. Robinson] the Chairman of the U.S. Delegation to the World Administrative Radio Conference of the ITU, Geneva, Switzerland September 24–December 6, 1979, [US] Department of State, Office of International Communications Policy, 1980 TD. Ser. No. 116.

175   It was the 1979 Conference which lowered the low end of the spectrum regulated to 9 KHz, allocating 9–14 KHz for radio-navigation on a worldwide basis. In the 50 years from 1927 developments had occurred at the high-freqency, not the low-frequency end of the radio spectrum range.

176   The Report of the International Frequency Registration Board (WARC–79 Doc. nos. 65–69 with relative addenda) was helpful for the work of 1979 as were major contributions from certain countries.

177   Rec2–1979 replaces and in part repeats Rec. No. Spa2–1 of the 1971 Space WARC to similar effect.

178   See also section 'The Geostationary Orbit', p. 387 below.

179   These are the cases of coordination of frequency assignments to an earth station in relation to terrestrial stations (art. 11.16–22) and the coordination of frequency assignments to a terrestrial station for transmission in relation to an earth station (art. 11.23–31).

180   INTELSAT systems are notified to the ITU by the USA, INMARSAT by the UK, EUTELSAT by France, INTERSPUTNIK by the USSR and ARABSAT by Saudi Arabia.

181   The data stipulated by Appendix 4 includes general characteristics as to the network, orbital positions and tolerances together with characteristics of the network in both earth-to-space, space-to-earth and any space-to-space relay, including frequency range, noise temperature, power, modulation characteristics (if available), bandwidth. The data required is highly technical and was made more detailed by the 1979 WARC to facilitate the whole process of coordination and accommodation between administrations.

182   Calculated for the appropriate coodination distance in terms of Appendix 29 to the Radio Regulations.

183   Similar provisions apply for the coordination of earth station/terrestrial service coordinations (art. 11.22) and terrestrial service/earth stations (art. 11.30).

184   These include administrative requirements to facilitate processing as well as information on the operator, frequency, date of use, identity and location of stations, class of emissions, antenna characteristics, modulation and polarization, noise temperature, link noise temperature and transmission gain, operational hours, coordinations effected and any relative agreements. In the case of space stations, orbital data and identity of the station is also required.

185   Final Acts of the World Broadcasting-Satellite Administrative Radio Conference held in Geneva, 1977, ITU 1978. I. Lonberg, 'The Broadcasting Satellite Conference'. *Telecommunication Journal*, vol. 44, 1977, pp. 482–88.

186   See Resolution No. Sat–8 'Relating to the preparation for an administrative radio conference for the detailed planning of the space services in the frequency band 11.7–12.2 GHz in Region 2', Resolution No. Sat–9 'Relating to the submission of requirements for the broadcasting-satellite service in Region 2' and Recommendation No. Sat–8 'Relating to the convening of a regional administrative radio conference for the detailed planning of the space services in the frequency band 11.7–12.2 GHz in Region 2' in the Final Acts of the 1977 World Broadcasting-Satellite Administrative Radio Conference, Geneva 1977.

187   I. Lonberg, 'The Broadcasting-satellite Conference' *Telecommunication Journal*, vol. 44, 1977 pp, 482–88, at p. 487.

188 The Vatican, for example, was allocated a superbeam which covers most of Italy. The Nordic countries, Denmark, Finland, Norway and Sweden were also allocated a superbeam.

189 Final Acts of the Regional Administrative Radio Conference for the Planning of the Broadcasting-Satellite Service in Region 2, ITU, Geneva 1983.

190 See section 'The Geostationary Orbit', p. 387 below.

191 The RARC itself had provided for the use of interim procedures on a provisional basis in the period between its own decisions and the bestowing of full legal authority on its proposals through the amendment of the Radio Regulations.

192 The band for transmission is 12.2–12.7 GHz, like that throughout most of Regions 1 and 3.

193 Arthur C. Clarke, 'Extra-terrestrial Relays. Can Rocket Stations Give World-Wide Radio Coverage?', *Wireless World* October 1945, pp. 303–8.

194 Cf. Convention on Civil Aviation, Chicago, 7 December 1944, 15 NTS 295; (1953) UKTS No. 8, Cmd. 8742; 61 Stat 1180, TIAS 1591, art. 1: 'The contracting States recognise that every state has complete and exclusive sovereignty over the airspace above its territory.' This is virtually the same language as art. 1 of the Paris Convention on the Regulation of Aerial Navigation, 13 October 1919, 11 LNTS 173; (1922) UKTS No. 2, Cmd. 1609. The USA never became a Party to the Paris Convention.

195 As to which see Chapter 1.

196 See C.Q. Christol, 'The Geostationary Orbital Position as a Natural Resource of the Space Environment', *Netherlands International Law Review*, vol. 26, 1979, pp. 5–23; his 'The Definition/Delimitation of Outer Space, the Use of the Geostationary Orbital Position, and the 1976 Bogota Declaration: Policies and Prospects', in his *The Modern International Law Of Outer Space* (New York and Oxford: Pergamon Press, 1982) pp. 435–546; A.G. Haley, *Space Law and Government* (New York: Appleton-Century-Crofts, 1963); R.S. Jakhu, 'The Legal Status of the Geostationary Orbit', (1982) 7 *Annals of Air and Space Law*, no. 7, 1982, pp. 233–51.

197 R.G. Gould, 'Protection of the Stationary Satellite Orbit', *Telecommunication Journal*, vol. 34, 1967, pp. 307–12 at p. 30.

198 See Christol, op. cit., at 453–5, material there cited, and from the *Twentieth Colloquium on the Law of Outer Space*, 1978; M.A. Ferrer, 'The Use of the Geostationary Orbit', at p. 216 and J.F. Galloway, 'Telecommunications, National Sovereignty and the Geostationary Orbit' at pp. 226–37. The UN COPUOS has also had the matter under consideration for years.

199 See note 197 above.

200 Cf. S. Gorove, 'The Geostationary Orbit: Issues of Law and Policy', *American Journal of International Law*, vol. 73, 1979, p. 444; K.G. Gibbons, 'Orbital Saturation: The Necessity for International Regulation of Geosynchronous Orbits', *California West International Law Review*, vol. 9, 1979, pp. 139–56.

201 Resolution No. Spa2–1, Relating to the Use by all Countries, with Equal Rights, of Frequency Bands for Space Radiocommunication Services.

202 Recommendation No. Spa2–1, Relating to the Examination by World Administrative Radio Conference of the Situation with Regard to Occupation of the Frequency Spectrum in Space Radiocommunications.

203 The 'Bogota Declaration' is Doc. No. 81 of the 1977 WARC–BS, the Broadcast-Satellite Conference 1977. It is reprinted, N. Jasentuliyana and R. Lee (eds), *Manual on Space Law*, vol. 2 (New York: Oceana, 1979–82) pp. 383–7.

204 Cf. Reservation 40 to the 1979 Radio Regulations by Colombia, Congo, Gabon, Kenya, Uganda, Somalia and Zaire; and Reservation No. 79 by the same Parties.

205 Resolution No. 4 of the 1979 WARC went on to provide for the expiry of frequency assignments for geostationary satellites, and for procedures for their prolongation and alteration.

206 See Statement No. 90, Final Acts of the Conference, Nairobi, 1982.

207 This goes along with the changes in the ITU in the realm of technical assistance mentioned above in 'Technical Assistance', p. 337.

208 E.D. DuCharme, R.R. Brown and M.J.R. Irwin, 'The Genesis of the 1985/87 ITU World Administrative Radio Conference on the Use of the Geostationary-Satellite Orbit and the Planning of Space Services Utilizing It', (1982) 7 *Annals of Air and Space Law*, 1982, pp. 261–81; M.L. Smith, 'Space WARC 1985: The Quest for Equitable Access', (1985) 3 *Boston University International Law Review*, 1985, pp. 229–55.

209 Cf. ORB–85, First Session, 'Proposals for the Work of the Conference' by the USA, Conf. Doc. 5 and Additional Proposals Conf. Doc. 30; by the USSR, Conf. Doc. 5 and Additional Proposals Conf. Doc. 30; by the USSR, Conf. Doc. 9; the UK, Conf. Doc. 18; Kenya, Conf. Doc. 20; China, Conf. Doc. 25; Brazil, Conf. Docs. 37 and 63; and India, Conf. Doc. 54. See also Patricia E. Humphlett, 'Use of the Geostationary Orbit and US Participation in the 1985 World Administrative Radio Conference', Congressional Reference Service, Report No. 85–105 SPR; and the FCC action cited next note.

210 Cf. *World Administrative Radio Conference, First Session on World Administrative Radio Conference in Geneva*, Hearing before the Subcommittee on Communications and the Subcommittee on Science, Technology and Space, of the Committee on Commerce, Science, and Transportation, US Senate, 99th Cong 1st Sess., 6 November 1985; and FCC Docket No. 80–741, which deals with US participation in both Sessions: 'In the Matter of an Inquiry Relating to preparation for the International Telecommunication Union World Administrative Radio Conference on the Use of the Geostationary-Satellite Orbit and the Planning of the Space Services Utilizing It', First Report and Order (1985) FCC 2d 976; Fifth Notice of Inquiry, (1987) 2 FCC Rcd 3843, Release No. FCC 87–151.

211 Cf. the Kenyan 'Proposals for the Work of the Conference', Conf. Doc. 63, at para. 2, particularly para. 2.3

212 See 'Letter dated 16 October 1985 from the Secretary-General of the International Telecommunication Union Addressed to the Secretary-General', COPUOS, A/AC.105/360, 6 November 1985.

213 Some countries involved made Reservations on the matter: see Conf. Doc. 361, statements 5 (Indonesia), 7 (Ecuador), 8 (Colombia), and Conf. Doc. 363, additional statement 3 (Somalia). Cf. Statement by the USA, Conf. Doc. 33, Statement no. 2.

214 World Administrative Radio Conference on the Use of the Geostationary-Satellite Orbit and the Planning of Space Services Utilizing It, First Session, Geneva 1985, Report to the Second Session of the Conference. Hereafter cited as 'Report'.

215 G.C. Staple, 'The New World Satellite Order: A Report from Geneva', *American Journal of International Law*, vol. 8, 1986 pp. 716–7; Conf. Doc. 363, Statement no. 2.

216 Cf. Conf. Doc. Nos. 1 (Kenya), 6 (USA), 12 (UK), 13 (India), 14 (Belgium, Denmark, Finland, The Netherlands, Norway, Papua New Guinea and Portugal) and 15 (Iran).

217 Report, ch. 3.1.

218 Report, ch. 3.2, 3.3.4.3, and 3.3.4.5.

219 Report, ch. 3.3.5 and Annex 1.

220 Report, ch. 3.3.5

221 I have above suggested such legal entities be given restricted membership within the ITU to help avoid this problem. See pp. 179–80.

222 See WARC–ORB–85 Conf. Doc. No. 83, in which the ITU Secretary General transmitted to the Conference the Resolution of the Ninth Assembly of Parties of INTELSAT calling for the Conference to bear in mind the system requirements of INTELSAT and similar organizations.

223 Cf. for example, WARC–ORB–85 Conf. Docs. 17 (Senegal), 36 (Cameroon), and 104 (Burkina-Faso).

224 Planning Principles, Conf. Doc. no. 216; Cf. Doc. 166, Consideration on the Requirements of the International Multilateral Intergovernmental Organization.

225 R.L. and H.M. White *The Law and Regulation of International Space Communication* (Boston, London: Artech House, 1988) pp. 216–20; G.C. Staple, 'The New World Satellite Order: A Report from Geneva' *American Journal of International Law Review*, vol. 80,1986 pp. 712–3, (note also the information in this article regarding politicking on both sides of the question).

226 The remaining part of para. 3.2.6 ensures that safety-of-life systems operating with feeder links in the FSS were included.

227 'Specific Characteristics of Multi-Administration Systems', Attachment No. 1 to the same titled BG–73–103, DG, 10 September 1987.

9 Developments and the Future

DEVELOPMENTS

The telecommunications arena is one of swift change and development. It follows that prediction and rune-casting is awkward. However, before turning to more general comment on the future we must indicate developments which have occurred since the foregoing chapters were written. All the areas previously dealt with have shown development, and will doubtless continue so to do. For reasons of space, we will concentrate on important happenings in the areas of two organizations, the ITU and INTELSAT, these being the dominant organizations in satellite telecommunications.

The ITU and WARC–ORB–88

The Second Session of WARC-ORB 1985–88 was duly held in Geneva from 8 August to 15 September.[1] I am informed that the conference was an efficient conference, workmanlike in its discussions, and relatively untroubled by points which had been fully argued at the First Session and at other ITU Conferences of recent years. The principles and methods recommended by the First Session 'to guarantee in practice for all countries equitable access to the geostationary orbit and frequency bands allocated to space services as well as the technical parameters to be used for planning' were generally complied with.

The Final Acts of WARC–ORB–88[2] contain a partial revision of the Radio Regulations and their Appendices, the bulk of which will enter into force at 0001 hours UTC on 16 March 1990 (art. 69). The all important Provisions and Associated Plan for the use of the geostationary orbit for the Fixed-Satellite Service operating in the radio frequency ranges it specifies will remain in force for at least twenty years thereafter (Appendix 30B [I]).

The provisions made by the Conference for geostationary questions are complex, making amendments to frequency allocations, revisions to coordination procedures and to those for notification and recording in the Master International Frequency Register, and changes in provision for certain shared radio frequency bands. Detailed regulation is made for the calculation of technical data necessary under various heads of the new arrangements.[3] New or modified definitions are also adopted.[4] A massive new Appendix 30A (Orb–88) to the Radio Regulations will govern the feeder links for the Broadcasting-Satellite Service Plans for Regions 1 and 3 and for Region 2.[5] Similarly a new Appendix 30B provides a a global Plan for the Fixed-Satellite Service containing both existing systems and allowing for further development. Each country has been allotted a nominal orbital position within a predetermined arc of the geostationary orbit, a band width of 800 MHz for up and down-links, a service area for national coverage and generalized parameters. Existing systems are integrated within the Plan, and there are procedures through which a new system is brought into effect and the allotment within the predetermined arc is converted into an actual assignment to a particular orbital position. There are special rules for the creation of subregional systems and also mechanisms for making a new allotment to any new member of the ITU.

There are many other important elements in the Final Acts of WARC–ORB–88 relating not only to satellite matters but also to the operation of the Master International Frequency Register, but one within our present interest and to which reference must be made is Resolution COM6/3.[6] This deals with the Multi-lateral Planning Meeting (MPM), the planning device recommended for use by the First Session of the WARC–ORB. The Resolution notes that there were already within art. 11 of the radio regulations elements of bilateral and multilateral consultations regarding coordination of space systems and networks. Building on these, and noting that cooperation and goodwill are often needed for the acceptable solution of such coordinations, the Resolution enshrines the MPM within the procedures for coordination for the fixed-satellite services in a variety of radio bands in the case where there are major difficulties in dealing with the matter under the normal procedures of art. 11. Administrations affected by the matter are to be invited to the MPM, but if any are unable to attend for any reason the ordinary provisions of art. 11 apply in its case. The MPM does not therefore displace the formal procedures. It does, however, provide alternative procedures through which bilaterally or multi-laterally, administrations can meet properly to discuss and negotiate. Administrations are urged to use the MPM in a spirit of international understanding and upholding the principles of equal rights and equitable access to the 'limited natural resources of the [geostationary orbit] and the radio-frequency spectrum.' In this way it is intended that the process of accommodation of interests through compromise, bargaining and mutuality can solve difficult cases. We may hope that this will happen, but note that the conference wisely decided that the Administrative Council should monitor the progress of the application of Resolution COM6/3, with power to recommend that the MPM be reviewed by a future competent conference, if difficulties make this necessary.[7]

WARC–ORB–88 therefore made useful and significant advances. Nonetheless there is one regrettable omission from the new arrangements. WARC–ORB–88 makes no special arrangements for the multilateral operational organizations such as INTELSAT or INMARSAT. It was apparently considered that these organizations, not being 'administrations' and members of the ITU, and because they had commercial thrust in their activities, could not and should not be given special consideration. Administrations involved with such organizations could arrange matters among themselves, including the use of their own allocated orbital positions for the purposes of the organizations. In addition some use could be made of the multilateral planning meeting (MPM) and the new procedures under Section II of the new Appendix 30B through which sub-regional systems can be introduced. We will return to this question below.

This is short of what is desirable.[8] The 'operational organizations' although in part creations of commercial competitors, are also treaty-based and are international organizations just like the ITU itself. INTELSAT in particular serves the international and domestic telecommunications needs of many countries. Recognition of that element should not be overlaid by a distrust of a commercial element – though one can see how governments, the members of the ITU – could so react because of the aggravated commercial and competitive emphasis which has been introduced into international telecommunications in recent years. But to understand is not to excuse: the role of the multilateral organizations, and their declared purposes, ought to be formally recognized within the ITU. That it has not may indicate how far the organizations have departed from the vision of their founders, and that there is a need for a return.

INTELSAT and separate systems

As discussed above, the US company PanAmSat was the first private telecommunications carrier to be dealt with through the procedures of art. XIV of the INTELSAT Agreement regarding the establishment of public international telecommunications services separate from that of INTELSAT itself, and the process was not without difficulty.[9] Although the proposed US–Peru service was not planned to use all the transponders on the PanAmSat satellite the consultation process took account of their full number. However, the approval given by

INTELSAT was restricted to the proposed US–Peru service, any extension of the service, including to other points, requiring a further consultation under art. XIV. The possibility that a trans-Atlantic service would later be proposed was therefore left open.

On 12 May 1988 COMSAT in its capacity as US Signatory to INTELSAT requested consultation of a US–UK service to be carried by the remaining transponders on the PanAmSat satellite. The UK Signatory's matching request was received on 13 May, and further data was provided by the US and UK on 24 May. West Germany requested that it be associated with the consultations on 17 June. Why the UK and Germany should now so act is attributable to a pervasive change of view on the virtues of privatization and competition, as we mention later. Here we discuss the INTELSAT response.

Technical coordination was a relatively simple matter, but clearly the proposed trans-Atlantic service raised the question of 'significant economic harm' in a form that had to be faced. The proposed service would be operating on one of INTELSAT's major routes. The recommendation of the Director General was that the INTELSAT Board of Governors should advise the Assembly of Parties that the new system be approved.[10] 'Significant economic harm' would not occur to the INTELSAT system, and the establishment of direct links would not be prejudiced. However the Director General also recommended that new consultation be required for material change in the technical parameters or operational scope of the network, for material extension of services or material extension in the use of the network beyond 1993.

By the time that the matter got to the Assembly of Parties in October 1988, Luxemburg, Sweden and Ireland had been added to the consultation process. The Assembly accepted the recommendation of the Board of Governors and the consultative process ended with only minor variations from the Director General's original suggestion. The Assembly specifically stated that the proposed use of the new network 'not interconnected to the public switched network' (an important caveat) will not cause 'significant economic harm' to INTELSAT. It also increased to ten years – to 1998 – the period for which the consultation was carried through. Use of the service beyond that period, or any extension or modification to the service will require further consultation.[11] Subsequently the FCC issued its authorization for the proposed service.[12]

This development is important, marking INTELSAT's coming to grips with a potential major competitor, which this time was not another international organization, but a private 'for profit' company. The route is a major telecommunications route. Under these circumstances one must ask: 'how does this affect previous discussions of the concept of "significance" in the phrase "significant economic harm" '? The matter is faced in the Contribution of the Director General to the Board's discussion.[13] That document rehearses the questions listed in 'Procedures for Non-Technical Consultation Under Article XIV(d) of the INTELSAT Agreement' (BG–64–80).[14] It finds that the new PanAmSat system will provide public international telecommunications services, which INTELSAT itself could provide, and which there is no reason to believe would not have been carried by INTELSAT. The effect on INTELSAT is then estimated in both the short- and long-term, and it is concluded that over the five year period dealt with by the Director General's paper, a maximum potential revenue loss of some 2.5 per cent of INTELSAT's global revenue projection would result from the PanAmSat trans-Atlantic service (para 36). The paper then observes that INTELSAT had not fixed a threshold figure at which such 'harm' became significant, but points to the Hinchman report on 'Significant Economic Harm'[15] as suggesting that it would be difficult to argue that less than a 5 to 10 per cent cumulative loss over a 5 to 10 year cycle was significant.[16] On that basis the paper did not consider the harm caused by PanAmSat to INTELSAT to be 'significant'. The cumulative effect of the PanAmSat case taken with other losses was estimated to total at some 9 per cent of INTELSAT's revenues (para 41), a figure within the 10 per cent indicator mentioned above.[17]

On these figures and on such reasoning one wonders when the 'threshold' of 'significance' of economic harm will be triggered, and what will then happen. The Hinchman Report[18] to which the Director General's paper pointed in respect of cumulative effects, also indicated (at p. 29) that a threshold of a 1 per cent limit for the individual effect of a separate system would be a useful *de minimis* threshold for INTELSAT to adopt. The PanAmSat service is reckoned in the Contribution of the Director General to have a maximum potential diversion of 2.5 per cent from INTELSAT global revenues (para 36), a somewhat higher figure than 1 per cent and presumably a much higher figure in regard to INTELSAT's trans-Atlantic revenues in general and in relation to the type of service to be provided.

The Thirteenth Assembly of Parties noted the Director General's Report which the Board had used in coming to its recommendations. It observed that studies on economic harm which it had requested at its previous meeting were in progress (AP–12–3, para 27). It has now, however, asked as a matter of urgency that the Director General develop for its own and the Board's consideration improvements in the methods of assessing economic harm and proposals as to guidelines for findings on 'significant economic harm'. These are to be presented to the next ordinary meeting of the Assembly. The Assembly also put it on record that any guidelines will have to take into account: a) INTELSAT's essential task of providing a single global system as part of the global networks providing telecommunications for all parts of the world, b) the economic effects of current and future separate systems on the ability of INTELSAT efficiently and competitively to perform that essential task, c) the importance which the markets that separate systems are aimed at may have for INTELSAT, and d) the need for a practical method accurately to assess the economic impact of individual separate systems and the necessity that terms be developed on which INTELSAT can readily deal with consultations on such questions.

INTELSAT's studies on 'significant economic harm' are therefore proceeding – but this takes us to the future.

THE FUTURE

The ITU

Let us begin, respectfully, with the senior organization involved in space telecommunications. It is remarkable how well the ITU, the second oldest international organization, has stood up to the demands and exigencies of the new facility. Although its procedures for adopting new regulations are, perhaps, rather slow in an area where technical development is quick, nonetheless the ITU's difficulties are less a matter of machinery and more a matter of those who make use of that machinery. The ITU is threatened. Its original purposes in radio are to provide an arena through which the rational use of the radio spectrum can be agreed through the device of allocation, and actual utilization safeguarded through the coordination and notification procedures. These are purposes necessary for physical reasons. However, politics irrelevant to these purposes have increasingly caused difficulties at ITU conferences of the 1970s and 1980s and impeded the Union's work. In addition there is the attempt by a numerical majority, although a contributing minority, to make the ITU a conduit for international aid to themselves. That attempt has been in measure successful, but has opened the possibility that the minority – the developed nations – may decide that the ITU in that form is no longer worth persevering with. The less developed countries are strident in their demands, and, as indicated in Chapter 7,[19] there is a danger that, as in other UN Agencies, they will harm the organization through attempting to use it for purposes for which it is not designed. Indeed, at worst they might wreck it, for it would be possible to form an alternative association in which numbers of the developed nations could secure their interest in minimizing harmful interference and so forth in sufficient measure to make that option attractive.

That said, I would hope that that step is not taken, and indeed that the forthcoming Plenipotentiary Conference in Nice in May 1989 follows the example of WARC–ORB–88, and is efficient, business-like, fruitful and reasonably harmonious. The ITU has done well, and we can but hope that sooner rather than later the hollowness of power without either responsibility or the means to make use of the opportunities of that power will lead the numerical majority in the ITU to wiser counsels. Perhaps the ITU should contemplate relating voting weight to contribution as the operating international satellite organizations do.

As indicated above, however, WARC–ORB–88 disappoints insofar as it gave no explicit recognition to the requirements of the multilateral operational telecommunications satellite organization. This is short of what is desirable. The 'operational organizations' although in part creations of commercial competitors, are treaty-based and are international organizations like the ITU itself. INTELSAT in particular serves the international and domestic telecommunications needs of many countries. Recognition of that 'public service' element of the organizations should not be overlaid by a distrust of a commercial element – though one can see how governments (the members of the ITU) could so react as they contemplate the aggravated commercial and competitive emphasis which has been introduced into international telecommunications in recent years. But to understand is not to excuse: the role of the multilateral organizations, and their declared purposes, must be recognized within the ITU. That it has not may indicate how far the organizations have departed from the vision of their founders, and that there is a need for a return.

Nonetheless, while recognizing the case that can be made to the contrary, I would repeat that full consideration should be given to the revival of the category of associate membership of the ITU.[20] It would allow a standing to be given to the intergovernmental organizations (which the international operational agencies are) within the councils and conferences of the ITU. That would better allow their peculiar requirements to be met, and tend to the proper use of space for the benefit of all mankind, and its better exploitation as the 'common heritage of mankind'. That, after all, is the thread which runs through the international treaties and declarations on the matter such as the Outer Space Treaty. We seem to be in danger of losing sight of these treaty obligations.

The satellite organizations

A major interest of the constitutions of each of INTELSAT, INMARSAT and EUTELSAT lies in the manner in which decision-making power is linked to investment and investment to utilization of the system. The curse of the UN system in general is that voting weight is unrelated either to contribution to costs, to size, or (bluntly) to importance. The only reference made to a criterion such as 'gross national product' is merely to calculate the 'assessment' laid on a country. This reflects the principle that each state is 'sovereign' in international law. But the concept of sovereignty was designed for international relations, not for international organizations. In the case of the 'operating' international telecommunications organizations a different scheme had to be adopted. Without that concession, the participants would not have banded together. As a result all three major organizations differ as to the provision they make for the particular interrelationships of states, of the PTTs as operators of telecommunications services, and of the agency itself as establishing and running the space segment of the system. But there is the common substratum that in all three cases cost is related to usage, which itself relates to 'importance'. That is good.

As far as the law is concerned, those who drafted the constitutional documents of the various international organizations should be congratulated. In the main the constitutions work well, and their emergent problems are not so much a matter of a fault of the various constitutional structures as an unwillingness on the part of some members to operate the structures with that degree of good faith and tolerance that any constitutional organism requires. In UK terms, the 'conventions of the constitution' are being put under strain in the ITU and in INTELSAT, and that is a pity. Much will be lost if unwise sectarianism is allowed to displace the basic ethos that used to underlie these organizations in their restricted respective areas. When President Kennedy affirmed the US intention to proceed to a global satellite system with access for all, and when the UN expressed similar views, there was an ethos abroad which may have insufficiently taken account of hard financial questions, but which resulted in an approach to the questions of a global satellite system which made things possible. One wonders whether, were the question to arise now, an INTELSAT would be conceivable, let alone possible.

Certainly one option is entirely out of court. There is no possibility that some general international organization be set up to operate international telecommunications.[21] It would be impossible to get the existing international communications carriers and their respective national states to agree to such a step. It would be quite impracticable to bring together the cable, radio and satellite arrangements and authorities, entirely apart from fears that such a conglomerate might be less than efficient. It is even impossible to contemplate the superficially smaller task of bringing together the satellite organizations.

But even so, we cannot stand still. We must continue to develop the organizations, and lawyers must continue to do their best to facilitate that development and avoid temptations to bring national practices and techniques into the international arena.

That said, economics will always remain a major factor, as will the speed of technical development. The international organizations must be sufficiently adaptable to cope with the swiftness of progress. INTELSAT has been attacked because its services are 'dated', 'unresponsive to public requirements', and 'unjustifiably highly priced' to the detriment of consumers. There may have been some truth in that, but the pressures that have been put on INTELSAT have had effect. The organization has been shown to be adaptable and its Agreements sufficiently flexible to contain the developments and demands of the last few years.[22] Those who proffer argument against INTELSAT as justification for they themselves being permitted to enter the public international telecommunications market have found that INTELSAT has become more efficient, and that the market they perceived is not as easy a place to make rich pickings as they had thought.

But the process of change will continue. In the technical realm the development of digital telecommunications will facilitate the growth of telecommunications networks. Satellites will be well placed to provide for the expansion which seems likely, although the operation of the new optical fibre cables will have an effect.[23] In business terms the growth of a number of private enterprise international telecomthe growth of a number of private enterprise international telecommunications carriers seems unstoppable. The 'privatization' of telecommunications within may of the developed nations will produce pressures on the existing international arrangements. Even the EEC's progress to the common internal market in 1992 has had its effects.

But the inspiration of INTELSAT, indicated in the Preamble to its Agreement, as in the 'constitutions' of the other agencies, was the provision of a service for all mankind. In the current *zeitgeist* that aim should be neither forgotten, nor downplayed. It is important that states should take seriously their international obligations under the treaties which are the constitution of the several space telecommunications organizations.[24] Further the PTTs and other communications carriers involved should properly market the internationally provided services within their domestic markets.[25] The organizations provide the space segment to the carriers: they do not market that to the end-users.[26] Given proper marketing, and a proper respect for their inspiration, the international space telecommunications organizations can continue to be examples of best intentions practically realized, and satellite telecommunications, while remaining a business, will retain the primary ethos of being a service.

Space law

As indicated, however, that will require that the organizations continue to be well served by a developed system of International Space Law. What will that law be? Will it be a truly international law, or will it be a national law writ large? I come from a 'small jurisdiction' – Scotland – where the law and the legal system has suffered through our having a large, insular and introspective neighbour. More recently Scots Law has been swamped with English and American concepts in regard to the exploitation of North Sea oil. I see these patterns of legal history reflected in the development of concepts for Space Law. Some of that explains why there are passages in this book which are short-tempered in discussing US attitudes, and I have not excised these lest this point be lost. But there is one other clearer warning from the past.

I note that the Law of England as applied to maritime matters through the Admiralty Court had significant impact, and became for most purposes international maritime law because of the dominance of the English forms of contract and of the English courts and arbitral procedures in settling maritime disputes at the time that international maritime law was evolving. History therefore leads me to suspect that US Law may become international space law – unless we foreigners are very careful, and determined.

By reason of its technical skills, its domestic market and its entrepreneurial attitudes, the USA is a major leader in space matters. That 'lead' may result in much US Law becoming the language in which problems are discussed and solved. Through such as the State Department and the Commerce Department, through the National Aeronautics and Space Administration, through the FCC, and through Congressional Hearings[27] and Congress's Office of Technology Assessment (the OTA) and the Congressional Research Service of the Library of Congress, through the legal sections of COMSAT, AT&T and other communications carriers and the large number of other US companies active in space matters, to say nothing of academic work, the USA is pressing ahead with developing legal concepts that can deal with the problems which emerge in space. US Patent Law and the National Aeronautics and Space Act have been amended to take account of developments on space platforms with the intention of providing patent protection and therefore an availability of ideas through the Patent Office in Washington,[28] and it is noticeable that INTELSAT patents are very likely to end up in the US Patent Office.[29] I feel that it is therefore likely that concepts developed in US law will increasingly be used in international forums to discuss matters of Space Law in general and telecommunications law in particular.[30] It may even be that questions of international telecommunications will be litigated before the US Ccourts. Certainly already the FCC has demonstrated no unwillingness to take under its scrutiny what others might construe as matters of state sovereignty.[31] I also note the willingness of various countries to submit evidence to the FCC in the proceedings just cited. It will be but a small step for US contracts and practices to dominate in legal thinking on practical matters of Space Law: then there will be a 'difficult' case which will be referred to the US courts, since they are best fitted to determine the meaning of US concepts. And the die will have been cast.

The possibility of US law becoming 'Space Law' in many areas is there, but remains only a possibility at present. It may not happen, but non-US commercial companies will have to help, both financially – for Space Law is not an interest which can command much funding in the current climate in universities – and through their being unwilling simplistically to accept concepts drawn from a foreign legal system. Familiar words, when used in a foreign legal contract, may not have the exact content one expects.[32] Those interested in Space Law must work out suitable concepts and their formulations in advance of problems. It may be that, just as Europe and others helped to form INTELSAT through banding together to reinforce their unwillingness merely to buy a service on a US-owned system, so by research and discussion, a truly 'international' space law may be developed in which telecommunications law will be a major component.[33]

NOTES

1   For earlier action see Chapter 8, Section 'WARC–ORB 1985–88', p. 393.
2   Final Acts Adopted by the Second Session of the World Administrative Radio Conference on the Use of the Geostationary-Satellite Orbit and the Planning of Space Services Utilizing It (ORB–88), (Geneva: ITU, 1988).
3   Eg. Annex 3 to Appendix 30A (ORB–88): 'Technical Data Used in Establishing the Provisions and Associated Plans and Which Should be Used for their Application.'
4   Article I of the Regulations has added to it definitions for 'Fixed-Satellite Service', 'Feeder Link'; 'Effective Boresight Area'; 'Effective Antenna Gain Contour'; and 'Steerable Satellite Beam' (the satellite antenna beam, not the earth station beam, is repointed). 'Deep Space' is re-defined as 'Space at distance from the Earth equal to, or greater than, 2×10^6 kilometres' – the first enshrining in treaty law of a definition of space in terms of a fixed height, and a sharp increase from the prior definition by which the orbit of the Moon was the inner limit of Deep Space.
5   For previous action see Chapter 8, Section 'Other Developments prior to WARC–ORB 1985–88' p. 381.
6   'Improved Procedures for Certain Bands of the Fixed-Satellite Service', Res. COM6/3, Final Acts of WARC–ORB–88, *supra*.
7   Resolution COM6/3, *ad fin*.

8   For prior comment see above, Chapter 8, Section, 'WARC–ORB 1985–88' p. 395.
9   See above, Chapter 4, Section 'Art. XIV(d) coordination: practice', subsection 'PanAmSat-Peru' p. 177.
10 'Article XIV(d) Consultation Concerning the Use of the PanAmSat Satellite Network to Provide Telecommunications Services between the United States and the United Kingdom, DG, BG–76–50, 27 May 1988, with Addendum 1, 13 June 1988.
11 From the information available to me I do not know whether these changes were made on the initiative of the Director-General, the Board, or the Assembly.
12 'In the Matter of Alpha Lyracom, d/b/a/ Pan American Satellite; Final Authority to Operate an International Fixed-Satellite System for Communication Between Specified Geographical Areas', FCC File No. CSS–84–004–P(LA), adopted 28 October 1988, released 7 November 1988. The satellite was launched on Ariane 4 in June 1988.
13 'Article XIV(d) Consultation Concerning the Use of the PanAmSat Satellite Network to Provide Telecommunications Services between the United States and the United Kingdom, DG, BG–76–50, 27 May 1988, with Addendum 1, 13 June 1988.
14 See above, Chapter 4, p. 168.
15 W. Hinchman Associates Inc., Report on 'Significant Economic Harm', Attachment No. 1 to 'Report on the Study of Significant Economic Harm', DG, BG–60–63, 15 August 1984. See Chapter 4, p. 167.
16 This appears to be an inference from the Hinchman report.
17 Reference is also made to 'Cumulative Economic Impact Tables for Article XIV(d) Consultations', DG, BG–76–49, 7 June 1988, showing the effects of prior Article XIV(d) consultations, and their potentials for the future. Some 6.4 per cent of the potential global INTELSAT revenues had already been 'lost' to systems consulted under Article XIV(d), of which approximately 3 per cent was attributable to systems involving the US and the UK. This paper went to the 13th Assembly as AP–13–20.
18 Op. cit. n 15.
19 See Chapter 7, Sections 'Finance' (p. 332) and 'Technical Assistance' (p. 337).
20 See above, Chapter, p. 179.
21 Cf. Grenville Clarke and Louis B. Sohn, *World Peace Through World Law: Two Alternative Plans* (3rd ed.) (Cambridge, Mass.: Harvard University Press, 1966), pp. 296–302, 'A United Nations Outer Space Agency'. Cf. also the International Seabed Authority set up under art. 156–88 of the Convention on the Law of the Sea 1982, (1982) 21 ILM 1261; (UK) (1983) Misc. No. 2, Cmmd. 8941: most major nations including the UK and US are not members of this treaty.
22 Cf. David M. Leive 'Flexibility of the INTELSAT Agreements', *INTELSAT News*, vol. 4 No. 3, September 1988, p. 2.
23 TAT–8, the first trans-Atlantic optical fibre cable, which runs from Tuckerton, New Jersey, USA and with landing points in both France and the UK, entered service on 14 December 1988. It can carry up to 40,000 simultaneous telephone calls.
24 It would also help if the incoming US Administration were to gather together the elements of government which deal with telecommunications and entrust them to a single cabinet level department. Cf. evidence of S. Doyle, *World Administrative Radio Conference*, Hearing before the Subcommittee on Communications and the Subcommittee on Science, Technology and Space of the Committee on Commerce, Science and Transportation, US Senate, 99th Cong. 1st Sess., 6 November 1985 at p. 60.
25 Some Signatories have interests which militate against a full commitment to the

purposes of the organization – for example, through having substantial investment in cables.

26 INMARSAT is somewhat different in its relationship with end-users, though it also technically acts through PTTs.

27 See all the various Hearings cited earlier in this book. See also *Space Commercialization 1985*, Hearings before the Subcommittee on Space Science and Applications of the Committee on Science and Technology, US House of Representatives, 99th Cong. 1st Sess., June–October 1985.

28 See *Patents in Space*, Hearing before the Subcommittee on Courts, Civil Liberties, and the Administration of Justice, Committee on the Judiciary, US House of Representatives, 99th Cong. 1st Sess., 13 June 1985; HR 4316, and 'Patents in Space', Report 99–877, 13 August 1986, to accompany HR 4316.

29 See Chapter 4, Section 'Inventions, Patents and Technical Information' subsection on 'Procedures', pp. 129 and 137.

30 Cf. the use of US domestic concepts by Orion Satellite Co.: above, Chapter 4, subsection 'International Public Telecommunication Services', and subsection 'US Separate Systems', pp. 159 and 182.

31 Cf. In the Matter of Regulatory Policies and International Telecommunications, CC Docket No. 86–494; Notice of Inquiry and Proposed Rulemaking (1987) 2 FCC Rcd 1022; Report and Order and Supplemental Notice of Inquiry, adopted 25 February 1988, released 25 March 1988, release No. FCC 88–71, 64 Rad. Reg. 2d (P&F) 976. Cf. also problems of the extraterritorial application of US law, notably that on commercial/military security.

32 From a different area of Space Law I would cite problems emergent in the legal underpinnings of remote sensing.

33 The establishment of the European Centre for Space Law in May 1989 though the European Space Agency is a very welcome development. It will help meld the European effort which is currently unduly fragmented.

Selected Bibliography

Alper, Joel, R. and Pelton, Joseph, N. (eds) (1984), *The INTELSAT Global Satellite System*, 'Progress in Astronautics and Aeronautics', vol. 93, New York: American Institute of Aeronautics and Astronautics.

Barty-King, Hugh (1979), *Girdle Round the Earth: The Story of Cable and Wireless*, London: Heinemann.

Bleazard, G. Bernard (1985), *Introducing Satellite Communications*, Manchester: National Computing Centre Ltd.

Christol, Carl. Q. (1982), *The Modern International Law of Outer Space*, New York; Oxford: Pergamon Press.

Clarke, Grenville and Sohn, Louis B. (1966), *World Peace through World Law: Two Alternative Plans* (3rd edn.), Cambridge, Mass.: Harvard University Press.

Codding, George A., Jr. (1972), *The International Telecommunication Union: An Experiment in International Cooperation*, Leiden: E.J. Brill, 1952: New York: Arno Press.

Codding, George A., Jr. and Rutkowski, Anthony M. (1982), *The International Telecommunication Union in a Changing World*, Dedham, Mass.: Artech House.

Cohen, Maxwell (ed.), (1964), *Law and Politics in Space*, Montreal: McGill University Press.

Colino, Richard R. (1973), *The INTELSAT Definitive Arrangements: Ushering in a New Era in Satellite Telecommunications*, Geneva: European Broadcasting Union.

Fawcett, J.E.S. (1984), *Outer Space: New Challenges to Law and Policy*, Oxford: Clarendon Press.

Haley, Andrew G. (1963), *Space Law and Government*, New York: Appleton-Century-Crofts.

Hills, Jill (1986), *Deregulating Telecoms: Competition and Control in the United States, Japan and Britain*, London: Frances Pinter.

Howell, W.J., Jr. (1986), *World Broadcasting in the Age of the Satellite*, Norwood, NJ: Ablex Publishing Corporation.

International Cooperation and Competition in Civilian Space Activities, Washington DC: US Congress, Office of Technology Assessment, OTA–ISC–239, July 1985.

Jasentuliyana, N. and Lee, R. (eds) (1979–82), *Manual of Space Law*, 4 vols., New York: Oceana.

Jasentuliyana, N. and Chipman, R. (eds) (1984), *International Space Programmes and Policies. Proceedings of the Second United Nations Conference on the Exploration and Peaceful Uses of Outer Space (UNISPACE), Vienna, Austria, August 1982*, Amsterdam; New York: North Holland.

Jasentuliyana, N. (ed.) (1984), *Maintaining Outer Space for Peaceful Uses*, Tokyo: The United Nations University.

Kildow, Judith T. (1973), *INTELSAT: Policy-Maker's Dilemma*, Toronto, London: Lexington Books.

Lay, S. Houston and Taubenfeld, Howard J. (1970), *The Law Relating to the Activities of Man in Space*, Chicago: Chicago University Press.

Leive, David M. (1970), *International Telecommunications and International Law: The Regulation of the Radio Spectrum*, Leiden: Sijthoff; New York: Oceana.

Levin, Harvey J. (1971), *The Invisible Resource: Use and Regulation of the Radio Spectrum*, Baltimore: Johns Hopkins Press.

Levin, Lon C., Novak, Sally J., Allen, Rosalind K., Frank, Carl R., Gannt, John B., Leibowitz, David E. (1986), *International Telecommunications Handbook*, Washington DC: International Telecommunications Committee, Federal Communications Bar Association.

Lewin, Leonard (ed.) (1979), *Telecommunications: An Interdisciplinary Survey*, Dedham, Mass: Artech House.

Lewin, Leonard (ed.) (1981), *Telecommunications in the US: Trends and Policies*, Dedham, Mass: Artech House.

McDougal, Myres S., Lasswell, Harold D. and Vlasic, Ivan A. (1963), *Law and Public Order in Space*, New Haven: Yale University Press.

McDougal, Walter A. (1985), . . . *the Heavens and the Earth: A Political History of the Space Age*, New York: Basic Books.

Mance, Osborne (1943), *International Telecommunications*, London: Oxford University Press.

Marsh, Peter (1985), *The Space Business*, Harmondsworth: Penguin Books.

Matte, Nicholas M. (1969), *Aerospace Law*, Toronto: Carswell; London: Sweet and Maxwell.

Matte, Nicholas M. (1982), *Aerospace Law: Telecommunications Satellites*, Toronto and London: Butterworths.

Michigan Yearbook of International Law, 1984: 'Regulation of Transnational Communications'.

Final Report of the President's Task Force on Communications Policy, Washington: USGPO, 1969.

Robinson, Glen O. (ed.) (1978), *Communications for Tomorrow: Policy Perspectives for the 1980s*, New York and London: Praeger.

Smith, Delbert D. (1969), *International Telecommunication Control: International Law and the Ordering of Satellite and Other Forms of International Broadcasting*, Leiden: A.W. Sijthoff.

Smith, Delbert D. (1976), *Communication via Satellite: A Vision in Retrospect*, Leiden and Boston, Mass.: A.W. Sijthoff.

Tomlinson, John D. (1945), *The International Control of Radiocommunications*, Ann Arbor, Mich.: Edwards.

Wallenstein, Gerd D. (1977–), *International Telecommunication Agreements* 3 vols., New York: Oceana.

White, Rita L. and Harold M. (1988), *The Law and Regulation of International Space Communication*, Boston and London: Artech House.

Index